ABOUT THE AUTHOR

THEODORE H. WHITE was born in Boston in 1915. After graduating summa cum laude from Harvard University in 1938, he was recruited to cover East Asia for *Time* magazine, becoming chief of its China Bureau in 1945. This experience would inspire his first book, *Thunder Out of China* (written with Annalee Jacoby). After serving for a brief time as editor of *The New Republic*, he edited *The Stilwell Papers*. In 1948, he went to live in Europe, and his experience as a European correspondent led to *Fire in the Ashes*, published in 1953. That same year he returned to the United States to work as national correspondent for *The Reporter*, and then for *Collier's* magazine. After the collapse of *Collier's* in 1956, White turned to fiction, completing two novels, *The Mountain Road* and *The View from the Fortieth Floor*, in the next four years.

At the time *Collier's* closed down, White was planning a story on "The Making of the President 1956" for the magazine; he revived the idea in the next election year, resulting in his most famous book, *The Making of the President 1960*, which was awarded the Pulitzer Prize for General Nonfiction in 1962. White went on to produce three more Making of the President volumes, covering the 1964, 1968, and 1972 campaigns. Subsequently, he was author of *Breach of Faith: The Fall of Richard Nixon*, *In Search of History: A Personal Adventure*, and *America in Search of Itself: The Making of the President 1956–1980*. He was the recipient of numerous awards in addition to the Pulitzer Prize, and was recognized for his documentary work with Emmy Awards in both 1964 and 1967. White died in 1986 at the age of seventy-one.

Also by Theodore H. White

Thunder Out of China
(with Annalee Jacoby)

Fire in the Ashes

The Stilwell Papers
(editor)

The Mountain Road

The View from the Fortieth Floor

The Making of the President 1960

The Making of the President 1968

The Making of the President 1972

Breach of Faith: The Fall of Richard Nixon

In Search of History: A Personal Adventure

America in Search of Itself:
The Making of the President, 1956–1980

THEODORE H. WHITE

THE MAKING OF THE PRESIDENT 1964

HARPER**PERENNIAL** POLITICAL**CLASSICS**

NEW YORK • LONDON • TORONTO • SYDNEY • NEW DELHI • AUCKLAND

HARPER**PERENNIAL** ★ POLITICAL**CLASSICS**

FIRST HARPER PERENNIAL POLITICAL CLASSICS EDITION PUBLISHED 2010.

Library of Congress Cataloging-in-Publication Data is available upon request.

ISBN 978-0-06-190061-7

HB 12.13.2022

FOR

H. CHOUTEAU DYER

(1935–1961)

CONTENTS

CONTENTS

FOREWORD TO THE 2010 EDITION

LIKE MODERN American politics itself, this book begins with an ending: Dallas, Friday, November 22, 1963. With a novelist's feel for character and drama, Theodore H. White opens his chronicle by putting his readers in the backseat of the presidential limousine carrying John F. Kennedy through the streets. "It was hot; the sun was blinding; there would be a moment of cool shade ahead under the overpass they were approaching," White wrote. Then came the gunshots, and the blood, and the terror. To John Kennedy, the heat and the sun no longer mattered. The president of the United States was dead.

It is rarely given to us to be able to mark turning points in the life of the nation with such precision, moments after which nothing was the same. In that tragic autumn, Dallas joined Lexington and Concord, Appomattox, and Pearl Harbor as clearly discernible landmarks in our historical imagination, an inescapable signpost.

White's Making of the President series, of which this is the second volume, is a landmark in its own right. In terms of the practice of American journalism and the writing of contemporary history, there is the pre-White era and the post-White era—eras that are as different from one another as silent movies were from talkies, or radio from television, or black-and-white from Technicolor.

That we take so many elements of White's writing style as a given, half a century on, is evidence of the magnitude of his achievement. The rendering of presidential candidates as fully human characters, with emotions, uncertainties, and insecurities; the painting of political consultants and managers as critical figures, plotting and polling and maneuvering; and the casting of campaigns as sagas in which the contenders must do battle with one another and with fate: Such is now the expected stuff of political coverage. In White's hands, beginning in 1960, the dry became the dramatic.

How did he do it? It seems simple now, but White was a pioneer in applying ancient storytelling devices—techniques and sensibilities at least as old as Homer—to history. Character, scene, dialogue, and story arc became weapons in White's arsenal as he lifted presidential politics from the sterility of newspaper copy to the stuff of legend. Before White, presidential campaigns were reported as though people had no sense that their rulers (and would-be rulers) were human; the dramatization of politics was for imaginative literature, not historical nonfiction. Shakespeare could plumb the anxieties of Henry V, and novelist Allen Drury might shed light on the darker corners of the capital, but their stories were confined to the safe playground of fiction.

White's genius was to deploy the skills of a novelist in the pursuit of journalism. Though it is less well known today than the celebrated 1960 volume (a big bestseller and a winner of the Pulitzer Prize), *The Making of the President 1964* resonates in our own time even more than its predecessor. The politics of race, the rise of the Republican Party (which would go on to win five of six presidential elections from 1968 to 1992), and Lyndon Johnson's outsized contradictions (which would lead him from a historic landslide to Vietnam) are all detailed in these pages, deftly and presciently.

Theodore White was both a maker and a mirror of the story of American journalism in the twentieth century. Born in 1915 in Boston and educated at Harvard, White rose to prominence as a journalist in Henry R. Luce's Time-Life empire. The newsmagazine genre Luce created with co-founder Briton Hadden in the 1920s transformed the straightforward coverage of newspapers into colorful narratives, complete with heroes and villains. For *Time* and, beginning in 1933, *Newsweek,* reporting the news meant more than passing on the who, what, when, how, and why. It meant conjuring up a pageant of the interesting and the important.

Such was White's world. Stylistically he was very much a creature of the midcentury newsmagazine—a medium that not only had its virtues (a sense of drama, a passion for politics, an eye for narrative detail) but also its vices, including an occasional Olympian tone that may strike modern readers as stentorian. Yet, in a way, he was also a forerunner of the New Journalism of Tom Wolfe, Gay Talese, and Truman Capote, among others who, later in the 1960s and 1970s, pushed the boundaries further, turning reportage into literary nonfiction.

The key reason for White's success was that he intuitively understood politicians. In his portrait of Lyndon Johnson, White diagnosed a condition he called "the politician's optic," in which the hostile language in any press story leaps off the page while the positive recedes. By the same token, even the slightest compliment to an opponent "swells to double-size capitals" in the politician's gaze. "This is an occupational disease of politicians," White notes, "just as it is for authors and actors, who similarly live by public approval or distaste." Johnson loved the press, or hated it, depending on what was being said of him. Asked at a press conference whether he worried about "overexposure," LBJ went into a slow burn, as White notes: "'Well, I strive to please,' Johnson replied, 'and if you'll give me any indication of how you feel about the matter, I'll try to work it into my plans in the future.'" He refused to speak to reporters again for four weeks.

Though politicians and their personalities dominate White's canon, he also had a historian's feel for the big picture. Here, for instance, is his encapsulation of the differences between the two major parties: "The Democrats believe in government—government as an instrument to do things. . . . The Republicans' impossible dilemma is that they have never sorted out properly what it is that government should do and should not do—and at what level. . . . They campaign, generally, *against* government; the Democrats campaign, generally, *for* government. The Republicans are for virtue, the Democrats for Santa Claus." Looking at politics from the end of World War II forward, it would be difficult to improve on that basic observation.

Civil rights and the conservative reaction to the end of Jim Crow rest uneasily at the center of the story of 1964. White was present at the creation of the forces that would, in the fullness of time, give us the Southern Strategy and Willie Horton. Barry Goldwater is depicted in vivid colors, but what White calls the "Goldwater movement" was, even in real time, of larger importance than the candidate himself. "The wordless resentments, angers, frustrations, fears and hopes that were shaping this force were something new and had welled up long before Goldwater himself took his Presidential chances seriously," White writes. Here is the rise of the West against the Eastern powers in GOP politics, the quiet but unmistakable shift of influence and clout from the ethos of Nelson Rockefeller to that of a man who spoke out for Goldwater just before the election in a hugely popular broadcast: Ronald Reagan.

Reagan's name does not appear in White's account of 1964 (and neither does that of a young man who ran, and lost, his first competitive race that year in Texas: George H. W. Bush), but the old Hollywood star who would become governor of California in two years' time is the greatest of what White refers to as the "seed-names of some entirely new era" that was coming to be in 1964.

In the end, the reason Teddy White matters is that he transformed the way many Americans thought about politics and politicians. He brought them to life, for better and for worse. Sometimes he was too soft, sometimes perhaps too harsh, in his series about those he invariably called "the men of politics." But he took what he covered seriously, because he believed American democracy to be a serious thing. Politics, for him, was not about gamesmanship and gossip, though both played their parts. The way we chose our leaders, he believed, shed light on who we were, and what we wanted, and what we believed in. No one in our time has done more to bring that process—that *drama*—to life. That is why we are still reading White all these years on.

Jon Meacham
October 2010

EVERY MAN who writes of politics shapes unknowingly in his mind some fanciful metaphor to embrace all the wild, apparently erratic events and personalities in the process he tries to describe.

Over the past few years the image of politics that has taken shape for me is that of an immense journey—the panorama of an endless wagon train, an enormous trek, a multitudinous procession of people larger and more confused than any of the primitive folk migrations.

There—ahead—lies the crest of the ridge, and beyond it, perhaps, the plateau or the sunlit valley—or danger. The procession stretches out for endless miles, making its way up the tangled slopes through strange new country, and it has been marching for a long time. The country the people enter is full of unknown dangers and fresh promise, where the only thing to be anticipated is the unexpected. In the procession there are those who trudge on foot, those who ride in the wagons, those who go by horse. On and on they go, some weary, some gay, some infuriated by the slowness of the pace, and others who insist the journey pause because the pace is too fast. In the long, endless file are the sturdy ones, the happy ones, the good hunters; and also the lame, the old, the trouble-makers, the stupid, the children too. They quarrel with each other as they go, or they make friends and form groups about their own section leaders in the rear. But none can leave the march for very long, or go his own way; for to leave means to be alone in the wilderness, or perhaps to fall into the hands of an equally savage enemy procession. All are bound to their own migration, their own wagon train, whether they enjoy it or hate it.

There—up ahead—approaching the crest, or already at the crest, are the leaders. From the crest the leaders can see whether it is jungle that comes next, or whether a rival procession may beat them to the watering place, or whether there is desert over the ridge. The leaders must judge how fast they must flog the march on, or whether they can let it relax because they are entering fair fields and good hunting land. The people behind and below cannot see as far as the leaders. Their feet hurt; the children are crying; or the old ones are sick. But they must march; they are compelled to go; they cannot choose the way themselves. Their only choice is the choice of leaders who will direct the way for them.

This, it seems to me, is politics.

Up there, at the head of the advance column, the leaders quarrel as bitterly among themselves as do the people behind. From their heights they have a wider view of the horizon. Not only that—but the scouts return continuously from advance patrols and explorations with fresh new information out of which the leaders must make their plans for tomorrow. They debate and argue over these plans, endlessly, day after day, sometimes violently. The problems that lie ahead, the prospects of the journey for the next day, should be the chief burden and responsibility of the leaders. But they have another and equal problem—the people behind, who want a holiday, who want to eat more, or want to be carried in the wagons. The only way the leader can get them over the crest is to pay attention to their aching feet and remember the sound of the old ones and the children crying. The leaders disagree continuously as to what course they should take, at what pace, at what cost. They know that much later they will be judged by some archaeologist's description of their route, their perception, their decision. But they know that now, right now, they can hold on to their leadership not by largeness of vision or logic of plan but only by the judgment and approval of those who follow so far behind below the crest. That judgment, they know too, will be made not by reason but by emotion; so the leaders must devote as much of their time looking backward to make sure that the people continue to follow as looking forward to find where they should go.

Such migrations have been going on throughout history—always accompanied by the struggle at the head, the debate about the route, the wailing children. Neither Orgetorix, who proposed to lead the Helvetii down from the Alps through the defile of the Juras onto the fields of Provence, nor Divico, who finally led them, saw that Caesar could so swiftly reinforce his single garrison legion with five more from over the Alps; nor did the people who trudged after Divico in good faith know that their women and children would so soon be caught and sold by Caesar into slavery. Nor did the Germans who gave their heart to Adolf Hitler ever even vaguely imagine where that parade would end. Nor did the Americans who chose Franklin D. Roosevelt, on his strange promises of 1932, know how swiftly he could turn the procession about and find a new way to the uplands once he spied them from a leadership position at the head. Nor did the British know what lay ahead when they gave their hearts to Churchill for his courage and found not death but liberty.

This conflict between what lies ahead and the needs of those on the march behind is the continuing problem of the working politician—and offers the stuff of art to the historians who tell it later.

This conflict perplexes men in politics at all times—the contradiction between what they must talk about to hold the hearts of their followers and what they see ahead and cannot talk of because so few will understand.

Once during the campaign of 1964 I was talking to Daniel P. Moynihan, Assistant Secretary of Labor, one of the most luminous and perceptive of the younger American officials in Washington. Moynihan (who is only thirty-eight) has himself come a long way within the procession, from the tenements of the lower West Side in New York to chief of Policy Development and Research for the Department of Labor, where his scholarly mastery of the data that define America's future is unmatched. Possessed at once of the vision to see forward and a memory that can recall the smell of cabbage cooking in a crowded flat, he was deeply involved in the politics of electing Lyndon Johnson in 1964. We were talking about the polls one day in October, when the verdict was already clear, and Moynihan answered a question of mine:

"What are the issues in this campaign? 'Issues' are talk about what's already happened or happening. Our poll takers bring back samplings of what people think about 'the issues'—about waste in government, or about Bobby Baker, or jobs, or civil rights, or Communism, or whether they're for or against the United Nations, or foreign aid, or Social Security.

"But these aren't the issues, really. Only a handful of people can see the advance issues. Can you explain that the greatest issue twenty years from now may be what's beginning in our knowledge of the human cell, and biology, and reproduction? Can you explain that we're beginning to be able to control our environment, maybe even change the weather—and discuss what we should do about it? Or can you talk about what we have to do to keep old people from growing lonely? Or can you ask them whether they think the purpose of industry should be changed from making *things* to making *jobs?*

"Maybe we're entering a new phase of government. Maybe the old legislative phase is coming to an end, the time when you passed a new law which set up a new bureau with a new appropriation to run new machinery. What lies ahead may be problems not answerable by law, or by government at all. But that's nothing you can discuss now in 1964— that's years and years ahead."

Historians commonly refer to those who lead the procession on to new terrain as "statesmen"; they stigmatize with the term "politician" those who spend most of their energies holding the people to the line of march. Yet I know of no great statesman who was not also a great politician; and I know few politicians whose dreams are not teased by what they can see ahead in the future.

This book, then, is about the procession as it moved in 1964 in America—from the crest of John F. Kennedy's vision, to the alternate routes offered from that crest by Lyndon Johnson and Barry Goldwater, to the final judgment of the people behind on which of the two was the better choice for leader. It concerns itself primarily with the politics of the year 1964—the quarrels of the leaders, the rivalries of the group chieftains, the riots in the streets, the bitterness of race clash. The story of what may lie ahead, which is also politics, I have left to the scouts, the scholars and the philosophers whose duty it is to define the future.

No one man can write any such book as this with any claim to authority. The movement is too vast for anyone to report it all—all the states with all their problems, all the ambitions of all the men, all the gossip around all the campfires, all the hurts and aches and greediness, all the meetings at which decisions were made. This book is the result of two years of one man's reporting and travels. And, even so, it would have been entirely impossible without the help of countless friends.

The first of these friends are the newspapermen and reporters who have been my companions over the years. The generosity of these men with their precious impressions and observations can be appreciated only by members of the brotherhood. I cannot list them all; but I would be remiss if I did not publicly thank: Hugh Sidey first, and then Murray Gart, Richard clurman and John Steele of *Time*; to Nick Thimmesch, Benjamin Bradlee and William Tuohy of *Newsweek*; to David Wise, Douglas Kiker, Richard Dougherty of the New York *Herald Tribune*; to Mary McGrory of the Washington *Star,* and Carroll Kilpatrick, Chalmers Roberts of the Washington *Post*; to David Maness and Edward Thompson of *Life* Magazine to Felix Belair and Earl Mazo of *The New York Times;* to David L. Wolper and Mel Stuart of Wolper Productions; to Robert Healy of the Boston *Globe;* to Charles Bailey of the Minneapolis *Tribune* and Philip Potter of the Baltimore *Sun;* to Scott Blakey of the Concord *Monitor,* Peter Lisagor of the Chicago *Daily News* and Walter Ridder of the Ridder newspapers; to Blair Clark and Bill Leonard, Louis Harris and Eric Sevareid, Mike Wallace and Walter Cronkite, all of CBS; Edward P. Morgan of ABC and Sander Vanocur of NBC. We live, all of us, in such simultaneous interdependence and rivalry that it is everlastingly incomprehensible to me how friendship manages to bind us together in affection and common cause. But, obviously, any such book as this would be impossible without the trust and help of such friends and companions.

It is more difficult to thank the men of government and politics. Their most truthful reflections are generally offered only in privacy, and their interests are best served if I offer my thanks to them privately. Nonetheless, the corps of public press officers, both in government and in campaigning, requires a public expression of my gratitude: Pierre Salinger, George Reedy and Malcolm Kilduff of the White House have been of enormous courtesy and help. So, too, always were Samuel Brightman, press officer of the Democratic National Committee, William B. Prendergast, formerly Research Director of the Republican National Committee, and William Perry of the Senate Periodical Press Gallery. Edwin Guthman of Justice and Robert Manning of State were exceptionally helpful. Robert McManus of Governor Rockefeller's staff, Tony Smith, Victor Gold and Paul Wagner of Senator Goldwater's staff, William Keisling and Jack Conmy of Governor Scranton's staff all did everything in their power to help make this book a responsible one. And Robert Jensen of Vice-President Humphrey's staff has performed for all of us correspondents far beyond the call of duty.

Lastly, I must thank my staff—which is Mrs. Constance Hellyer Corning, an ever cheerful, untiring young lady whose scholarship equals her dedication. Her passion for accuracy and good phrasing has given this book much of whatever merit it may have; and whatever mistakes it may contain, either errors of fact or interpretation, are my responsibility alone.

New York, May 1, 1965

THE
MAKING OF
THE PRESIDENT
1964

CHAPTER ONE

OF DEATH AND UNREASON

I T WAS hot; the sun was blinding; there would be a moment of cool shade ahead under the overpass they were approaching.

But the trip, until this moment, had been splendid. For the President was beginning, with this journey, the campaign of 1964—testing the politics of his leadership, and hearing the people clap in the streets.

He had just turned easily, but with grace and precision as was his style, to wave at the Texans who cheered him—when the sound rapped above the noise.

It was a blunt crack, like the sound of a motorcycle backfiring (which is what his wife thought it was), followed in about five seconds by two more; then, suddenly, the sniper's bullets had found their mark and John Fitzgerald Kennedy lay fallen, his head in his wife's lap.

There is an amateur's film, 40 frames long, twenty-two seconds in all, which catches alive the moment of death. The film is soundless but in color. Three motorcycle outriders come weaving around the bend, leading the black Presidential limousine; a gay patch of background frolics behind them—mint-green grasses, yellow-green foliage. The President turns in the back seat, all the way around to his right, and flings out his hand in greeting. Then the hand bends quickly up as if to touch his throat, as if something hurts. His wife, at this moment, is also leaning forward, turning to the right. Slowly he leans back to her, as if to rest his head on her shoulder. She quickly puts her arm around him, and leans even farther forward to look at him. Then, brutally, unbelievably, the head of the President is jolted by some invisible and terrible second impact. It is flung up, jerked up. An amber splash flicks in a fractional second from his head into the air. One notices the red roses spill from her lap as the President's body topples from sight.

That is all. But one who has seen the filmed action knows that though, technically, his pulse beat for another twenty minutes and some flow of blood went on, President Kennedy had ceased to be from the moment the second bullet entered his skull. He had died quickly, painlessly, perhaps even without consciousness of his own end. The faint recall of the President's wife a few days later may be the most accurate recapture of his sense of the moment—she remembered that, as he turned between shots, an expression of puzzlement, almost quizzical, crossed his face.

The time was 12:30 in Dallas—1:30 P.M. Eastern Standard Time, Friday, November 22nd, 1963. American history would be punctuated forever by this bloody date. Government would go on; but its direction, now, at this moment, was ruptured. All the next year and for years thereafter Americans would debate that direction.

The medical term later used for the President's condition on arrival at Parkland Memorial Hospital Trauma Room Number One was "moribund"—which means that all the skill the doctors showed in the next twenty minutes was only an exercise in medical technology. "Moribund" was a word he would have loathed. He had said once after the stroke which crippled his father, Joseph P. Kennedy, that he would not want to survive such a blow; rather he would die quickly, completely, at once. He was alive to the very end—entirely alive, totally alive, alert, searching and seeking. And, thankfully, the bullet that tore out the brilliant mind of the 35th President of the United States took his life with it on its way.

What was on that mind at precisely the moment it ceased to be we shall never know. He was in Dallas on the kind of mission only an American President can set for himself—its purposes stretching from the grubbiest roots of American politics to the endlessly distant reaches of outer space.

He was there, as we know, chiefly for politics. The President had insisted that Texas was going to play a "very vital role" in the election of 1964; but the state's Democratic politicians were quarreling with each other. The Lone Star State had, in the post-war years, lived through an industrial revolution, and new industries, new sciences, new cities were gathering together new kinds of people. The Republicans had in 1961 elected a United States Senator from Texas for the first time since Reconstruction—and the growing political unrest of the South threatened the Democrats' traditional hold on this, the greatest state of the Southland. Yet the Democrats were split between their liberal Senator Ralph Yarborough and their conservative Governor John Connally; these two hated each other, and the President, as their national political leader, had come, as Kenneth O'Donnell later testified, "to bridge the gap" between them.

Yet there was more than that to the trip. The President was trying to reach the people, too. He believed in showing himself to crowds around the country so that people could get the feel of their President, and he of them. He had been planning this trip for weeks now; Kenneth O'Donnell, the master of his tours, had arranged it, and one of his favorite advance men, Jerry Bruno, had prepared the way in this city, Dallas. Though Dallas had a bad reputation, the President had thought it was a reputation made by a few noise-makers; the average American in Dallas, he felt, was about like the average American anywhere else. Thus, he had been delighted by the throng in downtown Dallas—friendly, cheerful, respectful, "one of the heaviest I have seen in any American city," O'Donnell recalled later. It had been that way the day before, too, in Houston; and when the President had asked his companion, Dave Powers, what he thought about the size and wild enthusiasm of the crowds, Powers had quipped, "You're doing just about average—but there's a hundred thousand extra here for Jackie."

And there was a broader purpose even than this to his visit—for he was the leader of all Americans and he wanted, from Texas, to discuss outer space and America's role in a changing universe. It irked him that the Russians had outstripped America in the black and measureless reaches of space. The newest space center of the nation had been placed in Texas—so, from this platform, the President proposed to tell the nation how he hoped, soon, to loft the most powerful rocket ever built, the Saturn, into the sky, hoping thus to gain a lead in mankind's first primitive probing of the galaxies.

His mind may well have held thoughts about all these matters at the moment—as well as an instant crowd estimate. But there was so much more in that mind. All things. Care for the old and concern for the black; missiles and destruction and beauties and plazas; cities and personalities and budgets; peace and war and education and wonders. The range of the mind was immense, its reach dazzling.

The mind had become part of all Americans' thinking in the three brief years of his Presidency—their link with the future. It was as if, at that moment, the link had been ripped away. Yet it was not in response to his mind that America now reacted; it was to the man. For the man had loved America, and now, suddenly, Americans became aware of their answering love. It is with this instantaneous surge of emotion that the politics of 1964 began.

Within the thirty minutes of his passing, between the crack of the first shot and the intoning of Extreme Unction by a priest, the tragedy had been told all across America. The news struck the East Coast at mid-lunch, the Pacific Coast in mid-morning; it stopped all presses in their run; interrupted all TV screens; stabbed with its pain men and women in the streets, knotted them about TV sets and paralyzed them in postures first unbelieving, then stricken. Not until he was dead and all men knew he would never again point his forefinger down from the platform in speaking to them, never pause before lancing with his wit the balloon of an untidy question, did Americans know how much light the young President had given their own lives—and how he had touched them.

It was, for Americans, an episode to be remembered, a clap of alarm as sharp and startling as the memory of Pearl Harbor, so that forever they would ask one another—Where were you when you heard the news? They learned from radio, from television, from one another. Telephone switchboards clotted within hours. Some parents tugged little children directly to church; others lowered the flag over the green lawns of suburbia. Taxi drivers caught by red lights in city traffic passed the latest bulletin from one to the other. In New York a young woman was driving up Manhattan's East River Drive when her car radio reported that the President had been shot; when she reached the toll gate of the Triborough Bridge, a detail of guards had just left their stations and were lowering the flag to half-staff. She burst into tears, for now he was dead; and she remembers the drivers of cars in adjacent lanes sobbing, too. In Chicago a reporter on Lake Street noticed a passer-by approach a shabbily dressed Negro who held a transistor radio. "What's the news?" asked the passer-by. "He's dead," said the Negro, and then after a moment he gripped the stranger by the lapel and said, simply, "Pray, man." At Harvard, under the elms of the Yard where he had strolled as a student, no one spoke; on the grass under the elms only the gray squirrels stirred. A British reporter heard the bells in Memorial Hall begin to toll and saw the flags fall to half-staff. One student hit a tree with his fist; another lay on the grass on his stomach—he was crying. Uncontrollably, across the country, men sobbed in the streets of the cities and did not have to explain why.

It touched all men directly—but some more pointedly than others. Nelson Rockefeller, the first announced candidate to challenge John F. Kennedy, was lunching privately with Thomas E. Dewey in his personal dining room on 54th Street, seeking the former candidate's help to win the new nomination of 1964—when a Negro maid brought them the news, crying. Rockefeller issued an immediate statement and drove home to Tarrytown. Barry Goldwater heard the news in Chicago while on an errand of private sorrow; he was accompanying his wife and the body of his mother-in-law, who had just died, from Phoenix, Arizona, to Muncie, Indiana; he was supervising the transfer of the coffin from the airliner to a private plane at the airport when his nephew brought him the news. William Scranton was whispered the news as, in mid-sentence, he presided over a lunch in Harrisburg, Pennsylvania; he finished the sentence; said, "Gentlemen, the President is dead"; called for a minute of prayer; and left to go directly to the Pine Street Presbyterian Church. Richard M. Nixon had left Dallas that morning at 9:05 A.M.—about two hours before the President's arrival—and had arrived at New York's Idlewild (soon to be renamed Kennedy International Airport) at one P.M. He heard the news in a taxicab, at 21st Street and 34th Avenue in Queens, when the cab stopped at an intersection and a stranger, not recognizing Nixon, asked the cab driver if he had his radio on, for he had just heard that Kennedy was dead. Eisenhower was at a United Nations luncheon at the Chatham Hotel in New York and Secretary-General U Thant was in mid-speech when the announcement came. Eisenhower left immediately for his suite at the Waldorf.

The event was to shape the activity of all these men, intimately and deeply, for the next twelve months. Among themselves they must find and offer alternative leadership for the American government in the next year; and the memory of Kennedy would press on and ultimately quench the strivings of all of them in the election to come.

In Washington, men were even more directly touched; they hung now on the radio reports which, minute by minute, were relaying each fragment of happening in Dallas. His body was being wheeled out of Parkland Hospital. A bronze coffin was being brought. The body was being taken to Love Field, then hoisted into the Presidential plane, Air Force One. The plane was taking off. It was due in Washington by 5:30 or 6:00 P.M.

Men, women and children across the country reacted as emotion dictated. But in the capital every man whose sense of duty or love called was speeding to Andrews Air Force Base, where the plane was expected. Through the gathering dusk of a moist and balmy Washington evening —the sky broken, the gray cloud bars washed pink by the setting sun— the black government limousines, the private cars, the taxis, raced to Andrews Air Force Base. Yet the scene that remains longest in this writer's mind, as he hurried from New York to Washington to the base, were the people along the road. There had been no announcement of the route the President's body might take from the field. Yet all along the way, on every autumn hill, sitting in the brown fall grass, were the youngsters, teen-agers and adolescents, in clusters of five to ten—waiting for him to pass by. Their white shirts in the gathering dusk fleck the memory still;

but it is mostly their silence that remains, the bubble and laughter of youth stopped in sorrowing stillness.

Guards patrolled the gates to the base; other guards patrolled the approaches to the field. The servicemen stationed at Andrews, as well as wives and children, rimmed the wire fence in front of the operations shack and backed up beyond. From its roof and windows, from every cranny and vantage point, they watched. Only once before had I seen military discipline relaxed or dissolved as completely—when the Japanese boarded the U.S.S. *Missouri* to surrender in Tokyo Bay, and the sailors and marines boiled up from the bowels of the ship to hang from masts, turrets, radar webs or sixteen-inchers and watch the gathering of history below them. There was, now, this background of common people, all silent in the darkness, and in the foreground the glaring pool of light flooded for the television cameras. And within the pool of light was a gathering of what, until a few hours before, had been the leadership of the government of the United States.

Government is at once the most powerful and magnificent creation of the human race—and the most delicate and fragile. Here it was, in all its fragility, dissolved into these individuals stripped, momentarily, of direction. From the Cabinet there were Robert McNamara, the young Secretary of Defense; Averell Harriman of State, tall, grim and gaunt, last man in Washington to carry his vigor through from Franklin Roosevelt's New Deal to John F. Kennedy's New Frontier; here, too, were George Ball of State, Postmaster General Gronouski, Anthony Celebrezze, Franklin D. Roosevelt, Jr. Here at the airport already were the leaders of the Congress: Hubert Humphrey, red-eyed with weeping; Mike Mansfield of Montana, ghostly thin; Everett Dirksen of Illinois, the thin curls of his hair disheveled and glistening above his ravaged face in the light; the Speaker of the House, John McCormack of Massachusetts, next in line of succession to the Presidency, suddenly very old and clustered with several other Congressmen as if for comfort.

Apart from them, on the apron, were other, younger men of the President's personal staff: Theodore Sorensen, white of face, unapproachably solitary; Arthur Schlesinger and Ralph Dungan, bleak and somber; McGeorge Bundy, ice-white but contained in some unspeaking remoteness of his own; Fred Holborn, stunned, almost aimless in his moving. They sifted about from group to group as journalists, equally aimless, wandered among them, and within moments would leave one group, join another, then depart and mill about, seeking answers to questions they could not shape.

Government turned by itself as they waited; deep-rooted, orderly,

shaped by centuries of law and custom and tradition, the men who made
its parts responded to duty by instinct.

Government took precedence over grief. From Love Field in Dallas
the new President had telephoned his Attorney General to express con-
dolence—but also to ask for legal advice. Robert F. Kennedy had received the call through the wound and shock of his loss, asked for a
pause for consultation with counsel, replied within minutes that the Department of Justice urged, from a legal viewpoint, that the new President
take his formal oath of office there, in Dallas, before take-off.

Far out over the Pacific, at coordinates 2018 North, 1725 West, an
Air Force VC 137 had banked like a fighter plane at 35,000 feet altitude
and was now speeding back to the mainland at 625 miles per hour. It bore
among its thirty-three passengers more members of the United States
Cabinet than had ever joined together before in a single mission overseas.
Secretaries (of State) Dean Rusk, (of Treasury) Douglas Dillon, (of Interior) Stewart Udall, (of Agriculture) Orville Freeman, (of Commerce)
Luther Hodges, (of Labor) Willard Wirtz. Press secretaries Pierre Salinger and Robert Manning, White House advisers Walter Heller and Mike
Feldman had all chosen that relaxed moment of politics and foreign affairs
to fly to Japan for a meeting with that country's cabinet. Now, without
instructions, but knowing the government required them, they were
speeding home as fast as jets could thrust them.

At the Pentagon the military machinery stood at global readiness.
On the news of assassination, the Director of the Joint Chiefs of
Staff had immediately, suspecting a coup, warned all of the nine great
combat commands of the United States, which girdle the world, to hold
themselves in readiness for action. One of them, on its own initiative,
sirened its men to Defense Condition One, or combat alert. Within half
an hour the command was called to order and restored to normal readiness. In Pennsylvania, state troopers sped over the roads to throw a
guard around the farm of Dwight D. Eisenhower lest assassination be
planned for him, too.

Across the nation and across the world, no patrol flickered: no sub-
marine was called off station; no radio operator reported a failure of
monitoring or communication; in Vietnam the war went on. At the White
House, where the housekeeping administration had taken the occasion
of the President's absence to redecorate and repaint his office, service
personnel continued to replace his furniture and rehang his paintings as
if he were not dead. The Military District of Washington drew from its
contingency files the fully elaborated plans for ceremony and procedure
at the death of the Chief of State. Nearly all that night a group of seven
scholars in the archives of the Library of Congress were to research every

document on the funeral rites of Abraham Lincoln in 1865, so that
historic tradition might be joined to the formal military plan. Sometime in
the next few days the machinery of the Accounting Office, turning by
itself, brought in its final results too, crediting John F. Kennedy, President,
with pay up to November 22nd, 1963, for service performed, and paying
him that day for precisely 14/24ths of a day's work, death being assumed
as 2:00 P.M. Washington (Eastern Standard) time, rather than 1:00
P.M., Dallas time.

There is a tape-recording in the archives of the government which
best recaptures the sound of the hours as it waited for leadership. It is a
recording of all the conversations in the air, monitored by the Signal
Corps Midwestern center "Liberty," between Air Force One in Dallas,
the Cabinet plane over the Pacific, the Joint Chiefs communication center and the White House communications center in Washington. The
voices are superbly flat, calm, controlled. One hears the directions of
"Front Office" (the President) relayed to "Carpet" (the White House)
and to the Cabinet above the Pacific. One feels the tugging of Washington seeking its President—tugging at the plane, requiring, but always
calmly, estimated time of take-off and time of arrival. One hears the voices
acknowledge the arrival of "Lace" (Mrs. Kennedy) on the plane; one receives the ETA—6:00 P.M. Washington. It is a meshing of emotionless
voices in the air, performing with mechanical perfection. Only once does
any voice break in a sob—when "Liberty," relaying the sound of "Carpet"
to the Pacific plane, reports that the President is dead and then Pierre
Salinger's answering voice breaks; he cannot continue the conversation,
so that the pilot takes the phone and with professional control repeats
and acknowledges the message, as he flies the Cabinet of the United
States home in quest for leadership of the government of which they are
so great a part.

So, too, in uncertain quest for direction and purpose stood the rest
of the government of the United States on the apron at Andrews Base, a
shifting, tearful, two-score individuals, all masterless, no connection between them, except the binding laws of the United States, until Lyndon
Johnson should arrive and tell them in which direction he meant, now
and in the next year, to take the American people. Out of that direction
would come the politics of 1964—except that, whichever way he went,
he would have to start from the politics of the man now being borne,
dead, to this city.

No bugle sounded taps, no drum beat its ruffle, no band pealed
"Hail to the Chief" as John F. Kennedy returned for the last time to
Washington, the city where he had practiced the magic art of leadership.

It was seventeen years earlier that he had arrived there from Boston; and in the years since, his arrivals and departures had come to punctuate the telling of American history. When he arrived, the door of the plane would open and the lithe figure would come out to give that graceful wave of the arm which had become the most familiar flourish in American politics. There would follow then the burst of applause, the shouts and yells and squeals as he tripped down the stairs with the quick, light step that was his style.

He came this time in silence.

The faint shrill of distant jets, the sputter and cough of helicopters, the grunting and grinding of trucks on the base, the subdued conversation all made the silence even larger. A quarter moon, trailing a shawl of mists, had just made its appearance in the sky when Air Force One, the Presidential jet, silently rolled up the runway from the south. It had paused at the far end of the field to let Robert F. Kennedy come aboard. Now the pilot in his cockpit had sensed the hush—skillfully he stilled his motors, so that the plane glided surprisingly into the total glare of the lights and soundlessly came to a stop. It was six o'clock Washington time, only four hours, no more, from the moment the President had been pronounced dead.

One wished for a cry, a sob, a wail, any human sound. But the plane, white with long blue flashings, rested under the punishment of television lights—sealed and silent. A yellow cargo lift, whose enameled interior gleamed with white light, rolled as far as the plane and waited.

A door opened in the rear of the plane. A man appeared and for a moment it was as it had always been—the round face of Lawrence O'Brien peered out first. But O'Brien stooped down and, as he moved, lifted something. For the first time the ugly glint of the dark-red bronze coffin showed. Behind O'Brien was Dave Powers; and then Kenny O'Donnell. These men had followed John F. Kennedy, lieutenants in his quest for power, all the way from Boston to Washington over a decade, and from Washington about the country for three more years; they had carried his briefcases and his papers, his errands and instructions, his affection and trust, his soup and his clothes. This time, in last service, they carried the President himself.

They set the coffin down gently on the floor of the giant box. It jounced, then steadied, then began to settle quietly to the ground with its burden and those who accompanied it. An honor guard of six reached out their hands to receive the coffin on the ground. The coffin wobbled in the passing from group to group, and O'Brien's hand, almost caressing, reached out as if to steady a tumbling child; then, not being needed any more, the hand fluttered uselessly in the air.

There was still no sound, except the broadcasters pattering as quietly as possible into their microphones. The silhouettes on the edge of the lift parted and a slender woman in a raspberry-colored suit appeared, then hovered at the lip of the low floor of the box. The Attorney General, her brother-in-law, was there first to help, offering his hand to steady her down. He guided her into a gray service ambulance after the coffin had been placed inside. She tried to open the ambulance door, but her hand fell limply. Bobby leaned forward, opened the door, helped her in.

One watched and knew the nation was watching. One felt the group swaying imperceptibly after the ambulance. So many of these people had for so long been in the unquestioning habit of following immediately behind him on any tour, journey or expedition. His dead body and physical presence drew them after, by an invisible pull of obedience and love. It had been the custom of the reporters gathered here to gallop or sprint after John F. Kennedy wherever he moved in public. Yet now they only turned, held to the ground, immobile. The men of government were equally perplexed. In the foreranks of the shuffling group stood McNamara, Secretary of Defense, his jaw muscles flexing, otherwise rigid and unmoving; the others took lead from him, as they must. On the last heartbeat of John F. Kennedy, Lyndon Baines Johnson had become President of the United States, its 36th. For this new leader these individuals now waited, their silence still unbroken. The gray naval ambulance, its red dome light steady and unwinking, rolled off to the north with its burden, followed by four black limousines of family and personal staff. No one else followed. Then, after the passage of a few minutes, another door of the Presidential jet opened and Lyndon Johnson appeared.

The lights of the television cameras had gathered in front of the plane. Into these lights strode the new President and his lady. The new President spoke, assembled the advisers of his choice and departed by helicopter for the White House lawn. The cameras turned to the White House.

Somber, bleak and yet fully ablaze as it would not be again for a long time, the White House, in all its empty majesty, haunted the nation's vision that night. Lights played on the fountain of the north lawn. The chandelier over the great front entrance was dim but lit. Upstairs on the third floor the blinds and curtains were drawn, but one could see that here, where the President and his family had made their home, men stirred. The East Room, too, glowed through the shades as old friends prepared for the reception of the body.

On the front walk, along Pennsylvania Avenue, a reinforced police guard patrolled the iron palisade that fences the grounds. But it had no need to act. The White House was patrolled by the mourners—and all evening, until long after midnight, a subdued float of individuals, families and couples paced back and forth. And the autumn leaves, which begin to thin out in Washington in November, seemed particularly thick as they rustled and rattled and rolled over the sidewalk.

Late that night the same wind brought rain. It was to gush and drench the unhappy city all the next day, Saturday.

The spectacle of the next three days is so new to memory that to retell it falls impossibly short of still-fresh emotions. What will be difficult for historians to grasp, however, was that the ceremonies that followed were more than spectacle—they were a political and psychological event of measureless dimension. And in this event the chief servant was American television, performing its duties of journalism with supreme excellence.

Within minutes of the shot, American television was already mobilizing. In half an hour all commercial programs had been wiped from the air, and thereafter, abandoning all cost accounting, television proceeded to unify the nation.

When Lyndon Johnson arrived at Andrews Base, emergency crews of the telephone company and the networks had barely finished connecting the wires to the field installations, and as the plane arrived, they simply pointed their transmitters to the relay points in Washington—and hoped. By then, from all over the country, technicians were en route to Washington and the central headquarters of pooled TV coverage; by late Saturday over 1,500 men of all three major networks, 44 mobile cameras, 33 taping machines were established in the capital to feed the nation what its hungering emotions required. And beyond the Washington broadcasting pool stretched the uncounted mobile camera units and newsmen of the affiliates and independents from coast to coast.[1]

---

[1] The response of the American people to this effort was as unbelievable as the exertion itself. The Nielsen Audimeter Service has made available an hour-by-hour survey of audience response in the New York area (which, with 10 percent of the nation's television sets, is a fair sample of the whole). By 2:15 P.M. on Friday, within twenty-five minutes of the first report, the television audience had doubled, and by the time the President arrived at Andrews Air Force Base, 70 percent of New York (and presumably America) was at its sets. The next day, Saturday, almost half of all sets in America remained on for the full twelve hours from eleven in the morning to eleven at night. These peaks were surpassed on Sunday, when, for the ceremonies in the Capitol Rotunda, 85 percent of all television

The political result of this participation, of this national lament, was a psychological event which no practical politician will ever be able to ignore. Out of it began the Kennedy myth and the Kennedy legend. Learned doctors have already assembled to discuss the psychological impact on masses of people of this emotional outburst. It is best to summarize their thinking simply and report that the drama gave all people a sense of identification, and translated the majesty of leadership into an intimate simplicity of Biblical nature. There was in the drama of the four days all things to bind all men—a hero, slain; a villain; a sorrowing wife; a stricken mother and family; and two enchanting children. So broad was the emotional span, embracing every member of every family from schoolchild to grandparent, that it made the grief of the Kennedys a common grief. And this is the stuff of legend—a legend which, in the case of John F. Kennedy, has already begun to transform the man and will transform him yet further.

Beyond the beginning of the myth, one must stress another major political result of television's stupendous exertion. One has only to recall other assassinations, domestic and foreign, to remember the confusion, the bitterness, the wild rumors, the instability, the violence so often following the slaying of a chief of state so great. Ordinarily, television, by the sinful laws of its existence, must make artificial drama out of all normal political events in order to hold audience attention. But it achieved greatness in November, 1963, by reporting true drama with clarity, good taste, and responsibility, in a fashion that stabilized a nation in emotional shock and on the edge of hysteria. By concealing nothing; by sharing all; by being visible when their private natures must have craved privacy, Jacqueline Kennedy first above all, then the grief-stricken Kennedy family, then the new President permitted television to give strength and participation to the citizens. For this unconsidered and magnificent political benefit, television can take credit second only to Jacqueline Kennedy, the grace of whose leadership in the proceedings was the noblest of all tributes to her husband's spirit.

Saturday was a day of private sorrows and indoor concealment under Washington's drenching rain.

---

homes in America became live witnesses. And then, finally, for the funeral and burial ceremonies on Monday, the graph of audience attention reached its limit—a measurable 93 percent of all American television sets, the highest total in TV history, brought these ceremonies to what must have been more than 100 million people. And beyond these was yet another audience which no one has even attempted to measure—for electronics not only brought, live, to America the benediction of Pope Paul VI from Rome, but recast the American ceremonies live by the same Telstar to twenty-two foreign lands, among them Russia and Japan. Nothing like this had happened in the history of the human race—so large a participation of mankind in the mortal tragedy of a single individual.

Friday the wound had been so sharp and new that most emotions were numb—but Saturday was the day that men cried. Now their nerve ends were coming raw—and aching. Men who had, dry-eyed, performed their duties all through Friday night broke on Saturday uncontrollably. One would wander through the White House, seeking old friends for the solace of company and then, finding them, be unable to talk because, on sight, we wept.

A private Mass had been celebrated for John Fitzgerald Kennedy at 10:30 in the morning in the somber East Room. Thereafter the room was opened to an endless procession of his great and lesser friends, to all who were known somehow to have held his affection or respect. The enormous room, its polished floors gleaming, was hung with black draperies; its huge chandeliers were dimmed; in the center on a dais, draped with the American flag, rested the coffin. Five guards of the armed services attended it. A few feet south of the coffin were two prayer stools for those who wished to kneel and pray. Most people, however, chose to file by quickly in silence, occasionally crossing themselves, then disappearing out the ground-floor south entrance.

Many of those familiar with the White House filtered back to the official lobby in the West Wing—and there, meeting each other again, wept again. All that afternoon, as the rain came down soaking the city, the lobby smelled of dampness and of people. The Secret Service had by now concentrated its protection on the new President across the way in the Executive Office Building. And the White House guards, whether through sadness or kindness to men they recognized as old friends of the President, let them wander almost at random through previously barred corridors of the working West Wing of the White House. More open than ever before, the working wing was, however, emptier of purpose than ever before, too.

A wanderer had the impression of the macabre and the necessary all grotesquely confused. Already by Saturday afternoon the Oval Office where the President governs was changing. Looking in as one passed, one could see that it was being stripped—of John Kennedy's ship models and ship paintings, of his famous rocking chair, of his desk adornments, then of his desk, until by late afternoon only his two white facing sofas remained. Already the Kennedy personal files and cases, sealed the night before at 9:30, were being moved across the street to the Executive Office Building, as were all the belongings of his secretary, Evelyn N. Lincoln, and her desk and her working place.

In the confusion, however, a few men had begun to pull themselves together. Some time earlier President Kennedy had gone with his good friend Robert McNamara to visit the Arlington National Cemetery in Virginia, and they had lingered on a slope just under the Lee mansion overlooking the Capitol. Kennedy had been enchanted with the unfamiliar panorama of Washington as seen from that particular spot. McNamara now remembered this visit, and with Robert Kennedy, on Saturday afternoon, he trudged through the downpour to inspect the site. The Attorney General agreed this was the place for his brother to rest. So, on his return to Washington, he and Mrs. Kennedy persuaded the rest of the stricken family that there, in Arlington Cemetery, rather than in the family plot in Boston, should be the grave of the 35th President of the United States.

With this decision, the elaborate preparations for the pomp and ceremony of the next two days could proceed, and next morning, Sunday, brisk and blue, the skies cleared so that the nation could witness the pageantry of death.

All across the nation from Friday on, celebration had been extinguished, parties canceled, sports events (including the Harvard-Yale game) wiped out, Times Square blackened, night clubs closed. By now, Sunday, emotion had fused the entire country and mourning was national. In Washington it was a brisk November day, the wind chilly but not uncomfortably so, the breeze fresh enough to flutter the flags at precisely the proper ripple as the cortege set out from the White House, or stiffen them wherever they flew at half-staff.

There followed then those indelibly visual scenes that will be seen and heard for generations, as long as film lasts and Americans are interested in the story of America:

§ The cortege itself, the flag-covered coffin, the six matched grays pulling and tugging and tossing their manes as if unwillingly dragging a burden in a direction they would not take; then the riderless black horse, its silver stirrups reversed, prancing and willful, turning and twisting, yet held to course by its handler.

§ The dome of the Capitol rising in the distance; the arrival of the cortege at the Capitol forty-five minutes later; the President's lady, the haze of a black veil shielding her face, head held high, her children's hands in hers, moving soundlessly through tragedy. One came to search out her presence, not from that reflex that turns curious eyes to the grief of famous or beautiful people, but drawn by the quality of her strength and her presence—real in image, yet more than real. And the gesture as she stooped to kiss the flag that draped the coffin, repeated a moment later by Caroline, stooping, touching the flag to her lips in reverence.

§ And Washington that evening: the serpentine of people winding through the Rotunda to pay respect, a quarter of a million of them, from every corner of the Eastern Seaboard ("I didn't know my son cared for anything in the world," said one mother later, "and I was so proud when I learned afterwards that he and his friends drove down from Lehigh University and stood in line all night just to say goodbye"); one of the guards, who had stood ramrod-stiff and stone-faced at the coffin throughout his tour of duty, breaking into uncontrollable tears as soon as he was relieved and could march away.

Monday was for farewell; and, as Mary McGrory wrote, "Of John Fitzgerald Kennedy's funeral it can be said that he would have liked it. It had that decorum and dash that were his special style. It was both splendid and spontaneous. It was full of children and princes, of gardeners and governors."

The skies still held clear and blue, with a nip of frost in the air, and one million citizens sat or stood to line the long route from the Capitol to the White House to St. Matthew's Cathedral to the Arlington Cemetery.

Now began the sound that echoed and will echo forever in every memory: the sound of the *klop, klop, klop* of horses' hoofs; the sound of the muffled drums' cadence, 100 to the minute; the sound of funeral music occasionally broken by skirling bagpipes. This was the sound of Washington and the nation all day long as the screens showed the procession moving, then repeated the sight over and over again into the night.

For those who watched on television—and they were the ones who saw the ceremonies whole—the high passage stretched from the White House to the cathedral to the cemetery. At the White House the cortege paused and Mrs. Kennedy, President Johnson and other members of the family climbed out of the limousines that had brought them from the Capitol. Now, on foot behind the coffin, they proposed to walk the eight blocks to St. Matthew's Cathedral, where Requiem would be sung and the Church murmur blessings over the body before returning its soul to God. Behind the family trudged all the great of the world—and also the small and forlorn, unknown to fame, who had been his personal friends. Long and short, robust and frail, stooped and erect, they walked, kings and princes, queens and foreign ministers and generals. In the endless, spankingly smart military precision of banners, troops, guards, pomp, formation, step and parade-timing of the long day, these walkers, shuffling and ambling with neither pace nor order, reduced all glories and glitter to the stumbling pace of common mourners who must be shepherded on the way they go.

All this could be seen only by those who watched on television. At the center of the ceremonies in the cathedral, those who had come to be present at the Requiem, could only sit silent in their pews, rustling and shifting, until far off in the distance came the sound of muffled drums and above that, very soft, the first strains of the dirge.

The cathedral hushes, listening; the distant band falls quiet, and high in a corner of the cathedral the choristers begin, their pure voices singing Latin airs centuries old. When they pause, the muffled tattoo is louder, coming nearer; over the drums come the bagpipes, distinctly closer than the band before. After the bagpipes, those in the cathedral sing again, but the tattoo of drums is closer, always closer, intolerably, relentlessly approaching with the body. There is the fragrance of incense in the cathedral; Cardinal Cushing leads a file of prelates to the door to await the procession; now the band blares just outside—loud, immediate, insistent, *there*. It is playing "Hail to the Chief," but playing it slowly, with such a dirge-like beat that it takes seconds to recognize. Now the voices inside the cathedral, soft and mournful, contend with the brass of the band outside, one layer of music folding over another. The voices of song die away; the band hushes. Black-veiled, his wife comes, head high, leading a child with each hand. She is followed by her family; by the new President; and the church weeps softly.

An interminable few minutes is needed to seat the dignitaries of the great world; then comes the coffin.

And now the Mass begins.

The Mass is said by Cardinal Cushing, friend from childhood of the dead President, and the tall, gnarled figure chants with the harsh, dry inflection of the Boston Irish homeland. For those in the cathedral, this is the moment of passing. Century upon century of liturgy and faith have graven these Latin words in this form, have taught this high priest the melancholy drone that now fills the air, have instructed the singers in the sad melodies that now and then break the service. This is ritual —a poultice on the memory and on the hurt.

When prayer had implored God to take back the spirit of this man, John Fitzgerald Kennedy, it was left to those present only to carry his body back to earth.

At the grave site on the slope, in the sere grass and dry leaves of autumn, Irish Guards and Black Watch, Third Infantry troops and green-bereted honor guard of the Special Forces lined the grass plot to do honor. At ten minutes of three the Air Force flung fifty jets, one for each state of the Union, through the sky over the hill. The Presidential jet, Air Force One, followed instantly in a last salute. And then it was quick, all procedures timed meticulously: The casket, still flag-covered, was lowered to the ground. The Cardinal prayed. The guns thudded salute. The widow and the two brothers moved to the head of the casket and a rifle volley cracked the air. Then taps. Then the Navy Hymn. Then the flag of the coffin was stretched and snapped into its ceremonial triangle folds and given to Mrs. Kennedy.

She stood there a moment and someone gave her what appeared to be a rod; she touched it to the ground and a flare rose. She handed the rod to each of the brothers, and each in turn touched the burning rod to the flame that rose from the mound. For a moment the three of them grouped together—she veiled in black, and they slim and gaunt beside her—made a group atavism, calling to the blood, an echo of tribal life of primitive Gaels and Celts before Christianity gentled them. The brothers and the wife peered briefly at the orange tongue of flame, the wisps and fringes of black smoke, trying to postpone the moment of loss of the tribal chief. But they must accept it. And then she gave her hand to Robert Kennedy, who led her gently down the slope away from the casket.

It was a quarter of an hour later, after she had departed, that they lowered the casket holding John F. Kennedy into the earth; and that was the end.

Or perhaps not.

For no man there on that hillside but bore his mark—and would until the day he died. Here were all the rivals he had defeated in his race for the Presidency in 1960—Lyndon Johnson and Adlai Stevenson and all the other rivals: Hubert Humphrey, who held him, despite combat and contention, ever close in heart; Stuart Symington; Richard Nixon. The touch of John F. Kennedy rested on those who had been Presidents of the United States—Harry Truman and Dwight Eisenhower, who now at this place made friends after their angry separation and drove away together to have a snack at evening as old companions. As well as on those who aspired to be President of the United States in his place—Nelson A. Rockefeller with Mrs. Rockefeller, both of these usually smiling people somber and unhappy, pausing to talk to Averell Harriman, Rockefeller's predecessor as Governor of New York; the sad and composed William Scranton of Pennsylvania, upon whom, in these few days, men's attention had come to focus; Barry Goldwater, a perplexing man whom the late President had found personally delightful but whose philosophy he found absurd. Past Presidents, Future Presidents, Might-Have-Been Presidents, Also-Ran Presidents, President-Makers—all were here on this slope. But it was time to go, as the sun set behind the white columns of the Lee mansion, gilding the strange view of Washington one might see from this knoll—a Washington whose row upon row of government buildings made it a walled town in the distance, behind whose tawny parapets, like mysterious temples, rose the Lincoln Memorial and Washington Monument and the dominant Capitol, all as it had been before he came, apparently unchanged.

Yet it had changed. The changes were still invisible, not yet come to harvest.

For Eisenhower was the last of the "post-war" Presidents. The oldest President in American history had been followed by the youngest elected President. With Kennedy, the nation had begun to reach toward a new understanding of its own times; and it is over this change of understanding that we must linger.

All politics in 1964 would proceed from the new base of understanding that John F. Kennedy had established in the three short years of his leadership. The making of the President, 1964, was to take place in a new landscape of America. It was as if a starshell had soared into a dark sky and then burst, illuminating distances farther than one could ever see before. One remembers the light in the sky as a matter of romance; but one must examine the terrain below to understand how completely the perspectives of action had been altered. For one cannot tell the story of what was to come, either in 1964 or the years that will follow, without measuring the architecture and achievement left behind by the 35th President as legacies to his people and party.

We must, first, see John F. Kennedy as a protagonist in a drama of classic order, faced with two historic antagonists: the Communists abroad, with *their* dogmas; and the Congress of the United States, with *its* dogmas.

The dogmas of his antagonists made clear the quality of the protagonist. For John F. Kennedy, above all, was a man of reason, and the thrust he brought to American and world affairs was the thrust of reason. Not that he had any blueprint of the future, ever, in his mind. Rather, his was the reason of the explorer, the man who probes to learn, the man who reaches and must go farther to find out.

Reason held the inner circle of his court together. From the days of Ahasuerus to Charles de Gaulle, "The King's Ear" has always been the input center of political action. Over all the thousands of years of development among men toward civilization, two problems have permanently intertwined—first, how to bring the thinking of intelligent men to the place of power; and, next, how to enable power to impose a reasonable discipline on action. In Kennedy's time, The King's Ear was held

by men of learning and study, by as visionary a group of thinkers as
have ever held the ear of any chief in modern times. Without him these
thinkers would have gone their way in life as scholars, or commentators,
or administrators; bound to him, they established a new frame of American action in which his thrust and leadership gave purpose. At his
White House the dialogue of scholars was the atmosphere of political
decision.

Though immensely read (probably the best-read President since
Woodrow Wilson), he was no academician; yet he possessed the one
essential of the scholar—the ability to admit and learn from mistakes. I
remember one of my earliest meetings with Senator John F. Kennedy,
when I was still trying to understand him and criticized him for an early
attack he had made on two American scholars of the Far East as being
part of the web of Communist conspiracy about the State Department.
This had happened in his junior years as a Congressman, in the era of
Joe McCarthy. He shook his head in exasperation—but not at me, at
himself. "I was ignorant," he said, "I was ignorant; I was wrong; I was a
kid Congressman, and you know what that is—no staff, no research, nothing. I made a mistake—what else can I say?" And I remember talking to him one day much later at the White House, shortly after the fiasco
of the Bay of Pigs in the first Cuban crisis. He was depressed and melancholy and took on himself the blame for the disaster. "I was the last to
speak," he said. "It went around the table and everyone said 'yes'; but I
was the last to speak, and I said 'yes.' " Then he added, "I learned a lot
from it. It would almost be worth it for what I learned from it—except
that I can't get those twelve hundred men in prison out of my mind."

He was always learning; his curiosity was total; no one could
come out of his presence without coming away combed of every shred
of information or impression that the President found interesting—what
had the Communists told you in Asia, how fast could Communist gueril-
las march in a day, how had Jean Monnet made that national plan work
in France, what was happening in Harlem and the relationship between
Adam Clayton Powell and Ray Jones, who was going to play the lead
in a movie he was interested in, why did Adlai Stevenson capture the
imaginations of the eggheads but why did Irishwomen dislike Adlai so,
why was Lumumba able to make sense in the Congo, how was the promotion system working in the State Department?

And yet if there was no blueprint, he was a man not without direction. I never even once discussed religion with John F. Kennedy,
except in the practical political terms that made it a campaign issue in
1960. But religion was there—not in a parochial Catholic sense, but in a
profoundly Christian sense, of the kind now finding faint new sound in

Rome. It was a Catholicism neither of the sere and narrow Irish-Catholic tradition of Boston, nor of the amiable French tradition where the Church is a family inheritance that serves up baptisms, weddings and funerals. It was a Catholic faith more nearly kin to the new intellectual Catholic Left in Western Europe, out of which have come the Demo-Christiani parties, where the perspectives of a welfare state are joined to a sense of mercy for the weak and a disdain for the countinghouse men. In West Virginia, during his primary, when shown for the first time the dry relief rations given to unemployed workers, he gaped and said, "Do you mean people really eat this?" His juvenile-delinquency bills, his Medicare proposals, his interest in child health, his immense concern for family life all flowed from this sense of mercy and warmed the great bones of the ideas fashioned by American liberals. Many streams of culture and tradition joined in him: there was a touch of the English Whig from his reading of the eighteenth and nineteenth centuries; there was much of what Harvard had given him in his sense of history; and a good deal of the knuckle-dusting reflex of the Irish of East Boston and South Boston, where his grandfathers had led the Celts out of the slums to challenge the Saxons on the hill.

All of these instincts were led, of course, by reason. But what set him most apart from sedentary thinkers was his taste and zest for action. No one present in the old Boston Garden on the night of November 7th, 1960, the night before his election, will ever forget the howling mob of his home city hushing before him, and the candidate, so young, so pale, so worn at the end of his march, standing in the floodlight and pointing his long finger down at them and telling the home people in that clear voice and Boston accent: ". . . I do not run for the office of Presidency, after fourteen years in the Congress, with any expectation that it is an empty or an easy job. I run for the Presidency of the United States because it is the center of action, and, in a free society, the chief responsibility of the President is to set before the American people the unfinished public business of our country."

The Presidency was indeed the center of action in John F. Kennedy's day and he the protagonist of action. As the leader who had to act, he is best remembered for his deeds and personal decisions; but all these depended on another and often overlooked quality of Kennedy as President—his ability to find, direct and organize the talents of other men. As an executive, his taste in men was as fine as Franklin D. Roosevelt's. When he died, he left behind a team that was probably his greatest single gift to Lyndon Johnson. His Pentagon was, across the board, from McNamara through the Deputy Secretary to the Assistant Secretaries, the ablest civilian team to direct the Department of Defense since its founding—the equal, probably, of the great war team of Stimson-Knox-McCloy-Patterson-Forrestal. The revolution in civil rights could never have been achieved without the brilliant staffing of a Department of Justice that was simultaneously the scourge of the underworld, the trusts and the racists. (Kennedy had thought long before appointing his brother Robert to the post. Bobby had objected for days, on the ground that it would expose them to the charge of nepotism and "they would kick our toes off." Kennedy pondered the matter and then one morning called his brother and said, "You hold on to your toes— I'm naming you Attorney General today.") The choice of Willard Wirtz to succeed Goldberg, and the deputies Wirtz chose later, made the Department of Labor, for the first time, a powerhouse in American policy. Dillon and Heller made an outstanding—though at times quarrelsome—team in reorganizing the economy of America. Only State baffled Kennedy. He would defend State vehemently to outsiders; but, privately, it bothered him; its senior officers had grown up just before the war in a time when the doctrine of American policy was non-intervention, and they were trained not in action, but in observation. He would defend Rusk over and over again ("Bob McNamara has divisions he can move around; he can do things; but Rusk has to deal with one hundred different countries and you can't move them around the way you want to"); but, inwardly, he yearned for more action.

This, then, was the protagonist—restless, civilized, gay and witty, suffused in the talk of scholars, knowing his duty to translate their thinking to action. And the actions he took changed the landscape.

It is easiest to describe his action in face of the great overseas antagonist—the Communist world. It was in grappling with this adversary that he achieved dimensions larger than life.

In his struggle with the Communist world there is an artificial neatness that lends itself nicely to drama. One must begin with the first serious problem of his administration, the crisis in Laos, and his discovery that the power of the United States had shrunk to a spastic nuclear response. Americans could destroy the entire world with their bombers and their bombs—but they could not spare a single combat unit for Asian ground action and have any left for the greater crisis bulging in Berlin. America had become a muscle-bound giant that could only kill, inflexible and all-destroying in military response. There followed, then, the dismal errors and blunders and squalors of the Bay of Pigs—from which Kennedy learned that the art of war can never be left to administrators, whether generals or CIA—and the imposition of civilian rule on the Pentagon. There followed next the slow rebuilding of American military strength (the number of ready-to-go home land reserve divisions was to more than double in the first three years of McNamara's direction, and their flexibility and movability increased by a huge multiple).

While this strength built and flexibility returned, Kennedy had to live through the Berlin crisis and meet with Khrushchev in weakness. (When Khrushchev, at Vienna, explained to the President that the decoration he wore was the Lenin Peace Medal, Kennedy observed dryly, "I hope you do nothing to make them take it away from you.")

But then, finally, the strength was there, the new mobile divisions, the Strike Command, the cherished options and range of escalations, the responses open again to reasonable and civilized men who knew diplomacy can be effective only when backed with force. Thus, followed the second Cuba crisis of October, 1962, when the Russians were finally and openly faced down by superior force. Here, too, reason guided the strength as (while subject to domestic criticism which embittered him enormously) he made the Cuban confrontation a victory in which, nonetheless, some remaining dignity, some final escape from total humiliation was left to the withdrawing antagonist. Which, ultimately, led to the Matterhorn of the Kennedy achievement—the Test-Ban Treaty with Russia, which not only cleansed the skies of poison but, more importantly, cleared the way for that détente which points the way to peace.

War and peace were to be the dominant issue of the campaign of 1964; Barry Goldwater was to be destroyed by this issue. The threat of nuclear annihilation had been removed from the American mind, as from the Russian mind, and this the Americans would not permit Goldwater to raise again. It was only a partial peace that Kennedy left behind, as we shall see later, for Southeast Asia still festered. But it was the direction, the visible open movement and reversal of the post-war development, that was his greatest legacy: Peace—plus the two essential ingredients of peace: Power and Goodwill.

The second antagonist, at home, was entirely different—and so was the theme of his struggle. The domestic adversary was not the Republican Party, though in political talk Kennedy tried to make it appear so ("The Republican Party has made the word 'no' into a political program," he would declaim in the Congressional campaign of 1962). The domestic adversary was all the congealed past, all the native reluctances embodied in the Congress.

The theme of the struggle had been clearly spoken in his campaign of 1960—that America must move, that it must move from the postwar world into the newer world of tomorrow. How often have those who followed him heard him muse about tomorrow, when, in fifty years or less, America would have doubled her population. The schools had to

be built—*now*. The wilderness beauties of America had to be preserved
—*now*. The unskilled must be retrained—*now*. His scholars could describe quite clearly what he instinctively sensed: that the pace of change
in America was accelerating—industry changing, habits changing, tech-
nologies changing. Kennedy was no scientist—but he was a political
man in the highest sense. He knew that what science could uncover of
nature's wonders must be made useful, not destructive, for ordinary
people by politics. People in America in our time live so much longer
than they used to, and science can keep their feeble bodies alive so
much longer than they can afford, that the political man has no choice
but to pay attention to their growing torment and numbers (Americans
over sixty-five have jumped from 12.3 million to over 17 million in fifteen
years)—and thus Medicare. The air and the rivers are being polluted
by the wastes of our civilization; the cities sprawl beyond reasonable
government; loyalties and manners crumble under artificially stimulated
appetites. All these must be grappled with quickly. And beyond that—
space. There, too, America must move. This was his theme always—
movement and exploration, from his earliest campaigning in 1960 to the
day before his death, when, at San Antonio in Texas, he spoke of America's reach to the moon:

"Frank O'Connor, the Irish writer," said Kennedy, "tells us in one
of his books how, as a boy, he and his friends would make their way
across the countryside, and when they came to an orchard wall that
seemed too high and too doubtful to try and too difficult to permit their
voyage to continue, they took off their hats and tossed them over the
wall—and then they had no choice but to follow them. This nation has
tossed its cap over the wall of space, and we have no choice but to follow it." Kennedy, no more than the rest of his countrymen, knew what
lay behind the wall. But he insisted they move nonetheless, lest they be
trapped by the changes that would squeeze them against ancient walls
and crush our world to death.

Congress, his adversary at home, was always his most perplexing
problem—although the Communists were the most dangerous. Kennedy
entered on his first administration as a crippled political leader. His
Party had lost 20 seats in 1960, and its majority of 263 to 174 was only
nominal. Ever since the election of 1936, when the South first deserted
Franklin D. Roosevelt, any Democratic majority contained between 60
and 70 hard-core Southern Congressmen who would support a Democratic leader in foreign affairs but who went their own way on domestic
matters. On domestic matters, thus, Kennedy's leadership was always a
chancy thing, depending on two, three or five votes. Congress is the best

reflection of the structure of the United States; it is not a collection of fossils; it, like the country, is frozen by its past—Southerners with their fears, Negroes with their indignation, Catholics and Protestants divided by school controversy, farmers, lawyers, union men, suburban leaders, each man defending a constituency that has hardened in defense of its way of living, or hardened to attack. When one challenges Congress, one challenges all the American past.

And Kennedy challenged it all—on a dozen fronts. It was part of his philosophy. Once, as a junior Senator, in 1955, he spoke to me of his fight to raise the minimum wage; jauntily he shrugged off the certain prospect of defeat of his measure. "Sometimes," he said then, "you have to lose a few now to win a few later." His proposal was defeated in 1955—but six years later, in 1961, the $1.25 minimum-wage act was passed in the first session of his first Congress. And so he overwhelmed Congress, in the three years of his Presidency, with new bills, new ideas, new proposals.

It is customary, now, to refer to the moment of his death as the moment of his Great Deadlock with Congress. And, indeed, it was, for as he went on and grew more sure of himself, his pace and his appetite for change quickened with every month until, finally, Congress, digging in its heels, balked at the two greatest of his domestic measures. Yet one must note, beyond the defeats and reversals he suffered in Congress, how much new legislation was actually approved and passed into law —more than at any other time since the 1930's. Of 53 major proposals brought to the first session of the 87th Congress, Kennedy won enactment of 33; of 54 brought before the second session, Kennedy won enactment of 40; of 58 before the first session of the 88th, Kennedy won 35.[2] This legislation slowly changed not so much the shape and structure of the country as the country's way of looking at itself.

These, to be sure, were major domestic achievements. But the two greatest domestic legacies were those most dramatically deadlocked at the moment of his death. Those, however, were the ones that most directly

---

[2] Space prohibits anything more than the most fragmentary list of legislative enactment, but even the most partial list gives something of the flavor of his governing style: In 1961—establishment of the Peace Corps, the Alliance for Progress, Area Redevelopment Act, Community Health Facilities, Juvenile Delinquency Study Appropriation, Minimum Wage Increase, Water Pollution Act; in 1962—Reciprocal Trade Act, Communication Satellite Act, Manpower Development and Training Act, Educational TV, Drug Labeling Act, Poll Tax Amendment, National Institute for Child Health; in 1963—Nuclear Test-Ban Treaty, Mental Health Act, Maternal and Child Health Act, Higher Education Act, National Cultural Center in Washington, Aid to Medical Education, Air Pollution. And this is only the minimal listing of the most vigorous three years of legislation since the first term of Roosevelt.

influenced the election of 1964 and will remain to shape the future—
and we must look at them.

The first was the reorganization of the direction of the American
economy. Uninterested in any technical sense by interest rates, money
rates, depreciation rates or the Greek hieroglyphs of scholarly economists, Kennedy nonetheless let himself be led by his scholars to put into
practice the best thinking of the campus. What he was interested
in above all was movement—the United States, in its growth rate, must
not lag behind other nations expanding at a rate greater than America's.
He entered office with a Gross National Product of about $500 billion a
year in 1960, and a growth rate of 2 1/2 percent annually in the
previous administration. He left office as the National Product approached $622 billion in 1964 (a jump of 25 percent) and a growth rate
of 5 percent a year. In essence, what he did was simple—it was the
courage to understand, seize and apply such a simple idea that made him
leader. It required only the abandoning of the myth that a budget must
be balanced by specific calendar year in specific dollar terms; it required only that a President see the national deficit as that gap between
what was actually being produced and what could be produced if people and machinery worked at full capacity. If it required budgetary defi-
cits or tax cuts to close the gap between full capacity and under-
achievement, that was what government should do. This Kennedy did.
And in the great tax-cut bill of 1963-64 what was, in effect, a revolution in economic thinking was written into law. The figures, of course,
are as dusty and technical as the figures of McNamara's operations at
the Pentagon. But out of the figures emerged the dominant domestic
condition of the election of 1964—good times, prosperity, work, comfort on a scale no other civilization had ever known before. Abundance
—controlled abundance—was, thus, the first of his domestic legacies.

And the last, of course, was the Civil Rights Act of 1963-64. Of
this there will be so much to say in this book that it will be best to
let its full impact show later. Yet of all the motors in the campaign
and politics of 1964, none was more important than this. America is a
civilization that promises all men equality of opportunity, and, in his
Civil Rights Bill, Kennedy insisted that America finally try out the promise and see if it worked. He came late to this conviction. In the campaign of 1960 he had mused privately and with much perplexity about
the condition of American Negroes. Once he said to this reporter, "The
question always is—what kind of Negroes? Which ones are we talking
about? There's no other community in the country that has a gap so big
between its leaders and its men in the streets." But he had come finally
to the conviction that the only way to find out was to open every possible

opportunity for all Negroes as for all white Americans. Shortly after he introduced his Civil Rights Bill in the summer of 1963 he said to a friend, "Sometimes you look at what you've done and the only thing you ask yourself is—what took you so long to do it?"

These, then, were his legacies to the politics of 1964—Peace, slowly becoming real; Power, actual and at the ready; Prosperity, so vast as to be unsettling; and Equality for the American Negro, at whatever cost in adjustment.

To politics, as politics, during his Presidency he paid but little attention. It seems to be axiomatic that a great President must be a poor politician. This is because a President's time is too much taken with larger affairs. It is more important that one choose a good Ambassador in Paris, a good Chairman of the Council of Economic Advisers, a good Chairman of the Joint Chiefs than that one choose a good Federal judge in Wyoming, a good Customs Collector in New York, or manipulate the choice of a good mayor in Los Angeles.

Occasionally—sometimes wistfully, sometimes dramatically—the old political Kennedy would show. His blazing denunciation of United States Steel and Roger Blough for raising steel prices was inspired not so much by economic indignation as by political reflex; he thought he had had an understanding with United States Steel *not* to raise prices; they had broken one of those subtle agreements of honor which are binding on politicians but not on legal-minded businessmen. He thoroughly enjoyed the work of Averell Harriman in setting up a coalition government in Laos—"Old Ave," he said, "is putting together a balanced ticket in Laos, and it looks good." Late one afternoon in the White House he had a telephone call from a Southern Senator about the appointment of a judge; he rose from his rocking chair and went back to his desk and for a few minutes worked the buttons and the holds and the callbacks as he held the Senator on the line, conferred with the Department of Justice, conferred with the Senator again, conferred with the Department again. Obviously, the Department had checked out the nominee and objected. Shaking his head, Kennedy came back to the rocking chair and said, "The trouble with the politics of this administration is that the Department of Justice has got higher standards for judges than the President has."

His preoccupation with government, his neglect of the kind of politics that brought him to the Presidency, was a failing rather than a strength. It was evident in such blundering as caused the defeat of his proposal to establish a Department of Urban Affairs; it was evident in several minor domestic mishaps. But nowhere was this failing more apparent than in his conduct of foreign affairs. In foreign affairs he was guided by diplomats, and he dealt skillfully and, on the whole, with great success with Europe and Russia. But in the fermenting politics of Asia and Africa, where people vote with guns, he could never bring himself to see that the solutions lay in the kind of politics-at-the-grass-roots that had brought him to power himself. His greatest substantive failure was in Southeast Asia, where he groped year after year with names, with facts, with figures—but the politics of guerilla warfare eluded him, as they continue to elude American leadership today.

John Kennedy came, finally, in the fall of 1963, to turn his mind to the approaching politics of America, the Presidential choice of 1964. Earlier that year he had felt certain that his opponent would be Nelson Rockefeller, and when someone observed that Rockefeller liked him, he replied, "And I like him, too—but watch how we get to hate each other when the campaign gets under way." Later, after Rockefeller's remarriage, he felt certain his opponent would be Barry Goldwater. And he rather enjoyed that prospect.

He held but one meeting on the Presidential politics of 1964— on Tuesday, November 12th, 1963, ten days before his death. It lasted three full hours and was held in the Cabinet room in late afternoon, and, by the account of those present, was more an amusement or a matter of mischievous administration for him than a matter of high concern, as such meetings had been in 1959 and 1960. All his old teammates were there, with but one newcomer, Richard Scammon, Director of the Census. O'Brien and O'Donnell, Robert Kennedy and Sorensen, Ralph Dungan and Richard Maguire, brother-in-law Stephen Smith and John Bailey. John Bailey, to his own delight, sat in the austere Cabinet room across the table from the President, in the seat of the Vice-President. Vice-President Lyndon B. Johnson was not present at this first look at the politics of 1964.

There was little doubt that they would win. Therefore, certain longer-range matters than victory had to be attended to. The South was prime on the agenda, and as between the Sorensen and the Bailey-Scammon strategies for reducing the South's mischief-making in the Party, the President chose the Bailey-Scammon strategy—which was to reduce the strength of the South in the Democratic Party slowly (by a new adjustment of delegate seats at the national Convention to reduce Southern seats from 23 percent to 19 percent of the total), as against the Sorensen strategy which would have broken the back of the South right then and there in 1964. The problems of political housekeeping held them but little—Bailey would manage the campaign from the Democratic National Committee, Stephen Smith would be Executive Director. Registration would be stressed across the country. The President was seeking issues, some way of making clear in public what he was really trying to do. But no one had any adequate answer to that. The conversation lingered over how they could make the Democratic Convention interesting, since its choice was so certain; they discussed television possibilities; decided that the renomination of Lyndon Johnson as Vice-President would be staged on Wednesday evening and the renomination of the President on Thursday, when he would come himself. At one point the President observed that if he and his brother Bobby got into a public dispute, it might enliven things. But, beyond such wry comments, the President, sitting cross-legged on a cushion in his customary place, was more observer than participant. His mind was on larger things than politics. He had learned all that and was learning newer, bigger things. He closed the meeting at twenty minutes after seven and suggested they all get together in early January—he had a busy week and had to go to Texas the week following. Two weeks later he lay at Arlington.

So many things will be said about Kennedy in the future, and the myths are already so thick, that without doubt the man himself will soon be lost in the myth.

He was responsible for much of the myth himself, and his particular style was such as to captivate myth-makers, the men of words and phrases. He loved the sound of good English in his ear and the sight of beautiful things. He brought an appreciation of beauty into the White House and made elegance a quality Americans expected of their President; his taste was not for the angular and the abstract (though his doyen in matters artistic was William Walton, one of America's outstanding abstract painters), but rather for color and perspective; these perspectives will show in the future Washington, with its new plazas, its new esplanades and its new vistas, all of which he approved before he died. His wit and his phrasing will always entrance those who love a sentence with good cadence, as also those who love a good sense of timing, where the anecdote is rounded to perfection and then snapped off with a punchline, as a good glass-blower knows when to snap off his work. These will endear him to the myth-makers, who will find in his love of children and in the beauty of his wife much to work with.

The contemporary myths which were wrapped about him as he lived will probably fade, although the contemporary myths were probably closer to the truth than the newer ones. He was, indeed, and at times, ruthless—because judgment, at certain times, must take the cold cleaving act of decision and he could do it. He was, indeed, and at times, and in contradiction, indecisive—but generally not on issues, only on people. He would agonize over the firing of people and try to avoid dismissals until absolutely necessary; but this was because he believed that more could be achieved by kindness than by cruelty. With his own intimates, impatient, nimble, snapping his fingers, he could be abrupt; but with strangers no man was more quick with "thank you" and good manners, no man could be kinder in remembrances.

Of all these qualities, however, none will give the myth-makers more trouble than his original contribution to the politics of his time, the central quality to which we must return: his faith in reason.

Here, perhaps, the best light for the future will be shed by the act of the assassination. For John F. Kennedy was killed by a lunatic, Lee Harvey Oswald, who had momentarily given loyalty to the paranoid Fidel Castro of Cuba. And Oswald was, in turn, within two days, slain by another madman, Jack Ruby. It would have been easier for the American people to accept any enemy, any conspiracy, any plot, and then avenge John F. Kennedy. But what they had to face was an act of unreason, avenged by an individual act of obscenity. These were his enemies—as, in the changing society of America, the chief enemy of the good the change may bring is the undisciplined response of emotional unreason.

Normally, reason is an unattractive quality—it is a dull, gray gnawing at the conscience, or a cold denial of what warm indulgence and craven instinct invite. Kennedy, however, made the quality dance; in his time, poetry was in the White House and music, too. He added to reason the quality of courage, a courage that never failed him, from his captaincy of a patrol boat in combat in the Solomon Straits to the weekend of the Cuba missile confrontation.

In Washington, on the night of November 25th, the Monday of John F. Kennedy's burial, the crowned heads, the elected heads, the aging heads, the ambitious heads and the bureaucratic representatives of all the scores of governments with whom the United States maintains friendship or any other relations came to take a cup of sorrow and friendship with Lyndon B. Johnson. Friend and enemy alike, high and low, all were less and weaker than they had been four days earlier because the starshell had faded and they must grope again to find the will of the world's greatest power. One by one, they presented themselves to the new President, examining him, listening to his softly spoken words, asking themselves how he would continue—as millions of Americans asked themselves the same question on the same night.

CHAPTER TWO

LYNDON JOHNSON
TAKES OVER

THERE is no word less than superb to describe the performance of Lyndon Baines Johnson as he became President of the United States.

All the accounts of his behavior through the week of tragedy—his calm, his command-presence, his doings, his unlimited energies—endow him with superlative grace. Yet such stories limit the tale only to his positive deeds.

To measure the true quality of his take-over, one must consider not only these positive acts, but what did *not* happen. So much *might* have gone wrong—yet did not.

Only once before, with the killing of Abraham Lincoln, had a President of the United States been cut down so swiftly in so turbulent a passage of rushing history. And with the assassination of Lincoln, the government of the United States was torn entirely away from his pursuit of reconciliation, its open hates permanently frozen in the strata that divide Americans even today.

If Johnson had yielded to bitterness, or a Red hunt, or revenge; if he had faltered and let Congress or the Cabinet take leadership away from him; if he had not understood the nation to be in the beginning stage of a race crisis; if he had not understood the immediate tangle of international affairs; if he had been confused—if he had deviated by a hair from the course he was to choose in the next fourteen days—the country might yet be seething from the aftereffects of the Kennedy assassination. In retrospect there is a quality of inevitability about the course he did pursue. But historians insist that the inevitable is inevitable only *after* men have made it happen.

Lyndon Johnson had been up late the night before the killing, recounts a companion, relaxing until one or two in the morning in his room at the Hotel Texas in Fort Worth, unwinding from a happy day. A recent misunderstanding with the President had been cleared up, and that day he had managed a long, friendly conversation with John F. Kennedy. It had been a day of exuberance—the Houston reception had been unbelievably good, and a supposedly hostile Texas was showing its heart and smiling face to John F. Kennedy. All in all, the Vice-President felt just fine.

He could have had no more than six hours of sleep before meeting Kennedy in the Longhorn Room of the hotel for a contributors' breakfast, and then joining the journey and events that were to climax in the killing in the sun.

Then, suddenly, three shots; and he was slammed to the bottom of his car, buried beneath the body of his Secret Service agent, Rufus Youngblood, in the howling race to Parkland Memorial Hospital, not knowing what was happening. For the next forty-five minutes he paused, suspended in a transmutation of history in a suite for minor medicine and surgery on the first floor of the hospital while Secret Service agents closed around him, barred the doors, and shut the Venetian blinds so he could not be seen from the outside.

Lit by sterile overhead fluorescent lights, surrounded by surgical rolling beds, he waited.

As he waited, history waited on him. The staff of President Kennedy, convulsed by love and duty, attended the dying body. Sixty feet away, down the corridor, the Vice-President was alone with Mrs. Johnson until joined by several Texas Congressmen. The Secret Service, spread thin between the quick and the dying, tried to guard both lest a Constitutional thread be snapped.

Grotesque as it may sound now, the first thought of those who could think was: Conspiracy! Was this a *coup d'état?* Was this a foreign plot, a Communist strike, a native putsch? Were more marked for assassination? If so, whom? What was happening in Washington? Was all quiet around the world?

If one may separate a first strand from the skein that was to involve Johnson in high history, one should perhaps take Chester V. Clifton, military aide to President Kennedy, a man who loved the late President with all his heart—but who, though an extremely youthful Major General, had been fashioned by twenty-seven years of Army discipline. Someone must gather the threads, make contact with the national command center in Washington to find out whether this was, indeed, coup or conspiracy. Within minutes Clifton, on his own initiative, had reached the manager of the Southwestern Bell Telephone Company, clearing two lines from the hospital switchboard in Dallas direct to the White House and the Pentagon. Within minutes again, Clifton had dispatched the "walking courier" [1] from his station outside Trauma Room One, where the last pulse of John F. Kennedy was fading, to the minor-surgery suite where the next President was waiting. Within minutes thereafter, Clifton had linked the communications and telephone net from the Presidential plane, Air Force One, to the minor-surgery suite. And by 1:20, when Kenneth O'Donnell said to Lyndon Johnson, simply, "He's gone," the new leader was in contact with every ganglion of communication in the entire world.

From then, and for the next few days, and for weeks and months and years to come, all of Lyndon Johnson's moves would make these ganglia quiver. But his first move, instinctively, was to still them. For immediately after O'Donnell there came Malcolm Kilduff, press attaché of the trip, to ask whether he might announce the death of President Kennedy. Kilduff recalls Johnson saying, "No, Mac, I think we'd better wait for a few minutes; I think I'd better get out of here and get back to the plane before you announce it. We don't know whether this is a worldwide conspiracy, whether they're after me as well as they were after President Kennedy, or whether they're after Speaker McCormack or Senator Hayden. We just don't know." He mentioned the wounding of Secretary Seward in the Lincoln assassination, and the need, again, of getting back to the secure base of the plane at Love Field.

His next move, after discussion with friends in the cabin on the plane, was to make the ganglia quiver. The office of President was already his in substance; but it must be constitutionally, publicly, proclaimed. Thus he decided to take the formal oath there at the field before departure. Representatives of the press were summoned; thirty people filed into the hot and steaming cabin to hear him; an Army photographer caught the famous picture of the new President, hand upraised; on one side of him stood his wife—on the other, the gallant, blood-smeared widow of the dead President. Then, while the jets of the great plane howled, straining for take-off, Clifton threw the roll of exposed film to a waiting aide on the field so that, while the plane flew, the nation might have visual evidence of a new President in the first ceremony of office. And then, by a curt order of the President, the plane was in the air.

On the flight the party learned that there was no conspiracy; learned

---

[1] The walking courier is a mufti-clad military officer always within instant call of the President; he carries a plain briefcase called "the football," which holds the emergency plans and optional responses available to the President at any moment of warning of atomic attack.

of the identity of Oswald and his arrest; and the President's mind turned
to the duties of consoling the stricken and guiding the quick. One might
well try to envision him on this trip, for there is something essentially
Johnsonian about it. Of all men in public life, Lyndon Johnson is one of
the most friendless. Those who come in contact with him are accepted
generally as cronies, or partners, or supplicants, or men he can use—as
servants. But of real friends he has few, for, above all else, he lacks the
capacity for arousing warmth. Of all the things to which Kennedy was
born and which Johnson lacked—wealth, background, elegance—Johnson probably envied Kennedy most his capacity for arousing love and
friendship.

One has the picture of the President in the two-hour-and-eleven-
minute flight (the plane slowed down so as not to overshoot the preparations of reception at Andrews Base) sitting alone with his thoroughly
loving wife, Lady Bird Johnson; collecting his thoughts on the scratch
paper of Air Force One as other stunned men waited on his will,
then took his scribbled notes and coursed up the aisle to the signal center
of the plane, which flashed them as command to a Washington waiting
for the first pulse of new leadership. There emerges from all the stories
the picture of a massive, lonesome figure somehow—from under grief and
shock—making the machinery work and beat again, almost by instinct.
One of his oldest cronies has remarked that Johnson's instinct for power
is as primordial as a salmon's going upstream to spawn. This instinct, always there but now educated by experience and civilized by sadness, became a national resource in the days to come.

He arrived in Washington, his home and capital, at 6:00 P.M. By
then his orders had brought to the airport the men he needed most to
govern (he had originally asked for the entire Cabinet; been reminded that
most of the Cabinet were en route to Japan; asked that they turn back;
learned that they had already turned back and, in mid-Pacific, were approaching Hawaii and inquiring whether they should proceed to Dallas or
Washington; responded that they should proceed to Washington). It was
a truncated government, thus, that met him at the airport. But now, after
speaking to the nation over television the minimum, slow, decorous
words the bleak occasion required, he gathered three of them to him in
the helicopter. And the helicopter, its belly lights flashing white, then red,
took off on the seven-minute hop to the south lawn of the White House.

In the helicopter, the President sat in the big chair in the forward
compartment. Facing him, from left to right, were George Ball, Undersecretary of State; Robert McNamara, Secretary of Defense; Mrs. Johnson; and McGeorge Bundy, Special Assistant to the President for Na-

tional Security. Johnson broke for a bit on the short ride, speaking very slowly and with deep emotion, but disjointedly, about what had happened that day: the red roses Mrs. Kennedy held; how she would not change her clothes for the ceremony of swearing-in in the cabin, her bravery. And then of how "we" had to go forward with what we had; how he was counting on a few people the President had got together, particularly the three with him in the helicopter; how Kennedy could do some things better than he could—like putting together this kind of team.

Then, he was the President. Shifting gears, he asked Presidential questions. Was there any matter in any of their three areas on which he must reach a decision now, in the next twenty-four hours? Solemnly, each one in turn, starting with McNamara, answered No. Kennedy had left the nation with no urgent crisis that must be solved overnight. So that the President could now turn to the virtuoso performance of continuity which already, on the plane, he had outlined as his chief task.

He arrived at his old office in the dove-gray Executive Office Building (Room 174) at about a quarter of seven, and for more than two hours the visits and telephone calls rolled over him—visits from William Fulbright and Averell Harriman, telephone talks with Harry Truman and Dwight D. Eisenhower, a deputation of Congressional leaders, other telephone calls, a bowl of soup (the first meal since breakfast), more telephone calls. Then, at nine o'clock, home to The Elms in Spring Valley. A group of old friends gathered with him there—Bill Moyers, Jack Valenti, Horace Busby and his wife, Clif Carter, Dr. Willis Hurst—and in his sitting-room he sipped an orange drink, talked until about midnight, at one point saluted a portrait of his old friend Mr. Sam Rayburn that hung on the wall, then went upstairs to bed.

Johnson hates to be alone—his temper is not for solitary contemplation. So that those who are with him are usually audience to a piece of the thinking-out-loud that reflects the way the President ruminates. Now, after midnight, he called to his bedroom three of his guests who were staying for the night—Moyers, Valenti and Carter—to think out loud with them, planning schedules, pondering his own behavior on the morrow. Now and then his mind, sifting and resifting all the parts of the government, would stimulate a telephone call—Get me McNamara, get me the CIA. Every now and then he would pay attention to television and was watching television when he saw, at about 12:30 A.M., the plane bearing Dean Rusk and the Cabinet party arrive in Washington—which relieved him. It was after three in the morning when he dismissed his companions to their beds; and fell asleep almost immediately thereafter himself.

As he slept, it could be well stated that no man would ever waken to his first full day of the Presidency of the United States better educated in the meaning of that office, better trained in its mechanics, more artistically interested in its execution than Lyndon Baines Johnson.

Not since James Monroe had anyone entered the Presidency knowing so well so many predecessors in that office. Yet Johnson knew the office only as a critic knows the drama, only as an engineer knows the performance specifications of a plane. To be a pilot or an actor is something else again—a matter of background and instinct. And though Lyndon Johnson's background for more than thirty years—for more than half his life, since the age of twenty-three—had been the Federal politics of Washington, his roots and instincts were of the hill country in Texas.

We should look at both the roots and the background of this earlier life to see how profoundly they contributed not only to his performance in the next month but to the politics of the Presidency in 1964.

Men come to power in America from so many quadrants of geography, so many layers of its society, that each flavors the Presidency freshly with his own style, with the sound and manners of his region. If Harding gave the White House the coarseness of lower-middle-class, small-town Babbittry; and if Roosevelt gave it the quality of the Hudson Valley aristocracy; if Kennedy gave it the best flavors of Boston, from its Irish vigor to its Harvard polish, Lyndon Johnson was about to bring to the White House and the politics of 1964 the manners and style of rural Texas.

There are several kinds of Texas, even now as change sweeps over that state. There is the central, populated basin of Texas state politics, with its great centers of Austin, Dallas, Fort Worth, Houston; there is the Texas of the oil communities; the Texas of the chemical industries of the coast; the old cotton lands of the East; the new cotton lands of the West; and ranch-country Texas, dirt-farming Texas and several other Texases too.

Lyndon Johnson's Texas is hill-country, dirt-farming Texas and he is as native to this Texas as native can be in the United States. A John Johnson of Oglethorpe, Georgia, who had fought in the Revolutionary War, sired a son Jesse; and Jesse in turn made the wagon-train trek to Texas in 1846, the year after Texas joined the Union. Jesse's son, Sam Ealy Johnson (Sr.), the President's grandfather, moved on up to the hill country of Blanco County, still wild, still penetrated by Indians—some of them ritualistic cannibals—still open, filling with German settlers all clearing land, all shearing away the tall buffalo grass, eroding its soil. Sam Ealy Johnson left to fight the Civil War—and returned to fight Indi-

ans. Hill country is raw country, civilized less than a century; and about
the time that John F. Kennedy's great-grandparents were suffering the
enmity of cold-roast Boston for the early Irish pioneers of the big city,
Lyndon B. Johnson's grandfather was fighting Indians. As late as 1869
his grandmother Eliza hid from an Indian raid in the cellar of her
log cabin, infant baby in her arms, diaper tied over its mouth to hush it,
while Indians on the warpath looted her home, above, and scalped two
neighbors nearby.

Hill country is poor country. You drive out of Austin, the state
capital, west toward the Texas horizon, and slowly that horizon begins
to pucker into what the geologists call the Edwards Plateau. The best
word for the panorama is stark—it begins to roll, and rolls on, and continues to roll on for 200 miles in harsh, low hills, none higher than 600
or 700 feet. There is nothing here of the green or the snugness of New
England hills and valleys—these hills are primeval, hard, unending, and
tell only of more hills to come toward the west, and the desert and the
mountains. Today it is bleak land, with a beauty that comes only in
spring when bluebonnets flower and the grass grows sap-green. But from
the remaining clumps and wisps of bunch grass one can imagine the
tufted landscape as it must have been when the grass reached saddle-high
to a man on a horse and stretched over every hill and beyond, when the
buffalo roamed as they did until the settling of the Johnson family and its
neighbors. One can imagine it dotted with stands of cedar, of juniper, of
live oak and white oak. Only yesterday.

But the white man ruined it. He ruined it by overgrazing it with
cattle that cropped the grass to the soil and let the soil erode; he ruined it
by planting cotton and mining the soil of its nutrients, neglecting it. By
the time Lyndon Baines Johnson was born it was worn and poor—the
kind of country that bred agrarian radicals and Texas statesmen. And it
was rural Texas that sent to Washington the large-sized Texans of history—John Nance Garner and Mr. Sam Rayburn, Wright Patman and
Lyndon Johnson.

It was thus in hardship that Lyndon Johnson grew up. We know
little of his boyhood or his early life, for he has never told it all and no
major biographer has yet mined the rich ore of country life in Texas fifty
years ago. Normally, the story of Johnson's life is told in bucolic terms:
the barefoot lad, riding double on a jackass with his cousin to a country
school, hunting in the hills, watching lizards in the sun, remembering the
powder going off on the anvil on Armistice Day. But poor boys either
harden early or are destroyed. Hearing the politics of the gut-hungry
from his father's friends in the Texas legislature, he was one who
learned to scratch—hard. A hobo at sixteen, he wandered off in an old "T-

model" Ford with a few friends to California and came home almost two years later, as he says, "with empty hands and empty pockets." Hungry, brawny, standing in line in unemployment offices, waiting for a job assignment, footing a shovel with a highway gang for a dollar a day, dawn to dusk, he was finally persuaded by his mother to go back to school.

School was Southwest Texas State Teachers College. It took only three and a half years to finish college—including a year out to earn enough money to go back by teaching a primary-grade class in a rural school of predominantly Mexican-American children. At college, force of personality won him a place as secretary to the president. And then, on graduation, there followed a year of teaching public speaking to high-schoolers in Houston. After which came a political appointment as secretary to newly elected Congressman Richard Kleberg, which brought him in 1932, at the age of twenty-three, to Washington and the way of life that was to be his.

Washington has been his home ever since, and Washington politics his life. To understand Johnson, one must recall Washington as it was then, as it lurched from Hoover's Depression to Franklin D. Roosevelt's New Deal. All things were fluid—old ideas were being thrown out, new government structures built, torn down, renamed. Men were heroes one day, nobodies the next, lobbyists thereafter. In the high and creative years of the New Deal, all government seemed plastic and the mission of government itself changed. This, above all, was the formative period of Lyndon Johnson's thinking. Young, befriended by a group of youngsters like himself finding their first experience of manhood in this upheaval of order, Johnson acquired a vision of government that has never left him. To most of us, government is a massive structure, as permanent as the hills, the slow accumulation of generations of American experience and constructive legislation and, before that, of centuries of British experiment in the machinery of freedom. To men like Johnson who lived through the New Deal—as to the Russian Bolshevik generation that made the November revolution—government is a system of movable and replaceable parts.

As Congressional secretary to Kleberg from 1932 to 1935, Lyndon Johnson served a rugged mechanic's apprenticeship in the working of the House. He learned its politics too—aggressively winning for himself the title of "Speaker" of the "little Congress," an unofficial gathering of Congressional secretaries, an honor won from a man many years his senior. His ability and imagination led on to another job—appointment as Administrator for the State of Texas of the National Youth Administration. At twenty-seven, thus, he could return briefly to home country to make work the new machinery he had seen the Congress create. In a year he had helped 18,000 Texas boys get back to school, and another 12,000 to find work on government or private projects.

So much of what has been put forth by President Lyndon Johnson as national policy for a Great Society echoes so much of this early experience of Youth Administrator Johnson that one is tempted to linger over that record. The preservation of natural beauty was a program to be announced thirty years later as a mark of the Great Society to come—but in 1935 his Texas youngsters were already planting trees all up and down Texas highways to make them beautiful. (To this day the ugly billboarded approaches to Eastern cities and their cluttered roads annoy him.) The anti-poverty legislation of 1964 was far in the future—but his early war against poverty in Texas mobilized, with government money, thousands of youngsters for useful life who otherwise might have wasted in the despair of the depression. The use of government apparatus for political ends was also there, right from the beginning; from the youth corps he picked one of the poorest lads of a seven-children family, John Connally, who became his assistant from first term all the way through to the Secretarial level in Washington—until John Connally left to become Governor of Texas.

It was the youth corps—the youngsters, their families, the publicity, the leadership exposure—that projected Johnson first to the attention of President Roosevelt (Johnson called on the President at the White House for the first time at the age of twenty-seven) and then directly into politics. In 1937, after two years of directing one of the best-run youth administrations in the nation, he ran for Congress as an all-out, 100-percent New Dealer (he favored the Supreme Court packing proposal too) and won, in a field of ten, by nearly double the vote of his nearest rival. And with this, Lyndon Johnson was thoroughly and finally launched as a political figure in his own right, with a code that never thereafter changed.

If the frontier wrestler's code holds that "there ain't no holt that can't be broke," Lyndon Johnson's code held that "there ain't nothing government can't fix."

With the proper knowledge of the joints, pressure spots, personalities, laws, procedures and ambuscades of the Federal system, almost anything can be done by a smart, energetic Federal Congressman who cultivates his elders, works around the clock and knows how to bargain. As a demonstration of what government can do and has done for the American people, Lyndon Johnson can offer his own Tenth Congressional District of Texas, so sere and dreary and parched thirty years ago. Federal Housing Project Number One rises in Austin, Texas—as a grace of Franklin D. Roosevelt to his young Congressman. The New Deal's REA created the Pedernales River Electric Cooperative, which serves

Blanco County, Johnson City and the present Johnson ranch—and which remains one of the largest rural electric cooperatives in the country. ("The power companies said you couldn't put lines through those hills," Johnson is remembered as boasting, "but the REA, they showed them how to do it.") [2] The farmers of the district who curse the Federal Government get soil-bank payments, conservation payments, peanut allotments, cotton allotments. Like everyone else, they benefit from veterans' hospitals, and in the towns there is Social Security for all. Austin boasts four hospitals built or enlarged with Federal money under the Hill-Burton Act. Bergstrom Field, a major SAC air base, is one of the largest employers of labor in the area. The highways that now stream out through Texas, including Highway 290, which leads from Austin due west to Johnson's ranch, are built with 90-percent Federal money.

And the hill country is coming back. Federal laws and Federally encouraged science are restoring its prosperity; buckbushes (called "Roosevelt willows" locally) are replacing bull nettle and thistle on the eroded land; Bermuda grass is covering the bare soil; cattle are being raised at a profit. The people in Blanco County are fewer than when Lyndon Johnson first left for Washington, but they are healthier and more prosperous than ever before, as are all the people of the Tenth Congressional District and of Austin, its largest city.

The sour bubblings of Texas provincialism and jingoism, and the attitude of so many Texans that the Federal Government is their sworn

[2] The founding of the Pedernales River Cooperative is told in a rather affectionate manner, by those who know the career of Lyndon Johnson, as a marker in the Education of the Young Congressman. Johnson yearned to bring light and power to the homesteads of his district from the new Federal dams on the Colorado River; but the rules of the Rural Electrification Administration held that cooperatives could be financed only where population density was three per square mile—and along his native Pedernales, population was only 1.6 per square mile. Only President Franklin D. Roosevelt could waive these rules, and so young Johnson sought an interview with the President. Through the grace of James Rowe and Tom Corcoran, both then young White House assistants to the President, a ten-minute appointment was set up. But Roosevelt was in a discursive mood that day. He sat the young Congressman down in his chair, rambled on for fifteen minutes about matters of his own, then ushered the young Congressman out without letting the Pedernales River be mentioned once. With great difficulty Rowe and Corcoran then arranged a second interview for the young Congressman, and this time, well coached, Congressman Johnson was set. He entered the President's Oval Office carrying books, charts and maps and, without breaking stride, before sitting down, began: "Water, water everywhere and not a drop to drink; public power everywhere and not a drop for my poor people." Without pausing, he pulled out charts of power consumption and maps of his district, and for ten minutes, talking as only he can, overwhelmed the flabbergasted President. So amused was Roosevelt by the performance that, without bothering with further investigation, he simply called the Rural Electrification Administration and directed it to make the founding loan to the Pedernales Cooperative.

enemy, are thoroughly alien to a man like Johnson. For him the Federal Government is the greatest benefactor of Texas. Just as John F. Kennedy used Massachusetts as a staging base for national politics, all the while despising its internal politics, Johnson regards Texas politics as bush-league politics. Like Rayburn's and Garner's and Patman's, Johnson's stage is national—and Austin politics a dead end. Aloud, he mused to a newspaper friend of his several years ago that he could never see why John Connally left a good job like being Secretary of the Navy in Washington to go off down there to Texas and be governor. For Johnson, the Federal Government is the chief way of doing things—a vehicle of action.

No man has watched that vehicle in action over a longer period of time, or participated more directly in that action, than Lyndon Johnson. As a Roosevelt favorite, he was a favorite of all the younger New Deal movers and shakers—who still, now as they age, remain the most direct intellectual influence on him.

Under Truman, he made the transition to a Senate leader; and under Eisenhower, for six years as Majority Leader of the Senate, he made the wheels spin for the Republican President as loyally as he had for his Democratic chieftain. He had earned the Vice-Presidency and in 1960, if he could not be President himself, desired the second spot. But when, finally, it was offered to him at Los Angeles in a day of such farce, confusion and pandemonium as rarely occurs even at conventions, it was offered to him without dignity, without honor, without the great courtesies and bows that Johnson expects from other people. It was offered coldly, not as if he were the second ablest man in the party or a man fit to qualify for the Presidency, but in a manner that cast it as a hard deal, an index of Kennedy's need for a Southerner to balance a Catholic. The pressure, instant intrigues and back-room throat cuttings that went on from early on the morning of his choice on July 14th, 1960, until finally, against the urgings of most of his Southern friends, he accepted the designation, hurt his pride far more than any pre-planned slight. As he stood crestfallen, about to emerge from his suite at the Biltmore in Los Angeles to accept the high honor, his friend the late Philip Graham thought he and Lady Bird Johnson looked "as though they had just survived an airplane crash." As an example of the mocking comedy which surrounds such great decisions, the moment of Lyndon Johnson's choice as Vice-President was to be surpassed only by Johnson's choice of Hubert Humphrey as Vice-President four years later. (The historians are fortunate indeed to have had present on the 1960 occasion the late Philip Graham, publisher of the Washington *Post*, who left behind the most entrancing eyewitness account of the moment. It is published at the close of this book as Appendix B.)

What wounded pride Johnson brought to the campaign of 1960 must have been further wounded in the three years of his Vice-Presidency. Relations between the Vice-President and President Kennedy's staff were cold and suspicious. There, in the outer chamber of power, sat Johnson, a premature elder statesman, while in the inner chambers a group of young men—as young as he had been during the New Deal—who had bypassed his generation were experimenting with government as he itched to experiment himself. The humiliations inflicted on him by the Kennedy group were, almost certainly, unconscious on their part—and perhaps even more galling for being so casual and unthinking, as if he were a fossil decorating their mantelpiece as a reminder of olden times.

His hurts extended from the most frivolous indignities—Could he ride in the second car after the President? Could he come to a White House lawn party by helicopter?—to the largest matters. One man who sat with him regularly at National Security Council meetings recalls with vast admiration the discipline of Johnson's behavior—silent, listening, rarely talking unless asked to talk, his fingers locked and working together until sometimes the knuckles were white. Chafing in inaction when his nature yearned to act, conscious of indignities real and imagined, Johnson went through three years of slow burn, ceremonial formalities and international junketeering. It is to his credit that he was able to forgive and forget and later use so many of the men who had, in his opinion, so abused him.

This bitterness between Vice-President and White House did not, however, extend to President Kennedy himself. Occasionally amused by his Vice-President's manners, the President would enjoy the natural style of a Johnson who could bring a camel driver back from Pakistan as a national guest. "That's Johnson's luck," said Kennedy after that episode. "If I had tried it, there would have been camel dung piled up all over the front lawn of the White House." More often, and most seriously, the President made of Johnson, as much as any President can make of his Vice-President, a working participant in national affairs. Johnson and Theodore Sorensen alone participated in every senior council of state—Cabinet meetings, Congressional leaders' meetings, National Security Council meetings. And to make sure that no paper slip-up occurred, Major General Clifton was detailed, apart from the White House administrative staff, to make sure both that Johnson was always informed in advance of National Security Council meetings and that his whereabouts were always known. Through Clifton, again, Kennedy made sure that all military papers, all SAC briefings, all Joint Chiefs of Staff reports went, in copy, over Johnson's desk. At the State Department, Dean Rusk, with his outgoing courtesy and respect, made sure that all major State Department policies and papers were available to Johnson. For all this, Johnson, a man sensitive to every vibration of the web of personalities, was deeply grateful to Kennedy. He recognized Kennedy's personal care for his dignity, appreciated Kennedy's insistence that letters perfunctorily written by the Presidential staff to foreign chiefs of state for Johnson's foreign touring be rewritten with the appropriate dignity and flourish that the Johnson personality required. Once Kennedy stopped Johnson's jet, en route to a foreign land, at Otis Field, Massachusetts, and had Johnson helicoptered over to Hyannisport. It was important to Johnson that he be sent off to see foreign chiefs of state with the added impact of a recent talk with the President of the United States; and when the President came off his boat for a brief forty-minute conversation, Johnson said, "Well, I think we've stayed long enough for the record," and took his departure. "Thanks for dropping by," said Kennedy. On the way out to the airport, the Vice-President said to his driving companion, "When you go back, tell him I'll never forget how he's always been generous and thoughtful in every move to me."

Of all things Kennedy did for Johnson, none, however, was perhaps more instantly important on the weekend of November 22nd than a minor decision Kennedy had made months before. He had decided that in the secret and emergency planning for continuity of American government in the happenstance of a nuclear attack, Johnson should be given a major role. Through Clifton, who acted as White House liaison with the Department of Defense, all emergency operational planning was made available to the Vice-President in duplicate. These plans, envisioning all things from the destruction of all major cities to the bodily transfer of governing officers to an underground capital, included, of course, detailed forethought of the event of the sudden death of a President.

Because he had participated in all these plans, both panic and ignorance were already precauterized in the Vice-President on the night of November 22nd, 1963—he knew exactly all the intricate resources of command and communication at his disposal.

And beneath this lay the experience of a man who had spent thirty years observing the work of the Federal Government, while beneath that lay the instincts of a Texas hill-country boy.

Now it was up to him to act.

"Continuity" was the word in Washington within twenty-four hours of the swearing-in of Lyndon B. Johnson. It had become current within hours after Air Force One had landed at Andrews Air Force Base. It was to dominate Washington's thinking for the next month and a half. It was to steady Cabinet, White House staff, Congress and the nation.

When Lyndon B. Johnson first used the word is unknown, and testimony comes from so many sources that it is impossible to set the moment it was first spoken. McGeorge Bundy heard it on the White House lawn after the helicopter descended from Andrews Air Force Base. Dwight D. Eisenhower, who saw Lyndon Johnson the next day, told friends a few weeks later that the President was in a state of shock and kept repeating that it was up to him to carry out the Kennedy dream, the Kennedy vision. (Eisenhower himself urged the President to get to work right away on the Budget and the economy.) Pierre Salinger's description of the process, as he recalled it within three weeks of the assassination, ran about like this:

"It was about like taking over the driver's seat of a bus that had run up against a brick wall," said Salinger. "You had to get that bus started again, and you had to get it through that brick wall—but how? There's no time to find a new bus, or a new crew, so you have to use the bus as it is, and you have to use the guys who were running it, who know how it runs. He's got to keep them, all of them. In about thirty-six hours he'd buttonholed every one of them, and nobody says no to him—but some of them say they'll go with him and others say they'll stay as long as they can. 'I need you,' he tells them all, 'I need you more than John F. Kennedy ever needed you.' On Saturday he made one of the most eloquent speeches I ever heard, at the Cabinet meeting, pleading with them to stay. By Wednesday he had the full crew with him; but he knows they won't stay to help him drive unless he drives in the same direction that Kennedy was going. And that was his Congress speech—to tell them and the country that was the way he was going. That settled the question of direction. Then he was off; he was the bus driver."

This is, of course, the simplest way of telling the story. But it must be fleshed out with detail to show the phenomenal range of Johnson at his best, that virtuoso knowledge and understanding of all the myriad and mysterious interconnections between Executive and Legislative, between officialdom and community, between influence-makers and leaders and executives and symbol figures outside Washington who give America its system of orderly government.

What is amazing, really, and worth heroic examination, is the capacity of a man to function under shock—to work the wheels of his mind with logic when the emotions are numb, to sort out reasonable priorities of action under storm, shock and fatigue. Like a heavy-duty motor, Johnson functions best, most efficiently and most attractively under a heavy load. It growls as it goes, but the drive and capacity are the main thing.

The calendar of the new President in his first two weeks offered only a sterile and one-dimensional mirror of an unbelievable performance.

There was the civil-rights problem—the nation's most vivid concern, the most important unfinished business of Kennedy before Congress. In a seven-day period from November 29th to December 5th, one by one, he called to the White House every major Negro leader of the country: first, Roy Wilkins of the NAACP on the Friday following the assassination; Whitney Young of the Urban League the following Monday; Martin Luther King on Tuesday; James Farmer of CORE on Wednesday. Simultaneously he had called in all the legislative leaders to discuss civil rights; had conferred privately with Charles Halleck, the Republican leader in the House, to unfreeze the blocked bill; and had spoken to the nation in a fireside chat on Thanksgiving Day to urge its concern.

But simultaneously, too, he was dealing with staff—for he wanted not only the Cabinet to remain, he wanted the entire grief-stricken White House corps to carry on also. One of the grieving White House members directly concerned with the recruitment of quality, high-level, non-patronage personnel for government posts listened reluctantly to Johnson's plea that he stay to help the new President find the same kind of people he had found for Kennedy, then replied, "But, Mr. President—that's always been our weakness. We know scholars, intellectuals, politicians—we don't know the business leaders; you do and we don't." To which Lyndon Johnson replied with genuine sincerity, "But please stay—I don't know *anybody*."

And simultaneously, too, he was dealing with Foreign Affairs. He had amazed the Secretary of State, Dean Rusk, at the State Department reception on the Monday evening following the funeral by the range of his recognition of the foreign dignitaries and their particular problems—retrieval of faces, names and problems coming almost spontaneously from the many trips abroad and briefings in his three years of Vice-Presidency. And the morning after the funeral he had time, immediately, to receive at the White House the Prime Minister of Britain, the Chancellor of Germany and Anastas Mikoyan of Russia—then, in the afternoon, to call a full conference of Latin American chiefs of state who had lingered in the hope of meeting him. The meeting with Mikoyan is suggested by one of those present as the true beginning of his Presidency—a cold, correct, yet cordial talk with the shrewdest man of the Russian leadership, to let the Russians know that Johnson, like Kennedy, would balance an unquenchable desire for peace with an irrevocable commitment to firmness against pressure.

And, simultaneously again, he was reaching across the country, as if his mind itself were a switchboard winking with amber alert lights that should be checked for response. On his way home from the Executive Office Building the Friday evening of the assassination he had already detailed his personal assistant George Reedy, later Press Secretary, to put together the names of all the labor leaders in the country to whom he must speak personally in the next few days—and that evening he had telephoned Keith Funston of the New York Stock Exchange to thank him for closing the Big Board. As the governors dispersed from the funeral on Monday, their presence, too, suddenly registered on him and he summoned them to return immediately for a briefing at the Executive Office Building (Scranton was caught at the airport, about to board a plane, and turned back).

Editors and publishers, columnists and commentators received his calls. So did businessmen, and within a week they were streaming through the White House, one after the other, in a procession that reflected every shade and shading of power in the country. In the second week of his Presidency a record was made of what was publicly known of his activity that week: 90 visitors or groups to the White House, 250 recorded outgoing telephone calls, 10 major meetings, 10 major statements. And since the public record of the Presidency usually reflects only about half its activity, leaving out most internal staff transactions, the rest of the week's stupendous exertion must wait on publication of the White House log.

Above all, throughout this activity, there protruded his sense of the Congress of the United States. For it was there, in the Congress, that the visions of John F. Kennedy for America's future had been frozen in deadlock with the structure of American life as it is, hardened in the men the communities send to make their laws.

It was in Congress that Lyndon B. Johnson had spent his life, a Congress which he loved and understood perhaps better than any man in his time except his mentor, Sam Rayburn. He had, while John F. Kennedy lay still unburied in the Capitol, begun to think about what he must say to Congress; he had, on Sunday, detailed some of the best brains of the old Kennedy staff—Sorensen, Bundy, Galbraith—to draft the speech; added to his drafting board his old friend and companion of the New Deal, Abe Fortas, a Washington lawyer; added again two members of his personal staff, Bill Moyers and Horace Busby, to contribute ideas. And as he spun through his frantic activities he would return to this speech, his first appearance before Congress and the nation, again and again. All day Tuesday—as Sir Alec Douglas-Home, Ludwig Erhard, Anastas Mikoyan, Haile Selassie, Eamon de Valera, Ismet Inonu, Diosdado Macapagal and an assortment of Latin American dignitaries paraded through the White House to pay courtesy—he labored on his speech, changing words, phrases, seeking the right tone. It was a rare activity for him, for, normally, he had been casual and offhand about his public prose. Not until Wednesday morning was he satisfied. Then, when finally McGeorge Bundy and Bill Moyers brought him the finished speech, typed in outsize characters, at his home, he read it once more and as he lay in bed penciled in freshly those words that came from his own heart: "For thirty-two years, Capitol Hill has been my home," and was ready to deliver the address that made his personal policy national policy: "Let us continue."

What there was to continue was, indeed, the carrying out of the Kennedy dream. No less than fifty pieces of unfinished legislation lay before the stubborn Congress of the United States, which, split by its many divisions, had refused to process them into law. Major explorations of the future which Kennedy proposed to lead were trapped in committees, great and stupid, the directions clear but the will frustrated. They stretched from wilderness legislation to urban legislation, from foreign-aid appropriations to space appropriations. And of these, the two most important acts were most bizarrely ensnared. The first was the Civil Rights Act, a leap into the future which would bring the Federal Government to intervene in the most delicate and intimate personal customs of the nation, setting human rights above states' rights. The second was the Tax Bill, the enticing embodiment of a new theory of economics so seductive as to alarm every puritan conscience in both houses on the Hill.

Within twelve weeks of the Johnson Presidency, the Civil Rights Bill had passed the House of Representatives and the tax cut had passed the Senate. Aid to higher education; the Mass Transit Bill; the Poverty Bill—all were finally moving through committee. Foreign aid had been appropriated at a compromise figure which balanced State Department need against American exasperation as expressed in the House. By cajolery and pressure, by wheedling and promise, he had begun to chip away at the ice in Congress by an intervention more personal than any President's since Roosevelt. For most of Kennedy's New Frontiersmen, Johnson's actions were the best pledge of his commitment to their fallen leader. Others took a more sour view. Said one: "I don't think he's motivated by any greater desire than to please *The New York Times* and get a pat on the back from Dick Russell—both at the same time." Still others saw him as already motivated by the coming election in 1964 and a desire to write a record quickly. But the machinery was turning and the country was moving.

To all this, lastly, must be added the simple complexity of the Presidency, which now, for the first time, Lyndon Johnson must view as Chief Executive—who must decide matters beyond negotiation. No sooner had the Cabinet meeting of Saturday afternoon, November 23rd, broken up—while Kennedy's body still lay in the East Room of the White House—than Johnson must receive Adlai Stevenson. Stevenson's problem—abrupt, urgent, unpostponable—tangled every domestic and international problem of outer space. Stevenson must speak to the United Nations on Tuesday, to give America's policy on the orbiting of armed vehicles through outer space. Stevenson had one point of view; the Pentagon, with its defense view, had a contrary policy; the decision involved judgment on defense, on space, on the Russians, on peace. Stevenson outlined his policy. Johnson briefly ruminated about a similar position he had taken in a speech as a Senator in 1957, agreed with Stevenson—and told him to go ahead and put the words in the mouth of the President.

Sunday—while John Kennedy's body still lay in the rotunda of the Capitol—Johnson had to preside over the first decision-meeting on a matter of foreign affairs. Henry Cabot Lodge had flown to Washington on Kennedy's previous invitation to discuss the worsening state of affairs in Vietnam. A full-dress council (Taylor of the Joint Chiefs, Rusk, Ball and Harriman of State, Bundy of the White House, McCone of CIA) lasting nearly an hour had to be held, waiting on Presidential decision: which came, instinctively—Press on, increase the pressure.

Of all the overlapping, simultaneous, confusing streams of decision and thought, one, however, must be isolated and accepted as the background for all the more prominent activities suggested above. This stream we must seek in the Budget. For it was the Budget, above all the Budget, that made Johnson President in his own right and, for those who had the gift of imagination, roughly outlined the kind of President he meant to be. For the Budget, with all its billions, is only a multi-digit curtain that conceals the supreme decision-making process. Within its fat folds it encloses and buries the President's final measure of the nation's goals and activities. The first memorandum on the Budget had traveled downstairs from the office of the Director of the Bureau of the Budget, Kermit Gordon, to the first floor of the Executive Office Building early on Saturday afternoon, within twenty-four hours of the shot of the assassin. It was a one-page memorandum which made two points: it explained what the Bureau of the Budget did and, alerting the new President to the lateness of the year, insisted there was still time for Lyndon Johnson to make the next year's Budget *his* Budget.

By six o'clock that afternoon the new President had telephoned Kermit Gordon at his home and set an appointment for the next day, Sunday, at six o'clock. By the next week, the largest burden on his time, as it was to remain for the next month, was just this Budget. For a legislative foreman, which was Johnson's previous experience, the prime questions had always been: Who will vote for, who will vote against any given proposal? How much must be given to one bloc to get something for another bloc? But for a President the prime questions are: What are the merits? What are the priorities? What are the pros and cons? What shall be done with the resources of the entire nation? The President is the unifier—and the duty of the Budget Bureau Director is almost a priestly one. Eschewing politics and pressure, he must explain to the President what is buried in the figures, what they will or will not do, what the choices are in terms of exertion, time and national resources.

The Budget is a dreary and forbidding book, recommended to no one as casual reading. But from the Budget there rises, steaming, an emotion that colors in an indefinable way all political thinking in Washington. Indeed, it is not too much to say that the Kennedy deadlock was glued together by two ancient families of emotion on Capitol Hill—the one the emotions of civil rights, which had paralyzed the Senate for a century, the other the emotions on the Budget, which cemented to the racists men otherwise of goodwill. The emotions that rise from the Budget are all summed up in one word: "spending." (As: "We are spending ourselves to bankruptcy," or "We are spending ourselves to destruction.") Scores of Congressmen and Senators, Democrats and Republicans, are viscerally terrified that the unbalanced Budget will destroy the dollar, the life savings, insurance policies, and civilized life of Americans all together. The programs of social welfare are the particular targets of this emotion. Housing, education, research, beauty, all seem to open a frightful breach for more spending in the minds of men who see the Budget as a steamroller to flatten all Americans into an indiscriminate egalitarianism.

It is doubtful whether Johnson would word his thinking as precisely as this. But, as the most skilled master of Senatorial coalitions, he knew that part of the glue that bound the deadlock coalition together was the emotion that rose from fear of "big spending." If he meant to achieve the Kennedy vision, starting with the Civil Rights Bill and the tax cut, he must unglue the coalition. And in so doing, starting with the Federal Budget, he not only unglued the coalition fear of "spending," but began to rough out the shape of what later came to be known as "The Great Society."

Work on the budget started on Tuesday, the morning after the burial of John F. Kennedy. It continued for a full month thereafter—in sessions running from an hour to four hours every day.

The problems brought to Lyndon Johnson, cloaked in figures which he must translate into politics, touched all things.

§ The Rover project—a nuclear rocket program. The Rover project envisions exploration not of the lunar spaces of the moon but of the stellar reaches beyond the sun. It has no meaning to Americans for the 1960s, nor even for the 1970s—but for the 1980s. Yet if it is to be meaningful in 1985, it must start now. But at what rate? At what cost in scientific manpower, which is limited? At what drain of current resources? (Johnson's decision—Cut back.)

§ The Atomic Energy Commission. More missile-grade nuclear explosive was being produced than even the armed services needed. But how and where was such nuclear-explosive production to be cut back? At the gaseous-diffusion uranium plants of Oak Ridge (Tennessee), Paducah (Kentucky) and Portsmouth (Ohio)? Or at the plutonium plants in Hanford (Washington) and Savannah River (South Carolina)? (Johnson exploded: "I'm not going to make atom bombs into a WPA project." Johnson's decision: to cut back enriched-uranium production by 25 percent, close down three of Hanford's nine plutonium reactors, one of Savannah River's five. Later, reduction of missile uranium was cut by 40 percent, later yet by 60 percent from the production rate in 1963; and production of plutonium was cut by 20 percent.)

§ What about the MURA project—the high-intensity nuclear accelerator for the Midwest which would cost $170 million to build and $30 to $50 million annually to operate? The Midwest, muscle-bound by heavy industry, craved a new scientific center to give it leverage against the science-spawned industries of the East and West coasts. But scientists had advised the government that over the next ten years the future lay with high-energy accelerators rather than high-intensity accelerators such as the MURA proposal. (Johnson's decision—to veto the MURA accelerator and bear the political repercussions as best he could.)

§ Beyond these, all the other decisions couched in the Budget's figures. More agricultural extension services at a time when agricultural surpluses in America were swollen out of reason? A billion dollars for more river-valley projects to please Congressmen who had to please their voters in order to give a working Democratic majority in Washington? And a thousand other projects which Johnson examined in detail and savored both in figures and in the personalities they involved. His pencil worked over every figure; his cuts and decisions, with all the personalities to be hurt by decision, ranged from $170 million (for the MURA accelerator) to a picayune $10,000 (denying two extra secretaries for the staff of his chief economic adviser, Walter Heller).

Nor could any decision be made without pressure. No item of the Budget but touches millions of people or dozens of states, or, conceivably, can wipe out the jobs and facilities of entire communities. To all these pressures, with a magnificent sense of larger political strategy, Johnson responded with a resistance cloaked in every guile, skill and credit he had acquired in thirty years in Washington. Only once, so far as is known, did he yield. Robert McNamara of Defense had provided him with a ready-to-go list of economies in the Defense Department, obsolete installations that could be cut to fit the new "anti-spending" philosophy without hurt to the national defense. Among these was the Boston Navy Yard—which Johnson brought up at a legislative breakfast with his Congressional leaders one morning in early December, only to find the deacon-like Speaker of the House, John McCormack of Boston, boiling up almost instantly into a fast burn. To close the Boston Navy Yard, in John F. Kennedy's old Eleventh Massachusetts Congressional District, within a few weeks of John F. Kennedy's death was a truly unfeeling way to make orphans of John F. Kennedy's earliest faithful. Johnson bowed. It was to be a permanent retreat too, for a year later when other great Navy yards on the East Coast were wiped out by McNamara's accountants, the Boston Navy Yard, where *Old Ironsides* still berths, was preserved.

And yet, out of it all there finally emerged a political profile completely smothered in the gross figures of $97.9 billion, which in January became Lyndon Johnson's final Budget sum. What had been cut (about $2 billion) from the rough totals originally offered him had been cut most significantly from armaments: from the Defense Department and the Atomic Energy Commission. The few increases had taken place in the realm of American domestic need, in the field of human welfare—in health and medical research, for urban housing, for education, for the anti-poverty program. These were item-by-item decisions made according to the instincts of the new man in the Oval Office. But when it was pointed out to him that the contrasting cuts and rare increases made a new pattern, a new theme of government, he leaped on the new pattern of his own couture just like, says a member of the Budget process, "a fly on a June bug." Only later was it to become the theme of his election campaign, to be read both at home and abroad as the direction in which Lyndon Johnson meant to move his Presidency.

By mid-January, with continuity established, with a Budget prepared, with a staff fully functioning, Lyndon Johnson was obviously comfortable in his Presidency. At a lunch with several correspondents he re-

viewed his first few weeks and boasted of the Budget and its economies
as his major accomplishment. (He preened that he had eliminated 600
Treasury jobs that very morning.) From his left rear pocket he pulled
out three folded sheets of paper. One bore the $122 billion in original
requests from all government agencies; another, the amounts finally appropriated; a third, the new obligational authority involved. He was
proud of how he had handled Congress, and how he was swapping jobs
and favors for votes with Anderson of New Mexico and Bennett of Utah.
With obvious sincerity he said that he would give everything he had not
to be in the job that he was now in. But he reminded his listeners that he
was now in the job of President, and he could not run away from it and
shut the door and "hope that Mama would come and look for you."

By the middle of January the Presidency of the United States was
Lyndon Johnson's Presidency, brilliantly and effectively conducted. But
the honeymoon by then was over and the country was becoming aware
of other dimensions of Lyndon Johnson's personality and character.

These other dimensions of the Johnson personality were sensed
much more slowly than the political dimensions of Lyndon Johnson, for
they could be transmitted only by the ripple effect that characterizes certain kinds of news out of Washington. Even those who live in the nation's capital can only imperfectly describe this "ripple transmission."

News of personalities; anecdotes, episodes and scandal; the vital
overtones of alarm or anticipation begin in Washington sometimes with
no more than an overheard snatch of conversation, or a calculated leak
by telephone, or an often anonymous tip, or the impulsive outburst of a
government official when his emotions break through reticence and the
tone of his voice becomes more important than the facts he recites.
Washington boasts the finest press corps in the world—hundreds of energetic, devoted, scholarly and restless newsmen, as sensitized to a shred
of gossip or whisper of fact as radar to a distant cloud. Neither they nor
anyone else can describe the way their radar picks up gossip. But the
collective radar of the corps is part of the atmosphere of government.

Among these men, for years Lyndon Johnson had been a folk character, the man who dominated any conversation about the Hill—at once
respected, feared, admired, a huge yet often comic figure. His almost
daily briefings in the Senate when he was Majority Leader, his hunger for
recognition, his salty phrases, his virtuosity in negotiation and compromise, his professional dexterity as a lawmaker had enlivened Washington
conversation for decades, much as Casey Stengel's performances enliven
the conversation of baseball writers. But as President he now lived in the
White House in majesty. Had majesty changed the man?

The only way to learn was to hatch the answer by daily attention to the domestic life of the White House. Now, to the casual tourist the White House is the most open of all the great mansions of state in the world. A million tourists a year wander through it, and for certain hours every morning it is an absolutely public place. Not so, however, to those commonly known as "White House observers." For them the White House is a tantalizing place, tantalizing because news is more solid when it is gathered here, more important, yet more elusive, than anywhere else in the world.

For the men known as "observers," the White House begins in the business lobby of the West Wing—a huge room, in the center of which is an enormous table of Philippine mahogany, normally festooned with coats, cameras and news gear in an untidy pile; on the walls, some of the uglier paintings of American art; about the sides, black easy chairs and oversized black sofas for lounging. Just off to the left is the multi-cubicled, dingy, unbelievably overcrowded Press Room; and next to the Press Room an office of the Secret Service, which protects the President.[8]

Normally, the large lobby is neutral ground. Here visitors to the White House—dignitaries (domestic and foreign), recognizable Congressmen and officials, mysterious callers—sit and wait for their appointments while White House officials occasionally flit through and newspapermen mingle with them, hoping to strain a shred of news from the procession. Foreign chiefs of state enter by this lobby, the television cameras that show the nation the evening news are posted outside this lobby, occasionally the President himself will stride through, escorting some toweringly important figure, and drop a few words. Congressional leaders find it a convenient sounding board, which, if they strike it correctly, will echo their remarks to the entire nation. Newsmen who work here—and some old-timers would work nowhere else in the world—have the illusion, and occasionally the reality, of being at the center of things.

It is the conflict between this illusion and this reality that makes the White House beat so nerve-racking. Almost anything in the world can happen at any moment: wars may be announced, cancer may be conquered, Cabinet officials named, scandal erupt. If any of these things happens, the world will learn of it first from the men apparently lazing in the big black easy chairs. But then again, days may go by with nothing

[8] In the week this book was sent to press, President Johnson decided to enlarge the press room and redecorate the lobby described above. Newly lit and painted, its ugly oil paintings replaced by fresh American art, the lobby has seen the removal of its old furnishings and their replacement by glistening green leather armchairs and sofas in the style called "Dallas Modern." It is a gayer, pleasanter place, but the atmosphere and gossip remain the same. I have left the description of the original text, for this was the way it was in 1963 and 1964.

announced, no news made beyond the pronouncement of National Pancake Week, while all the world knows that matters of history are being transacted just a few feet away behind the desk in the lobby where sit, normally, one of two pleasant receiving ushers, Roger A. Hallman or Samuel Mitchell.

Both of these gentlemen are genial, soft-spoken and courteous—but they are symbols of separation. Behind their desk are two flags, and on the wall between the flags, the Great Seal of the President of the United States. Two doors open through this wall into a U-shaped corridor which rings the entire outer lobby and which, within, opens to the offices where the President conducts his business. Only the one wall and the two doors separate the traffic on the interior corridor from the outer lobby and the press. But when the President so chooses, this thin wall can be stouter and more impenetrable than the Great Wall of China. Here the Secret Service stops everyone not known to be a member of the staff or Cabinet or to have an appointment. If secret business takes place, members of the Cabinet and private visitors can enter by the side entrances or the south lawn and disappear unknown, unobserved and unreported. A President who chooses to conduct his private and public business in secrecy can cut the press off from news flow as effectively as the men who direct operations from the Kremlin.

The lobby, thus, is the first receiver of two kinds of information: hard news of state, whether formal statement of President, press secretary, Cabinet officer or visiting dignitary; and informal information, the shreds of gossip, whisper, rumor and extrapolation from silent, downcast or troubled visitors who may hurry through the lobby simply because it is closest to the street—or because they wish to drop a pebble into the news pool from which the ripples will reach, ring by concentric ring, all across the country. Rumors will speed from the newsmen here to the senior bureau chiefs, and from news bureau chiefs to the parlors and cocktail parties of Georgetown, where senior press and senior government intertwine socially in the evening. They will speed from Washington to publishers' homes around the country, they will creep into gossip columns and surface later in magazine stories, and slowly, increment by increment, after weeks or months of delay, the information of impression seeps into public opinion.

As the driving pace of events and Presidential activity finally slacked off in December of 1963, hard news again began to be amplified with gossip from the White House lobby, from Georgetown and from the Hill:

Lyndon Johnson had not changed—except to become more so.

There was, first of all, the vanity of the man. This was not a vanity of person so much as an obsession with self; and an obsession with self and self-performance so deep as to recall all the insecurities and awkwardness he had first brought with him to Washington from the hardship and reaching of his past. As a young Congressman back in the Roosevelt days, he was remembered as one of the best, most vigorous and earthiest conversationalists of the younger thinkers who were then remaking America. But he could go to a dinner party, talk like a man possessed of enormous force, then, when the conversation passed to someone else, droop his head, doze off in weariness as others talked, then come awake and seize the conversation by main force and carry it back to himself.

A correspondent who had known Johnson well before the war returned after the war and was visiting in Mr. Sam Rayburn's office when Lyndon Johnson dropped in to visit the Speaker. After Johnson left, the correspondent observed how he had changed in the war years, and Mr. Rayburn said, "Lyndon ain't been the same since he started buying two-hundred-dollar suits." As Senate Majority Leader and as Vice-President he would grow furious at indignity or neglect to himself in print—he could denounce in ugly terms an old friend who had reported on the enormous gold cuff links he wore with his shirts; or dress down like a top sergeant the eminent head of a broadcasting bureau who had been forced to cancel a broadcast with him.

As President he had not changed. On the evening after the ceremony of John F. Kennedy's funeral, as the State Department prepared to tape for posterity the international reception in the Benjamin Franklin Room, Walter Jenkins telephoned to State that the President wanted the broadcast to go out live—and make sure that the camera took only the left profile of the President. As December wore on and his enormous exertions became customary and his daily doings faded from the place of eminence on the front pages, he became at once irritated, embittered—and almost beseeching.

To his press officers and the press officers of the major Cabinet divisions he could be tough. Before going off for Christmas week, he assembled the senior press officers of Defense, of State, of Justice and half a dozen other departments at the White House, kept them waiting for forty-five minutes in a small chamber, then came to give them a brisk four-minute dressing down: he was going off to a reception with some important officials, and all he had to say was they better get on the ball. The White House had been on the front page with only one story that week—the lighting of the Christmas tree, and he had done that himself. He was going down to Texas now and they had to let Pierre (Salinger) have as many stories as possible so he could release them down there. Then, as a parting shot, the President added that he'd been checking the Budget and the government was spending almost a billion dollars on people like them and they better start earning it.

To newspapermen, during this period, he could be more kind and buddy-like than any President before him. Publishers and publishing executives, newspapermen and writers trooped through his office for lunches and dinners as frequently as Congressmen. On his first holiday at Christmas, in Texas, he decided to come up all the way from the ranch, sixty-five miles away, to the Austin hotel where the White House press was isolated and down his drinks with the boys at their New Year's party. And on the way back from Texas on the Presidential plane, courting them yet further, he told them that he was going to make big men out of all of them, they were gonna get along fine together; if they played ball with him, he'd play ball with them—and he didn't expect to see it in the papers if they saw him with one drink too many under his belt.

When, after a few weeks, this method did not seem to be improving his press relations, he went into his sulks, in the alternating cycle of courting the press and being angry with it that was to continue throughout the campaign of 1964. For Lyndon Johnson and the press were, and remain, natural enemies. This enmity can best be explained by something we describe as "the politician's optic." A politician can read a sentence about himself which begins, say, "Brilliant, effective, statesmanlike John Doe was momentarily confused as he rose the other day . . ." And the adjectives *brilliant, effective, statesmanlike* shrink to fine print, while the adjective *confused* swells out of the page into poster-size letters. If, on the other hand, a politician reads a dispatch about his enemy which begins, "Richard Roe, dirty, testy, bumbling solon of the old school, was in a civilized mood yesterday as he . . . ," then the adjectives *dirty, testy, bumbling* shrink to pica-size print and the single adjective *civilized* swells to double-size capitals. This is an occupational disease of politicians— just as it is for authors and actors, who similarly live by public approval or distaste; but in Lyndon Johnson this condition of the politician's optic was, in the early months of his Presidency, aggravated to an unprecedented extreme. And with this condition Washington newsmen had to live.

Then there were his manners, which hardly seemed to improve as he made the White House his home.

Socially, Washington is dominated—both in the press and officialdom—by graduates of the Eastern schools, with a system of courtesies, gravities, protocol and grace that are almost a binding code on the Executive branch of government and those who write about it. Lyndon Johnson's manners are those of the earth; and as much as the earth gives him the strength and the range to deal with all kinds of Americans from the high-executive level to the construction gang, so the earth has left its mark on his style and speech. It was not so much the speech, nor the mispronunciations, nor the simplicity of style ("Come on over here, honey, I'm kinda hard of hearing"—to a visiting lady) that began in December to filter into Washington conversation as the more earthy and barnyard idiom of a President who, when he normally relaxed, could reach the vernacular of the stable with more native ease than any President of modern times. Over-precious and over-sedate in manners, Washington was pleased that the President no longer speared meat at the table with the boardinghouse stab he once showed as a legislator—but winced to observe that his table manners were still largely country-style. Everyone enjoyed the story of the three New York broadcasting executives who had come to Washington in December with such solemnity, at the President's invitation, to lunch at the White House—and then, in amazement, been summoned to strip and join the President in the swimming pool after lunch, where they splashed, all four naked to the skin, in the water. Such stories, rippling out of Washington in embroidered exaggeration, helped to lift some of the all-pervasive gloom and balance such thoroughly resented farm-husbandry habits as cutting off the lights at the White House at night (for a saving of $2,000 a month, he said) and the throttling off of government limousines.

All agreed that these manners were entirely different from the manners and style of John F. Kennedy. Some found the manners of the new President simply coarse; others found them an expression of sturdy American plain living. It was best to think of them, however, as American-Rabelaisian. Those who had known the President as Vice-President and Senator, particularly those who had visited him at the ranch in Texas, had always returned with a much warmer picture after seeing him there. At the ranch all men were equal—Johnson sat at the head of the table and all members of his staff, from secretary to high official, dined with him in equality. As a host he was considerate of every detail of his guests' comfort, entertained them endlessly with earthy stories and rambling reminiscences, fed them copiously, kept them up until late at night until his vitality and hospitality exhausted them. But in the White House such manners were new, and it would take some time before Washington grew accustomed to his Rabelaisian private speech and style.

More disturbing, as the honeymoon phase of Johnson's public relations wore away, were the tales of his irascibility—an irascibility extending almost to cruelty. Here was a President who yelled and snarled—and whose closest private staff served him with an odd mixture of devotion and outright fear. He would snarl at the telephone operators at the White

House switchboard and berate them until one burst into tears; he could be harsh and threatening to the members of his Secret Service protective escort; he would repay old scores—and did. One startled member of the Kennedy group observed, as much in sorrow as distaste, "It isn't that he's mean to important people—he's mean to his servants." Later another Kennedy loyalist, having become accustomed to the drain of Johnson's appetite for work, and to the strain it imposes on others, said, "It isn't that Johnson abuses people—he simply dehydrates them." It was for his own personal staff, however—the old Johnson loyalists of previous years —that Johnson reserved his most exhausting impositions. Generally, the President treated the carry-overs of the Kennedy regime with utmost courtesy, understanding sensitivity and unfailing good manners—except for one episode.

The episode is worth noting for the light it sheds on both Johnson and the delicate tempers of men still in shock shortly after the assassination. On December 5th, an afternoon meeting gathered in the President's office a panel of government officials who were to discuss with him the ceremonies of conferring the Medals of Freedom on distinguished Americans for the year. It was pointed out to him that Kennedy had planned to make a forty-five-minute ceremony of the presentation, and as they talked the assumption of the group was, of course, that Johnson would do the same because this was the Kennedy plan. At which Johnson burst out that he was tired of people telling him he had to do this or that or another thing because it had been the most important thing in the world to John F. Kennedy—it would take him fifteen years to do it all. He wouldn't give the ceremony more than ten minutes for television—the American people were watching Jack Benny at that time. His manner was abrupt and harsh; George Ball, Undersecretary of State, was visibly upset. Theodore Sorensen left the room. The next morning the President telephoned Ball to apologize for his loss of temper. And a month thereafter, but for other reasons, Sorensen resigned from the White House.

The story is apropos—as are several others—for the light it sheds on a man under strain. So gargantuan are Johnson's energies, so wide his capacities, so spectacular was his initial performance that, like all good things, they were too easily taken for granted. Nor, where sympathy should be naturally forthcoming, does Johnson often receive natural sympathy.

Here was a man under enormous strain, of curious manners but of passionate devotion to his country, loving both the country and his government, seeking loyalty and disloyalties in an inherited staff whose hearts, he knew, still belonged to John F. Kennedy.

But he was President now. And he could wander through his White

House, and for weeks—until February—the pictures that hung on the walls of the offices of the White House were not pictures of himself, the President, but pictures of John F. Kennedy that no one had the heart to remove. Almost a usurper in his own mansion, sensitive to the fact that in his every move he was being compared to Kennedy, he had, nevertheless, to carry on government in his own manner. This was the period when his harsh, almost brutal treatment of his own people reached a peak. A man of great masculine physical vigor—a burly, husky specimen six foot three inches tall—he could outdrink almost anyone around him (Cutty Sark), and this was the brief period of his heavy drinking. Nor did the drinking bother those who knew him well, for no one has ever seen Lyndon Johnson befuddled by drink. It worried people concerned for him because it led his extravagant energies on to exertion even beyond his own physical capacities. Johnson tends to feed on exhaustion, driving his own body and nerves far beyond any reasonable capacity of human work, and alcohol has a tendency to invite a man to exert himself beyond the restraints that reason sets. For a man who had recovered from a serious heart attack caused by overwork only eight years before, this effort to stabilize, single-handed, the course of American government in the aftermath of assassination could be dangerous—not only to him but to the country, if a second catastrophe occurred and the Presidency fell to seventy-two-year-old Speaker John McCormack or the Senate President pro Tempore, Carl Hayden—eighty-six.

Yet there he was, performing beyond any normal human capacity; performing flawlessly as President, though less well as a human being; suspicious of those around him, yet at the same time trying to forgive; unable to translate himself to the idiom by which Kennedy had made the nation listen; conscious of his own style and resentful when comment was made on it in the press; and burdened by duty.

One can only see him as a man who receives a battlefield commission. As a President his style was new; but his commission had come not from the academy; he had been commissioned from the ranks, under fire, and if he brought barracks-room language and a top-sergeant crudeness to his tasks, he nonetheless knew those tasks better than any American interested in his country could have reasonably hoped. It was for this that John F. Kennedy had chosen him—as much as for his political value on a ticket headed by a Harvard Catholic.

By the end of January, one could begin to measure the new President.

He was, first, no man of words, not given to verbalizations or abstractions. When he thought of his America, he thought of it either in primitive terms of Fourth-of-July patriotism or else as groups of people,

forces, individuals, leaders, bosses, lobbies, pressures that he had spent
his life in intermeshing. He was ill at ease with the broad phraseologies,
purposes and meanings of civilization. Problems had to be brought to
him in the concrete, to deal with and solve. He did not like long papers,
or the supplementary papers of detail and speculation that John F. Kennedy had enjoyed; he liked to be briefed "to the gut"; and, usually, he
wanted the paper short while the author sat in the room with him to
answer questions. He did his homework as President, reading until late
at night, but not the way John F. Kennedy had done his homework.
Under Kennedy, meetings of the highest councils were attended by men
who assumed that the President was thoroughly informed and had done
all his prior reading; such meetings took on the cast of an appellate
court, with the search for alternatives going on in the presence of the
President, who would choose the final alternative. Johnson was frequently bored by long meetings; he called meetings for information, to
get facts, to educate himself. Yet while he hurried and beat upon all his
staff to perform their tasks, he would not be hurried himself. He had
the true executive sense of deliberation. Officials who would urgently insist they needed his signature on a document that very day were told
they should have thought of that three weeks ago or five days ago—or
simply that it was their hard luck. He did not like to be overconfused
with too many alternatives and options; information had to fit into his
own retrieval and information system.

Next—he was a man who believed in laws. So, too, had Kennedy—
but Kennedy had looked broadly and grandly out on the future of the
country with an intellectual perception of American life so sweeping as to
see Congress only as an obstacle (which it was) to the things that must be
done. But Johnson's perception of American life ran, not *above* the Congress, but *through* the Congress. The law is sacred to Johnson, however
odious the cooking smells on Capitol Hill. For Johnson, there must be a
base in law before action can begin, and Congress must sanctify the law.
It is certain, for example, that Johnson does not believe laws alone can
solve the racial crisis of America—but he does believe that until a base
in law is set on the statute books, no action can begin which will clear
the way for conscience to function.

And, finally, it became obvious that Johnson, for all the blurred
outlines of a man who had spent thirty years in compromise and bargain-
ing, was at heart a Texas liberal, not a Texas conservative. "Liberal" is a
word whose edges have been worn as dull as the word "conservative."
But one could discern in Johnson's entire career, and now in his first few
months of Presidency, a native grass-roots definition of liberal—something for everybody, you get yours, and he gets his, and we all share

around what there is to share. It is a practical code, harder and thus more durable than the word "liberal" as defined by campus scholars or Northeastern trade unions. Enemies of Johnson have tried—and all through 1964 tried intensely—to define this liberalism with a gloss that makes it read: You get yours, and he gets his, and I, Lyndon Johnson, get mine too. Yet it is a code which Johnson applies not only to personal life, but to all American domestic affairs, and which, extrapolated into international affairs, bears the seeds of peace. By the end of January, 1964, the course of legislation in Congress and the major decisions in the Budget had made clear what the political Johnson was all about.

It is difficult to say when the politics of the Presidency actually resumed. It was several months before Johnson became an active participant in the mechanics of his re-election, but his own experience had taught him that the best way a President can present himself for re-election is by doing the job well. A moratorium of thirty days after the assassination had been accepted by all politicians, Republican as well as Democratic. There followed then a confused period in which the President was half hallowed as successor to the martyr, and half human as Lyndon Johnson, subject to criticism.

Arbitrarily, then, every chronicler must make his own definition of when the politics of 1964 began. The final week of January, 1964, is as good a choice as any. In January of 1964 the simmerings and sputterings of the unstable world, at home and abroad, began to press Lyndon Johnson to his own new decisions on the front pages. Zanzibar's minuscule stupidities and convulsions raised the problem of Africa. General de Gaulle decided to recognize Red China. Students had rioted in the Panama Canal Zone and four U.S. soldiers had been killed there. For the first time the press insisted that public attention be paid to the Bobby Baker scandal, and the President felt constrained to discuss crisply the gift of a stereo set that Bobby Baker had made to him. Sorensen had resigned on January 15th. Now, in the last week of January, another Kennedy man quit—Arthur Schlesinger, whom, on assassination weekend, the President had begged to remain.

And much more important, though inconspicuous, happenings took place in the last week in January which were to foreshadow the politics of 1964. In New York City, discussions between the Board of Education and the extremists of the civil-rights faction over integration of the New York school system broke down—to be followed shortly by the first school boycotts in the nation's greatest city, capped finally in the summer by the bloody riots that announced a new menace to the domestic tranquillity. Another American plane was shot down on the approaches to

Berlin. In Vietnam the taste of possible disaster became material as General Duong Van Minh, the CIA's choice to rule Vietnam after the killing of the Diem brothers, was himself thrown out of office by a fresh coup installing Major General Nguyen Khanh (who would in turn, in ten stormy months, be thrown out by another coup which was predecessor to yet another as the Southeast Asian front worsened).

And in this last week of January the Republicans, restive under the unnatural moratorium on politics imposed by tragedy, turned to the task of defeating Lyndon Johnson in his first try at his own national election. All across the nation, at a series of GO-DAY dinners on January 29th, Republicans convened to begin the assault on the White House which would open the politics of the year and climax on November 3rd, 1964.

CHAPTER THREE

THE REPUBLICANS:
RENDEZVOUS WITH DISASTER

I T was impossible to foresee, as the calendar turned the pages of the new year, just how great would be the Republican disaster of November to come. But the forecasts were already dark.

All through the three transition years of Kennedy leadership, as the nation moved from the postwar world into the unnamed present, the Republican scenery had been alive with the restless comings and goings of men stalking or seeking a Presidential candidate. These ambitions, conspiracies, bustlings and explorations, which had begun as early as October of 1961, were, of course, entirely normal. But what was abnormal, even before the death of John F. Kennedy, was the chaos that kept surfacing among Republicans more and more often as 1963 wore its way to the end. The bitterness was of a new order of intensity; and it had begun within weeks after the election of 1960 had closed the truce of the Eisenhower years.

The election of 1960 could have been called by no man a Republican disaster. The tiny, almost invisible margin of Democratic victory had made it the closest run in American history. But the very narrowness of the defeat had brought on Richard M. Nixon the most violent abuse of both wings of his Party—if he had just made this speech, they chanted, or taken this position, they complained, or adjusted his stance that mite, said both his enemies and former friends, victory might have been his. Thus, instead of credit for a hard-fought near thing, Nixon, always a loner, found himself patronized and scorned by left *and* right of his Party, both of whom felt that his rejection of *their* advice had booted away the Presidency. Neither side could forgive the solitary Californian his solitary effort to apply the same stickum that Eisenhower had applied in his Presidency over the growing civil war in the Party.

This civil war in the Republican Party of the United States is one of the more fascinating stories of Western civilization. Historically, its solution may determine whether any nation in the Western world can adjust the traditional system of open enterprise to public need or whether enterprise, in the classic Western sense, must perish because the Republicans have let the word become a fossil. The rift is deep, a subject for a philosopher of government more than for a reporter. Yet one should touch, briefly, on the difference between America's two great parties before plunging into the Republican dilemma.

The Democrats believe in government—government as an instrument to do things. Thus they can promise all their client groups with complete sincerity that, when elected, a Democratic government will do something to help each of the groups. These promises are often contradictory; and when Democrats win, they find their promises all too frequently paralyzed in a Congress of feuding Democrats whose public contracts defy each other. Yet, nonetheless, the Democrats are the party that guarantees government will do things.

The Republicans' impossible dilemma is that they have never sorted out properly what it is that government should do and should not do—and at what level. Perhaps no political party should be asked to make an orderly philosophical sorting out of governmental responsibilities. But the Republicans' largest client group, business, wants the government to do nothing—to leave it alone; other client groups, chiefly inspired by the old Protestant ethic of individual salvation, feel morally that the individual must be left master of his own destiny. In the years of their governing of the United States, the Republicans have, of course, immensely added to the reach of government activity, and more than honorably acquitted themselves of great tasks thrust on them. From the establishment of the Department of Agriculture in 1862 to Eisenhower's Federal Highway Act of 1956, the record of Republican achievement is one of massive acceptance of government responsibility.[1] But the Republicans do not campaign on such issues. They campaign, generally, *against* government; the Democrats campaign, generally, *for* government. The Republicans are for virtue, the Democrats for Santa Claus. These are the rules of the game, implacably stacked against Republicans, and in their attempt, year after year, to solve this dilemma the Republicans confuse one another and the nation, and, in defeat, distill a bitterness among themselves greater than their bitterness against Democrats.

[1] See *The Making of the President—1960*, Chapter Three, for a brief review of the Republican Party.

By the fall of 1963 this bitterness within the Party had reached a condition of morbid intensity, and battle lines were more clearly drawn for struggle than at any time since the Roosevelt-Taft cleavage of 1912.

One must set the stage carefully to see what was about to happen. Each of the two sides had ranged the other and had given the other a name of contempt: on the one side were "the primitives" and on the other "the Eastern Establishment." Both names are corruptions of reality. But they are convenient shorthand for describing a twenty-year period of politics, from 1940 to 1960, during which the Eastern Establishment dominated the national machinery of the Republican Party and its major governorships, while the primitives dominated its Congressional machinery and set its national lawmaking posture.

One must start, in 1964, with the apparently dominant Eastern Establishment—with what it appears to be and what it actually is. From beyond the Alleghenies, the Eastern Establishment seems to inhabit a belt that runs from Boston through Connecticut to Philadelphia and Washington. Its capital is New York, a city shrouded in a blur of symbolic words like "Wall Street," "International Finance," "Madison Avenue," "Harvard," *"The New York Times,"* "The Bankers Club," "Ivy League Prep Schools," all of which seem more sinister and suspect the farther one withdraws west or south.

Technically, of course, the capital of the Establishment, New York, can easily be described as the executive center of the Western world. Here is the center of the cobweb of world finance; here is the greatest pool of investment money on the globe; here is the interlocking machinery of the idea-and-word business. Here are the image-makers, the idea brokers, the dream-packagers. Such executive services as the Establishment offers out of New York are rarely available to New York City or State; it uses New York as a base from which it intervenes around the entire world and across the nation. More than London ever was, it is a headquarters town for sophisticated expertise. The Persian economy was reorganized by men working out of New York; the Franco-British cross-channel tunnel scheme was revived by two young New Yorkers; the world center for the war against cancer lies in a four-block area on Manhattan's East Side; Robert A. Taft's victorious campaign against the "intruding" CIO in Ohio in 1950 was masterminded by public-relations talent who produced some of Taft's best speeches from an office in Rockefeller Center.

In politics, however, it is best to look at the Establishment capital in New York as a rather large village or a peculiar kind of community which has brought together, by an indefinable selection process, a group of men with the same kind of worries, families, little quarrels, private

pleasures as other men elsewhere, yet at the same time set apart from
other men by their enormous executive ability, superior responsibilities
or great wealth.

There are several neighborhoods in New York which the Eastern
Establishment—if there is one—makes particularly its own. It dwells,
generally, on the east side of Manhattan in the Perfumed Stockade that
runs from the East Nineties south to the East Fifties and from Fifth
Avenue east to the river. There is, indeed, no greater assembly of executive ability, inherited wealth and opinion leadership in all the world than
is domiciled in this Perfumed Stockade. Four Rockefeller brothers live
here, several Whitneys, three Kennedy families, six Harrimans, half-a-
dozen Strauses. There are Sulzbergers, Bakers, Millikens, Clarks, Sloans,
Pierreponts, Roosevelts—old names without number—but also an equal
number of more important new names. Their children go to the same
half-dozen Manhattan private schools (all excellent) whence they emerge,
go on to boarding school, and hope to attend proper Ivy League colleges.
They meet at dances, charities, art festivals, dinners as neighbors. It is by
no means a closed community; but raw money cannot buy entrance. It
may be said of old New York that its members resist change of their own
ways, but welcome change-makers in their midst. Fearful of appearing
themselves in the newspapers except at birth, marriage or death, they enjoy the company of those who do make news. Persons who make a mark
in the world—be it in industry, politics, art, science, writing—gradually
find acceptance in the Perfumed Stockade. And then discover they enjoy
it. Those who come to New York to live when great corporations summon them to headquarters stay on to watch their children grow up and
marry. And, as they age, they find there companions of their own high
level of interests, with whom they live out happily their fading years.

Washington is only an hour away by shuttle plane, and Paris or
London only seven hours. The ambience of conversation and contact here
is one of great affairs. All the United Nations embassies cluster in or
around the Perfumed Stockade; eminent visitors from abroad or from the
hinterland linger longer here than in Washington; many linger and stay.
Three former High Commissioners to Germany live here (none native to
New York); two former Ambassadors to England, fourteen former Cabinet members of various administrations (five of Dwight D. Eisenhower's last Cabinet, including Richard M. Nixon), and sub-Cabinet
members and generals in countless numbers. The community inter-
meshes a span of American life that reaches from the brightest young
playwrights of Broadway (if successful) to such aging patriarchs as Herbert Hoover and Douglas MacArthur (born, respectively, in West

Branch, Iowa, and Little Rock, Arkansas), both of whom chose to see
the end of their days at the Waldorf Towers.

The fact that this community is suspect to the broader nation beyond the Alleghenies comes generally from the strange way in which the
Establishment fathers of families earn their generous living or shepherd
their fortunes and corporations. And this suspicion is generally summed
up in two phrases: "Wall Street" and "Madison Avenue," which can
always rouse a healthy growl at any meeting in the hinterland.

Both of these symbol phrases are so generally misread that they
deserve some clarification.

What is essential to grasp, first, about Wall Street today is how
much it has changed—reflecting the nation's changes over the past
twenty years. There are no longer any individual titans of finance in New
York. The Chase Manhattan Bank, whose President is David Rockefeller (the Governor's brother), is, indeed, the last great bank influenced
by an individual family fortune. But the Morgans, the Bakers, the Selig-
mans, the Schiffs no longer dominate any bank. Instead the great banks
and insurance companies are managed, as most great corporations are
today, by skilled, carefully selected managerial personnel who, as executives, are paid generously but who lack private fortunes of their own and
worry about the tax bite and the college bill just as do all other Americans. Predominantly, they are intelligent, hard-working men—of small-
town origin. The Chairman of the Chase Manhattan is George Champion
—from Normal, Illinois. The president of the First National City is
George S. Moore—of Hannibal, Missouri. The chairman of the Manufacturers Hanover Trust is R. E. McNeill, Jr.—of Live Oak, Florida;
and the president of that bank, the scholarly Gabriel Hauge, is a
preacher's son from Hawley, Minnesota. The chairman of the board of
the Chemical Bank is Harold H. Helm—from Bowling Green, Kentucky,
and the president, William S. Renchard—from Trenton, New Jersey.
The chairman of the board of the Morgan Guaranty Trust, the most
elegant symbol of all American high finance, is Henry C. Alexander—
from Murfreesboro, Tennessee. One could go on—but these are the five
largest banks in the greatest financial center of the world.

If anyone is part of the Eastern Establishment, then these men are
part of it. They came there not by inheritance, however, but by hard
work and luck, and remain subject to dismissal for failure—hired managers of contractual rivulets of savings which, under their direction, ac-
cumulate into the fantastic financial pool of fluid capital that makes
New York the world's banker. As managers they perform a service function—they make credit available to the great industries when need calls
for expansion. But they can no longer, as they did fifty years ago, dictate

the decision of industry. General Motors, American Tel & Tel, U. S. Steel make their decisions about the geography of American investment, at home and abroad, by the cold logistics furnished by their own figurings. Then they use the great banks as a home owner uses a savings-and-loan association—to help him do what he has already decided to do—and shop for the best interest rate where they can find it, or conscript capital from insurance companies or out-of-town balances.

There is another change, too, in the power and character of these New York banks. As the country has changed since the war, so has its financial geography. Swelling population, swelling industry, the shift of energies and the gushing of regional wealth have created great rival centers of finance in America since the war. Los Angeles, San Francisco, Detroit and Dallas have been added to the great traditional centers—New York, Philadelphia, Chicago and Boston—as sources of capital financing. New York's share of the banking resources of the nation has shrunk from 25 percent at the end of World War II to 12 or 13 percent today.

Wall Street can no longer command. Yet it still leads. The Wall Street men do not meet collectively at The Bankers Club, nor can they be gathered any longer in a cozy group in J. P. Morgan's study to turn on or off the spigots of credit in America. Yet they set the climate. They know that a threat to the British pound is as much a threat to Liverpool, Texas, as to Liverpool, England. They can see the meaning of gold outflow long before anyone else. They must finance overseas investments and discount bills of foreign trade in every currency in the world. They are neither internationalists nor liberals. But they must live with the world as it is—as it is described to them daily in the flow of information from every capital in the nation and around the globe: information on personalities, economics, markets; information public, secret and diplomatic. A dogmatic man could not survive as head of a Wall Street bank, nor could an isolationist. Only in this sense are they liberal. Republican almost to a man, the Wall Street bankers and corporate executives must learn to live in a world of constant adjustments. They are, to be sure, "businessmen," but they can be compared to Midwestern industrialists or a Phoenix, Arizona, merchant only in the distant sense that one can compare an elephant and a sheep as herbivorous animals.

Republicans also dominate the other great enterprise of the Establishment—the idea factories of "Madison Avenue"—but they must share this domination with a highly vocal component of Democrats and a subordinate staff of "creative" or editorial talent largely Democratic.

Madison Avenue is even more difficult for strangers to understand than Wall Street, even though Madison Avenue toys with the spirit and the dreams of strangers across the land in a way far more intimate than Wall Street. "Madison Avenue" is more than a single lane; it is a twenty-block-by-four-block area in which all the communications in America are gathered into the single most complicated switchboard of words, phrases and ideas in the world. All three national broadcasting networks operate from here. So do both major wire services; so do 90 percent of the great book-publishing houses of America; so do most of the major magazines—*Time, Life, Fortune, Newsweek, The Saturday Evening Post, McCall's, The Ladies' Home Journal*—as well as *The Reporter* and the *National Review*. The hundreds of thousands of people who work here (and who drink, chat, gossip and swap schemes or fancies with one another) are without doubt the largest community in the world that lives by wit, words and communication. Several facts must be signaled about them: First—since they live by words, they deal constantly in abstractions, in phrases, in intellectualizations that make a house jargon, a neighborhood dialect as unfamiliar to the rest of America as the automobile talk of Detroit or the oil talk of the Petroleum Club in Dallas. Secondly—since they live so closely packed, their interconnections form a switchboard for all transmission of American ideas and fashions, and anyone who lives in New York thus has immediate access to the largest megaphone and finest brainwashing system the world has ever known. Lastly—except for the local New York newspapers, the commanding summits of this world are occupied by Republican small-town boys quite similar to those who dominate Wall Street—by men like Henry Luce (born in Chefoo, China) or Gardner Cowles (in Algona, Iowa) or Norton Simon (of Fullerton, California) or J. Wes Gallagher of the Associated Press (of San Francisco) or the late Roy Howard (of Indianapolis, Indiana) or David Sarnoff (of Uzlian, Minsk, Russia) or Leonard Goldenson (of Scottdale, Pennsylvania)—or the *Wall Street Journal's* editor, Vermont Connecticut Royster (of Raleigh, North Carolina).

All these leaders in Wall Street, in Madison Avenue, in the Perfumed Stockade (Republicans and Democrats alike) share a great concern for America. They and their law firms (where the old native New York blood runs bluest) know they must get along with government; they know that though they influence government, government influences them even more; good government, understanding government, is to them a prime condition of their own prosperity. They are concerned with the condition of American life as much as any men anywhere in the country—and more aware than any other group of its infinite complexity.

If one could choose a single institution which illustrates how profound and important is their concern for American policy and destiny, one might choose, say, the august Council on Foreign Relations, set center in the Perfumed Stockade at the corner of 68th Street and Park Avenue in the old Pratt mansion. The Council counts among its members probably more important names in American life than any other private group in the country—not only ex-Presidents, ex-Senators, ex-Governors; not only executives at the summit of all the great banks and industries headquartered in New York; but scholars, writers and intellectuals too. The meetings of the Council are deeply, profoundly concerned with government and America's role in the world. Its roster of members has for a generation, under Republican and Democratic administrations alike, been the chief recruiting ground for Cabinet-level officials in Washington. Among the first eighty-two names on a list prepared for John F. Kennedy for staffing his State Department, at least sixty-three were members of the Council, Republicans and Democrats alike. When, finally, he made his appointments, both his Secretary of State (Rusk, Democrat) and Treasury (Dillon, Republican) were chosen from Council members; so were seven assistant and undersecretaries of State, four senior members of Defense (Deputy Secretary of Defense, Comptroller, Assistant Secretary for International Security Affairs, Assistant Secretary for Manpower), as well as two members of the White House staff (Schlesinger, Democrat; Bundy, Republican).

I cite the Council on Foreign Relations not to emphasize the centralization of the Establishment, but to emphasize its brooding concern for America's larger position in the world which is the atmosphere of higher New York executive life; and because this sense of national responsibility is so rare in any country.

For the last fact to be stated about the Eastern Establishment is that it is *not* centralized. It shares a common code, indeed. But it is divided among Republicans (predominantly) and Democrats; and the dominant Republicans are themselves split and torn by groups, factions and personal antipathies which give to no one man, no one name, no one bank absolute and unquestioned leadership—a fact that was soon to become evident in 1964.

It is the dominant Republican element of the Establishment that concerns us here—for, as the curtain rose on the events of 1964, it was these men who apparently held the power in their Party. For an unbroken twenty years the Establishment of the East Coast had dominated the conventions of the Republican Party; in any floor fight it had been able to prevail by finding beachheads of Midwestern Republican Governors (Stassen for Willkie in 1940, as floor manager; Youngdahl for Eisenhower in 1952) about whom other Midwestern groups could be led to rally. The Establishment Republicans were content to let the primitive element write the record of the Party in Congress, if they could choose the Executive candidate and thus set the image of the Party. In 1964 they were to find, too late, no single Midwestern progressive to join them and frustrate Goldwater.

Their hegemony dated back to 1940, in the world of yesterday, when they had compelled the party to name Wendell Willkie as Republican nominee by a combination of every kind of pressure the publishers, broadcasters and bankers could collectively exert around the country. Through the provincial banks, then in trailing-strings to New York banks, the Wall Street men could mobilize for their Indiana favorite almost inexorable pressure on local businessmen who controlled delegates. (One Eastern political observer later reported, "I bet on money—not just any kind of money, but old money. New money buys things; old money calls notes.") From 1940 on, under the foremanship of Dewey, then under the Presidency of Eisenhower, the Easterners persisted in the illusion that they still could, when the chips were down, control the nominating mechanism and thus the image machinery of the Party. Yet as the years wore on, the Easterners were almost unaware of how the country was changing. They knew professionally that great rival banks were growing up in the West and Far West, that great new industries were surging with a power they could not control. They could spot population shifts and market shifts, for that was their business. But they could not grasp that, in the world of politics, faces and forces change far faster than in the world of business; voters' loyalties are more fickle than those of a board of directors. The men who had imposed Willkie in 1940, won the nomination for Dewey in 1944 and 1948, frustrated Taft and installed Eisenhower in 1952, still met as neighbors or companions in New York—and still thought their power of veto over the Party was undiminished.

Yet the Party was changing. It was changing at home in New York, where Nelson Rockefeller, never a favorite of the Establishment, was beginning to grow larger and larger. It was changing, too, in the West, where new names of which they had never heard were rising. It was changing most dramatically with the echo of a man named Barry Goldwater, who gave them the *defi* proper, an outright challenge, a taunting enmity which never, even in their wars with Taft, had they experienced before.

Slowly, as they came into the fall of 1963, they became aware that if Barry Goldwater meant what *he* said, and Nelson Rockefeller meant what *he* said, then the Party was on collision course. Even before the assassination, as this became clear in downtown New York and in the Perfumed Stockade, the more vigorous leaders of the old Dewey-

Eisenhower team began to study the situation to see what could be done to avert the disaster that lay ahead.

For months, in their uncoordinated gropings, the old team had vested their hopes in the old lion of the Party, the hero who had never known failure—Dwight D. Eisenhower. It was impossible, agreed the few remaining veterans of the old triumphs, to stop either Rockefeller or Goldwater without a leader; and only the old leader had a name which was still magic. Yet the old leader, now in retirement, found politics even more distasteful than he had in his prime.

A genial man throughout his life, Dwight D. Eisenhower has always acted best when duty was clearest. Given the clean assignment to destroy the Nazis or armor the West, he could perform such historic master-pieces as the invasion of France or the organization of NATO. But poli-tics was a murky world, alien to him. Even after the eight years of the Presidency, he could lament in 1964 (after the San Francisco conven-tion) that "I was never trained in politics; I came in laterally, at the top. In the service, when a man gives you his word, his word is binding. In politics, you never know." [2]

His attitude to politics in 1963 was still as fresh and honorable, but as simple, as that of a teacher of high-school civics; he felt that with good open discussion of issues and candidates the American people would always reach the best choice. The cookery and jiggering and fix-ings that go on in practical politics always annoyed him. So did the con-stant importunities, the demands made on him since his departure from the Presidency, the efforts of so many to grab and use his name as a factional label, to bend his great reputation to serve what other people considered the need of the Republic or the Party. In July, 1964, on his transcontinental train trip to the San Francisco convention, he amazed an old newspaper friend by suddenly quoting verbatim and with enor-mous feeling a passage from Stephen Vincent Benét's *John Brown's Body,* in which Lincoln is caused by the poet to say:

> "They come to me and talk about God's will
> In righteous deputations and platoons,
> Day after day, laymen and ministers.
> They write me Prayers from Twenty Million Souls
> Defining me God's will and Horace Greeley's.

[2] A former close working associate of the great war hero read this passage, disagreed, and commented, "I don't think Ike really found politics distasteful; in fact, I think he was beginning to like it more and more. But his motor reactions to it tended to be confused and contradictory."

God's will is General This or Senator That,
God's will is those poor colored fellows' will,
It is the will of the Chicago churches,
It is this man's and his worst enemy's.
But all of them are sure they know God's will.
I am the only man who does not know it.

And, yet, if it is probable that God
Should, and so very clearly, state His will
To others, on a point of my own duty,
It might be thought He would reveal it
Directly, more especially as I
So earnestly desire to know His will."

All through 1963, deputations and platoons of politicians had been buzzing through his farm at Gettysburg, telling Eisenhower what was God's will for the Republican Party and what he must do, whom he must support, how he must use his strength. Yet Eisenhower would not move.

In his own mind, Eisenhower knew but few men who, in his opinion, would make great Presidents. Had any of them sought his help, he would have bestirred himself and taken active leadership. The war hero listed them thus: His first choice was Robert Anderson, who had been his second Secretary of the Treasury (but Anderson did not want the job on any terms; later, in 1964, he actually supported Lyndon Johnson). His next choices were two war companions, General Alfred Maximilian Gruenther and General Lucius Clay; both of these are men of extraordinary ability; but Gruenther was ailing, Clay had no lust for the Presidency, and both were realistic enough to know they stood no chance for the nomination. For any of these three, Eisenhower would have saddled up and battled. And there was a fourth choice: his own brother, Milton Eisenhower. At Eisenhower's birthday party that year on October 15th, a gathering of his old Cabinet and political friends in Hershey, Pennsylvania, he had put Milton on the dais, seen to it that Milton was the only speaker of the evening besides himself. Friends of the old President were discreetly exploring the support for Milton around the country with the old President's knowledge and encouragement.

But Milton, as far as most of the New York group went, was out. ("It would have destroyed the brother issue," said one. "How could we hammer Jack and Bobby Kennedy if Ike was running *his* brother for the Presidency?") Moreover, Milton would not risk political exposure; he shrank from it; he showed no aptitude for political visibility.

Not being able to command the nomination of any of these four, Eisenhower had contented himself in August, 1963, with listing ten eminent Republicans who, in addition to Goldwater and Rockefeller, he thought were Presidential timber. He hoped the good civics of the Republican Party would choose among them in open discussion.

This attitude did little to satisfy the downtown New Yorkers whose operations on behalf of Eisenhower himself in 1952 had borne little resemblance to civics as taught at West Point thirty-five years before. They knew they could not attack Nelson Rockefeller at home in New York without tearing the state Party apart; they knew that the nomination of Barry Goldwater might permanently wreck the national Party. If they were to create a new candidate, then the fall of 1963 was already very late; and nothing could be done unless Eisenhower took an active role.

Eisenhower was due in New York to accept an award at Columbia University on Thursday, November 21st, 1963. The following Saturday afternoon they hoped to gather with the General and make him see reason: that Milton could not be nominated, that Goldwater and Rockefeller would split the Party, that they must find a new candidate—and he must lead. Lucius Clay was to be at the meeting. So, too, was Herb Brownell, Dewey's long-time political chief of staff. Dewey himself was reluctant to be officially a part of the Rockefeller opposition. ("You have to remember," said one of the group, "that Dewey's been a New York Republican all his life and he can't oppose a Republican Governor of New York—not until Rocky's out of it. Besides, he has his law business to consider. He doesn't handle any Chase Manhattan business. But still, he'd have to talk to his partners before taking a public stand.")

Thus, on assassination weekend the old Eastern Establishment considered the situation already desperate. Dewey, as we have seen, was lunching with Rockefeller when the news of the assassination came. Eisenhower was a guest at a U.N. lunch at the Chatham when he heard. And, suddenly, the tragedy erased politics: Eisenhower flew to Washington to meet Johnson; there was no Saturday meeting.

For a few weeks more several friends continued to explore the possibility of making a candidate of Milton Eisenhower before reporting to the General that it was out of the question. But by then it was January, and the Rockefeller and Goldwater forces, full steam up and flags at battle staff, were heading for the first clash in the White Mountains of New Hampshire.

The crew that Nelson Rockefeller took into battle in New Hampshire was one of the most elaborate ever to enter political war in America in modern times.

Tempered by the experience of its 1960 clash with Richard M. Nixon and beefed up with new talent, the Rockefeller staff of 1964 was a thing of splendor. Its ambassador-at-large and closest man to the candidate was George Hinman, who had become a veteran of national politics without losing the gentility of manner which had won him so many friends in 1960.[8] Hinman had added to the ambassadorial staff for national contacts the young and extremely promising Robert R. Douglass. Dr. William Ronan, of the Governor's Albany staff, had found a way of combining his professor's knowledge of political theory with an administrator's knack for cutting to the heart of things. Charles F. Moore, once vice-president of the Ford Motor Company, had volunteered as strategist of public relations as much out of determination to stop Barry Goldwater as out of simple gaiety of spirit.[4] A new and labyrinthine research department was directed by Roswell Perkins. Robert McManus, the press secretary of 1960's effort, had by now acquired an almost encyclopedic knowledge of the important newsmen and news outlets across the continent. Thomas E. Stephens and Anne Whitman of Eisenhower's former White House staff offered connections to old Eisenhower loyalists around the country. Carl Spad, later to be New York State Republican Chairman, controlled schedule. Professor Henry Kissinger of Harvard had replaced Emmet Hughes as chief theoretician on foreign affairs, and Hugh Morrow prepared speeches. Capping them all was one of the outstanding campaign directors on either side, John A. Wells, senior partner of the law firm of Royall, Koegel and Rogers, and a splendid administrator. Beyond these were second-string members of the Rockefeller staff in such quality and such numbers as to staff several campaigns.

Lavishly funded, high-minded, shrewdly deployed, enthusiastic and tough, this team had been put together over three years of effort. The first and initial survey of the Presidency had taken place within a few weeks of the election of 1960, at the Rockefeller grounds near Tarrytown, New York, in the playroom on the Rockefeller estate. Successive and irregular meetings in the executive dining room of the Radio

---

[8] The Rockefeller people had lost, of course, their chief theoretician of the 1960 campaign, author Emmet John Hughes, who had quit the Rockefeller service to become a columnist. It should be noted that they still suffered, however, from Eisenhower's bitterness at Hughes's memoirs of the Eisenhower years, the devastating *The Ordeal of Power*. Eisenhower's ire at Hughes's deed spilled over into a smoldering dislike of Rockefeller, who he felt had permitted or authorized it.

[4] Having lost with Rockefeller in 1964, Moore went back to his native Massachusetts and campaigned at the grass roots of Cape Cod, insisting that citizens *must* take an interest in politics. In the spring of 1965, having left Presidential politics behind, he was elected selectman of the town of Orleans in the largest vote of the town's election history, by a margin of 1039 to 432.

City Music Hall had followed as an informal strategy board surveyed the information from around the country and its developing mood. The gubernatorial campaign of 1962 had been won by a margin large enough to keep Rockefeller the dominant contender for his Party's nomination. By the spring of 1963 a campaign biography for publication in early 1964 was being written by Frank Gervasi. And by the fall of 1963 the planners had begun to grapple with their two major problems.

The first was the nature of their hero.

Their hero was one of the wealthiest men in the world; he was one of the most stubborn men in the world; he was also one of the most high-principled. And he was rough. His enemies called him, quite simply, the most ruthless man in politics. But what in other men would be simple arrogance was in Rockefeller the direct and abrupt expression of motives which, since he knew them to be good, he expected all other men to accept as good also. He could name his cousin (Richard S. Aldrich) City Councilman of New York over the protest of all New York City Republicans at this bit of nepotism, simply because he *knew* Aldrich to be the best man. He loved government—he delighted in the complexity of problems, as problems; he loved mauling them, taking them apart, putting them together again. His campaign speeches in the early primaries were to drive some of his associates to despair by their earnest, didactic quality; when a questioner at an open meeting would ask about a matter of public policy, Rockefeller would respond with a cascade of fact, figure and detail as if the questioner did *indeed* want to learn as much about government as Rockefeller could tell him.

When he said that he put principle above party, he meant it. The Rockefeller family, for example, has abolitionist roots that go back through a century of American life. His great-grandfather Spelman had run a station of the Underground Railroad in Ohio before the Civil War, to help runaway slaves get to Canada; his family had endowed the first Negro women's college (Spelman College) in 1876 before it was fashionable to send white women to college, let alone Negro women. In all, they had given approximately $80 million to Negro causes and institutions. Rockefeller would as soon repudiate these roots as the American flag. But his counselors moaned that even an abolitionist running for President would not go as far as Rockefeller went. If there were to be any hope of peeling off Southern delegates from Goldwater at the convention, why, *why* did he have to encourage the Negro revolution of 1963 in the South just at that time? Two gifts of $5,000 and $10,000 each to Martin Luther King's Southern Christian Leadership Conference were openly on record. But what was not announced was even more upsetting: a loan of $20,-000, arranged through the Chase Manhattan Bank, to the firebrands of the revolution, the Student Nonviolent Coordinating Committee. Principle demanded it, and principle committed Rockefeller. Principle had similarly demanded a three-year stress, as Governor, on the politically unappetizing project of fallout shelters. He had felt it was up to him to protect New York.

Principle defined, politically, the type of candidate the Rockefeller men had to run—a hard-line foreign-policy man, committed to stiffening America against Communism and increasing its military wallop. Yet, at the same time, a man who supported the Test Ban Treaty and the United Nations. He was for the most rigidly balanced Federal Budget and sound fiscal policy; yet he was for Medicare, and in his state budget, aid to education had doubled in his six years of office; state college scholarships had tripled. It was difficult to enclose Nelson Rockefeller in any neat political pigeonhole except that he yearned for the post of President, that he was supremely convinced that he could direct the problems of the United States, and that his talent, energy and experience made him obviously one of the rare men fully equal to the great task. John F. Kennedy had once observed that if it had been Rockefeller rather than Nixon he faced in 1960, Rockefeller would have won. Rockefeller shared that opinion—his problem as he saw it was not how to manage the United States but how to get the Republicans to nominate him and let him try.

Which brings us to the second problem of the Rockefeller campaign—his personal life. For one can no more discuss the Republican politics of 1964 without dealing with Nelson Rockefeller's divorce and remarriage than one can discuss English Constitutional development without touching on the stormy marriage of Henry II and Eleanor of Aquitaine.

It would be good if the private lives of public figures could be sealed off from their political records, and their leadership discussed as an abstract art in the use of men by other men. The politics of an open democracy, however, dictates otherwise. Men and women both vote, and they choose a leader by what they catch of his personality in the distortion of quick headlines. Yet the private lives of public figures are as three-dimensional, as complicated, as unyielding to interpretation by snap judgment as the lives of ordinary people. And the divorce and remarriage of Nelson Rockefeller offer a classic example of an immensely complicated tangle of personal tragedies distorted by quick summary into oversimplified scandal and blame. They warrant exploration as much to show how greatly public report distorts, as for their shattering impact on the politics of the Republican Party in the seeking of a candidate in 1964.

One must draw back and view the critical episodes in the tale of Nelson Rockefeller's divorce and remarriage against the background of his growth—against the backdrop of the Pocantico Hills estate at Tarrytown with all its four thousand acres, the old brownstone mansion on 54th Street in New York, the splendor of sea, rock and surf in the Rockefeller retreat at Seal Harbor, Maine.

It is the estate at Pocantico Hills that gives the clearest impression of the isolation and separation so characteristic of all the Rockefellers—as well as the near-paralyzing effect the Rockefeller fortune has on those who approach too casually its field of force. Only forty minutes from Manhattan, off a winding road in Westchester County, surrounded by a low fieldstone wall, the estate stretches away to the Hudson, so hidden from the public eye that the hurrying motorist will miss the gate unless forewarned. Behind the wall stretches some of the greenest and loveliest land anywhere in America—low, rolling hills, perfectly planted yet not manicured, that come to a crest in a Renaissance mansion built by the original John D. Rockefeller shortly after he transplanted his family here from Ohio in 1884. It is a beautiful mansion, yellowing now with mellow age, with grottoes for children, a loggia, a terrace, a swimming pool. From the terrace on the far side one looks out over the Hudson River as it winds majestically down from the north, with all its freight of the American past, before it is squeezed into the angry present by the Palisades. Both sides of the valley are equally green with grass and forest, and as one gazes down in enchantment on the broad-flowing river, it is difficult to imagine sorrow or anger or any ordinary human concern penetrating this paradise.

When Nelson was growing up, the Rockefeller children lived in the city during the week, carefully exposed to the proper schools; but on weekends and on vacations they would enter the shelter of the great estate where all the world could offer was theirs—sheltered and guarded from any turbulence. Playrooms and skating rinks, gardens and swimming pool on the estate all provided an ease protected by an outer privacy that was a fetish of the entire Rockefeller clan. To this estate Nelson Rockefeller returned in the summer of 1930 with his bride, Mary Todhunter Clark of Philadelphia. His parents had not been highly enthusiastic or encouraging about this marriage so soon after Nelson's college graduation; Nelson was very young (twenty-one); they themselves had known each other over six years before their marriage. But Mary Todhunter Clark was of the finest Philadelphia Clarks—intelligent, shy, as withdrawn, as jealous of her privacy as the most reclusive of the Rockefellers. She soon won their complete affection.

What happened within the marriage of Nelson and Tod Rockefeller must be a matter of private speculation. Five children were born of it. But somewhere in its long course, difficult to fix in time, it ceased to be a marriage of love. Some time after the birth in May, 1938, of the twins, Michael and Mary, the marriage ceased to be a true marriage—and remained thus for the last eighteen years of their married life, a fact well known to close friends. But during that period and for the sake of the children, both agreed to continue to remain together as a family and cherish the children while they were growing up.

It was during the war years that the separation matured. For Rockefeller is an activist, fretful and frustrated when not in action, and the approach of war in Europe called him down to Washington, where he found an excitement in government life which he could not find in the splendid privacy which his wife so prized within the Rockefeller estates. Going down again and again to Washington, returning again and again to the estates in Pocantico and Seal Harbor, Rockefeller found that the walls chafed—as did his marriage to a woman who chose not to share the excitement of political and social life beyond those walls.

When, in 1958, Rockefeller finally burst out of the traditional restraint of family to enter the governor's race in New York, what had become an empty marriage long before, bound only by love of children, became slowly a prison. Campaigning in America is done with wives; wives are on public display; the code calls for their participation, however unrealistic this code may be. Tod Rockefeller is a highly intelligent woman; she shared her husband's liberalism in politics. But her taste was rather for good causes than for people in the raw. Meeting people was a chore for her. She did try, however, and she tried hard. But her lack of relish for the pressure of people, as politics constantly throws them into a governor's circle, was not only evident publicly; it was more deeply voiced privately—for the first Mrs. Rockefeller is a lady of no uncertain opinions and no hesitation at reiterating them in the private presence of her husband—or his staff. Her attitude, which was an irritant from the beginning of Rockefeller's political career in 1958, ended by creating a bitterness in a marriage which love had long left and of which only the form remained.

In November of 1961 it was announced from the family office in Rockefeller Center that Nelson Rockefeller was about to be divorced. American politics can accept divorce: for every four new marriages each year, one old marriage breaks up, for there is no civilized way of imprisoning people in the agony of forced partnership. Divorced candidates get elected and re-elected in American life; and even after his divorce Nelson Rockefeller was re-elected in 1962 by the margin of 529,169 votes (and would have surpassed his 1958 margin of 573,034 had not the newly formed Conservative Party peeled off 142,000 traditionally Republican votes).

Remarriage, however, complicates even more the political problem —which brings us to the unhappy Murphy family.

Among those who lived in the gravitational field of the Rockefeller power was Dr. James S. Murphy, a distinguished scientist of the Rockefeller Institute, whose father before him had been included in the Rockefeller family's inner circle. Dr. Murphy—called "Robin" by his friends —was married to Margaretta Fitler Murphy, like Tod Rockefeller a Philadelphian. Tall, with fawn-colored hair, healthy, dressed not so much fashionably as wholesomely, she had thrown off from girlhood a smiling outgoing quality that had won her the nickname "Happy." Meeting her, one thought of tennis, not nightclubs; soap, not perfume; sailing, not sports cars. Radiant, warm and handsome, she appeared to be as serene a specimen of American woman as a good upper-class American family can produce.

But she was not serene—nor was her husband. Both were children of broken and unhappy marriages. Margaretta's father, an alcoholic, had separated from her mother when she was a child. Robin Murphy's father and mother, too, had separated when he was young. Margaretta and Robin had married when she was twenty-two, in December, 1948. And in the lives of both the central gravitational force became the Rockefeller family.

The first to favor Margaretta Fitler Murphy was the elderly father of Nelson Rockefeller, John D. Rockefeller, Jr. Robin brought Margaretta to Seal Harbor as his father had brought Robin as a child. The young bride filled a void in the life of the solemn and aging John D. Rockefeller, whose own wife, Abby Aldrich Rockefeller, had just died in 1948. The lonesome head of the Rockefeller clan saw the young woman almost as a daughter; she had just married the son of his own friend who had just died. She walked in the woods with him, cheered him, made his days brighter.

The favor of the patriarch of the Rockefeller family was not to be taken lightly. Robin Murphy wanted to be a medical scientist—the old man saw to it that he was posted as a junior scientist to a medical research project at the Rockefeller Institute in San Francisco, from where he returned with his family to summer at Seal Harbor. The promising young scientist wanted to come back to New York. And so David Rockefeller, then charged with the family's responsibilities in matters concerning the magnificent Rockefeller Institute, capitol of American biological research, saw to it that young Dr. Murphy was transferred to the New York headquarters. It was also David who later arranged that the Murphys be given the rare permission to build a house of their own on the Rockefeller acres at Pocantico Hills. So ordered, however, is the beautiful, landscaped estate that a new house must be planned so as not to break the skyline but to fit into the gentle curve of the hills and the valley. And among the Rockefellers the one most interested in housing and architecture and art is Nelson. Thus one has the picture of Nelson Rockefeller—who had known Happy Murphy and her husband at his brother David's Seal Harbor house, on sailing and swimming parties along the Maine coast—as he discussed with her the plans, designs and drawings of the house in which she would live.

It could not be hidden from Nelson Rockefeller, himself caught in an unhappy marriage, that the young woman was similarly caught. Scientist Robin Murphy was intensely jealous of his wife, and there had been frightening scenes. She had sought psychiatric help as to what she should do to preserve and rebuild her marriage. She did not want her children to grow up in a broken marriage, as she had herself. But finally she had been advised by her psychiatric counselor that it would be hopeless for her to try further—within the framework of her marriage to Robin Murphy—to do for her children what a conscientious mother wants to do for her children. She must leave him.

Thus—two unhappy marriages.

Thus—two unhappy people trying to escape.

And thus Happy Murphy was there in 1958, as a political volunteer, when Nelson Rockefeller first tried to break out of the walls of privacy to run for Governor of New York. Politics is exciting, and in its excitement Happy Murphy found a new and engrossing world. By 1960, at the Republican Convention, she was a working member of the volunteer Rockefeller staff at the Sheraton-Towers command post in Chicago from which was directed the Rockefeller clash with Nixon that climaxed in the Compact of Fifth Avenue.

Yet the Presidential try of Nelson Rockefeller in 1960 was only a preliminary. He knew then that he would run again in 1964. There was the politician—and the man. The man felt he could not campaign for the Presidency if he had to relive the public sham of private life which he had endured in 1960. He would break clean; and so he told his wife late in the fall of 1961. He was hoping to marry Happy Murphy.

But Happy Murphy's problem was the final complication of the political problem, and it was to echo and re-echo through every turning of the Republican struggle of 1964—from supermarket rally of housewives to smoke-filled gatherings of politicians: Happy Murphy was the mother of four children, whom her husband refused to yield. Should she stay penned in a marriage which her medical adviser and counselor told her had no hope of success—or should she try to get out for the sake of her children? And, very importantly, for the sake of her own health as well. This last consideration—her health—was causing her physician more concern than any other among the problems he and she were trying to solve.

All divorces are ugly; they often harden and make unreasonable antagonists of the finest of human beings; and they violate common sense and good taste most when children are involved. The cleanest of motives, the frankest of statements become punishing wounds when marriage partners fight for emotional reasons beyond their own recognition. Both Robin and Happy Murphy loved their children; their own love was at an end; how does one separate? Particularly if both, remembering their own childhoods, feel that the children should have the benefit equally of father and mother, equal access, equal kindness, equal control. But such delicate thoughts do not work out in a court of law, where matters must be written down in legal terms.

A friend explains Happy Murphy's thinking thus: "There she was, trapped. In prison. Loving Nelson—and loving the children. She feared her husband and she also felt that somehow the children would suffer from it. I don't think she knows herself yet whether she did right or wrong. But I believe I know what her thinking must have been. Although she realized the custody agreement placed the children, technically, in her husband's hands, it called for her to have the children with *her* at least half the time; she expected that Robin Murphy would want to work out with her —and in the best interests of the children—an arrangement fair to all. On at least two occasions, before the divorce decree was granted in Idaho, she talked with Robin Murphy on the phone with a view to reconciliation. But both times he refused to discuss the matter with her. Later, after her marriage to Nelson, it became abundantly clear to her that the original agreement was not working out satisfactorily for anyone, least of all for the children. That was why she went to court again in the summer of 1964; and though the custody agreement wasn't reversed, the situation has improved, following the new court action, to an extent that she can now work more effectively and confidently for the health and happiness of her children."

This hidden, private and many-sided anguish was, of course, going on untold, or lay ahead as the Republican voters sampled or savored their choices in the spring and summer of 1963. Yet voters had to react to what they read, or what was told them. And what lay in the mind of all four members of the two marriages could not be spoken either baldly or delicately in public. The voters must react to hard news: as they did on May 4, 1963, when Nelson Rockefeller and Happy Murphy made known that they had been married that day. Then followed the coarse but inescapable reporting of the nuptials and their aftermath: Nelson Rockefeller at his Venezuela ranch with the glowing and smiling Happy Rockefeller in blue jeans and sports shirt. There followed the immediate gathering of the Hudson River Presbytery to censure as a "disturber of the peace" the unfortunate pastor who had married the two. In Chicago the Young Adults for Rockefeller disbanded in anger. Ex-Senator Prescott Bush of Connecticut, a long-time friend of Rockefeller's, branded him a destroyer of American homes. And from supermarket to sewing circle to parlor to cocktail lounge there rose almost instantly the henyard clucking that every correspondent was to hear for a year thereafter wherever he tried to read the feminine psyche: "I ain't going to vote a woman into the White House who left her children."

If Goldwater was later to be hung on the "bomb issue" and "Social Security," Rockefeller was to be hung first on the "morality issue." One was to hear more of "morality" in the campaign of 1964 than ever before in American life, but Rockefeller was the first to suffer from a virtually uninformed public discussion of his life.

What, indeed, was the morality issue here? Said one of Rockefeller's friends, "Here were two marriages already actually broken, already destroyed and at an end. Here were two people in love. Should they marry publicly or have a clandestine love affair? A man of Rockefeller's wealth can have a love affair on any terms he wants. But he didn't want Happy that way. He wanted it open, she as his wife. He kept holding off hard delegate commitments until he could announce his remarriage—so that they could make up their minds with the facts on the table."

At least two of his most intimate advisers—George Hinman and Emmet Hughes—pointed out to the New York Governor that if he married again he would be putting his chances of nomination at an extreme risk. So did Robert McManus and William Ronan when they learned later. To which Rockefeller replied—so be it. He would not give up the woman he loved, even for the Presidency. He would stand openly with her—and morality demanded that he make the matter public, as directly and as quickly as possible.

It was a decision frankly faced, morally accepted—but politically perilous. People will forgive a politician they love almost any sin—as witness James Michael Curley, Huey Long, Adam Clayton Powell, Jimmy Walker and a score of others; in matters of romance, particularly, they will forgive him almost any peccadillo, as everyone who has worked on Capitol Hill knows—so long as the peccadillo is not flaunted. But the frank and open acceptance of a new marriage was a breach

with the general indulgence of the hypocrisy of politics, and in Rockefeller's case was particularly hazardous. Rockefeller occupied few hearts
as a politician; he had earned his strength by winning respect for his
talents as a governor, his ability to solve problems and administer affairs. As a politician, he appealed to the mind—not the emotions. No
counter reserve of emotion was available to him as now the emotions
of millions of women across America were engaged against him.

To measure the impact of the divorce on the Rockefeller campaign, one must go back to the spring of 1963, before the remarriage.
By that spring all the Rockefeller machinery was on stand-by. Directed
from George Hinman's corner office on the 56th floor of the RCA building in Rockefeller Center, its reach covered the map. All of New England—practically sewed up. The Middle Atlantic States—the same.
The Midwest—good (Ohio poor, but Michigan, Minnesota and Iowa
very promising). In California the chief problem was how to keep the
bandwagon jumpers from jumping on too early. Washington good; Hatfield in Oregon to be counted on; Smylie of Idaho friendly. Even in the
South there were pockets of friendship.

The horizons of the Rockefeller staff in early 1963, as they surveyed the nation, had not been, of course, entirely cloudless. There was
John F. Kennedy in the White House, and by early 1963 Kennedy was
entering his phase of mastery. The Republican nomination of 1964 was
Rockefeller's for the taking, thought the unknowing, younger Rockefeller men in the spring; the only questions were: Could any man beat
Kennedy in 1964? And if Rockefeller dodged Kennedy in 1964, how
could he later claim the nomination in 1968? As they read the portents,
their critical problem was simply whether to accept the nomination or
pass it off on someone like Scranton. All waited on word from Rockefeller, and meanwhile the machinery purred, ready to go.

From Hinman's office, as one peered over the spires of New York
in early 1963, the Rockefeller prospects had never been brighter—except for what Hinman privately knew and would not reveal to callers:
that Nelson Rockefeller meant to marry Margaretta Fitler Murphy as
soon as the New York State legislative session ended in late April.[5]

---

[5] From Hinman's corner office a knowing visitor had as good a view as any
of the Eastern Establishment. One could peer down catercorner at the corner office
of Louis Harris on the 33rd floor of what had once been the Time-Life building.
Harris had conducted similar surveys for John F. Kennedy in 1960; and now, from
his new offices, he was undertaking the massive survey of precinct-vote analysis
for CBS which was to revolutionize election-day reporting eighteen months later.
Harris, who had never met Hinman, could have told Hinman what the impact of
remarriage would be on the campaign of 1964, statistically, by ethnic group, by
income bracket, by religion. But Hinman already knew in his heart. So, too, could
have reported the previous occupant of Harris' new corner suite—Henry Luce, the

With the news of the remarriage, the Rockefeller operation staggered. And when, by late May, the Gallup Poll had taken its first reading, the results read even worse than Rockefeller's staff had expected. Goldwater, previously the choice of only 26 percent of Republicans as against Rockefeller's 43 percent, was now the choice of 35 percent as against Rockefeller's 30 percent.

None of the conventional gambits of politics could now work. Rockefeller's power in the Republican Party had never rested on his liberal politics. It had rested on his sock at the polls, the fact that he could get votes. In July, struggling for a comeback, he fired a bruising volley against the extremists of the Republican Party, a volley provoked by the rowdy tactics of the right-wing Young Republicans at their June gathering in San Francisco; but it misfired. His trip to Europe in September, to limn his silhouette as an international statesman, likewise brought no apparent political gain.

There was, for Rockefeller, no other court of appeal but the people. That had been Kennedy's route in 1960—to appeal directly to people at the primaries, over the heads of politicians. This is why primaries are important. But to appeal to people he must be an open candidate. Thus, on November 7th, two weeks before the assassination, Rockefeller announced from his chambers in Albany that he was an open candidate for the Presidency—and flew off to New Hampshire to begin his primary campaign there.

It was still, to all men except Rockefeller, a hopeless proposition.

But two weeks later John F. Kennedy died. I saw Rockefeller about ten days after that, and he was in a reflective mood. He was as cold and analytical about his own candidacy as if he were examining a case exposed for surgery. The chances were slim. No one, he felt, would vote for him unless a national crisis forced them to do so. The new Mrs. Rockefeller did not want him to run for the Presidency; no one, in fact, wanted him to run except himself. If the people needed him, they would take him. He had been impressed by the reaction to him not only at the Kennedy funeral ceremonies in Arlington National Cemetery, but also at the

---

founder of *Time* and *Life*, who had moved a few blocks into a new building he owned in partnership with the Rockefeller family. Harris, in turn, had moved here from the corner suite of the 49th floor of the Empire State Building, owned by Roger Stevens, the most important Broadway entrepreneur, and previously captained by Alfred Emanuel Smith, a former New York candidate for the Presidency, a few floors up from the law firm of which Robert Ferdinand Wagner, Mayor of New York, was a partner, and Wagner— One could go on with these interconnections, but that way madness lies. I cite this only to show how intertangled is the maze of the Eastern Establishment, even to a New Yorker—and how much more so to the primitives out beyond the mountains.

funeral of Herbert Lehman two weeks later. He had walked the half block down Fifth Avenue from Temple Emanu-El, where the services had been held, to his brother Laurence's apartment, and people had turned to clutch at him and shake hands—extraordinarily, as if he were in midcampaign. The Kennedy assassination had shocked the country; Rockefeller felt it wanted leadership—strong, visible continuity of leadership. But did it want him? Was this public turning to him an ephemeral reaction or something deeper? And since it was obvious that the politicians did *not* want him, he must, as Kennedy had done, show them his muscle at the polls, in the primaries. This was the only way. There were three critical primary appeals to the people—in New Hampshire on March 10th, in Oregon on May 15th, in California on June 2nd. He must fight in all of these—so he was off to barnstorm New Hampshire.

The Eastern leaders watched Rockefeller's headstrong course in consternation. They were convinced he could not win. But so long as he put himself forward as their champion against Goldwater's primitives, no one else would enter the lists; they insisted that this would split the Eastern forces. Many did not like Rockefeller—some disliked him personally for what they thought was his arrogance; others disliked him simply for his family name.[6] But now, added to their dislike was the bone-sure feeling that he would lose—and, in losing, forfeit the nomination of their Party to Goldwater. Many who had not dared to oppose Rockefeller before now found, in his remarriage, a convenient pretext for open opposition.

Yet who else was there? It was too late to create another Willkie out of nowhere, and the choices were few.

The first name to come to mind was that of Richard M. Nixon. But Nixon's name brought a wrinkling of the nose in Manhattan. He was acceptable, yes. But his campaign of 1960 had left so many internal scars in the Party that it would be difficult to make him even a compromise choice—Rockefeller froze at mention of Nixon's name, and Goldwater had contempt for him. Besides, Nixon had run for Governor of

---

[6] In downtown New York sixty years of experience with the Rockefellers have divided the financial community bitterly; the commonest word applied to the Rockefeller business people over the years has been "ruthless"—cold, hard, scrupulously honest but tough men; and the Rockefeller group has as many enemies in the financial community as it has friends or allies. Under the newer leadership of David Rockefeller and the family's chief investment counselor, J. Richardson Dilworth, the adjective "ruthless" is, in fact, no longer applicable. But the old feuds persist, and in 1960 many of the financial leaders downtown had delighted in raising money for Nixon's national campaign as an enjoyable nose-thumbing at the Chase Manhattan and Rockefeller people who traditionally, over decades, had raised the big New York money for Republicans.

California in 1962 and lost, thus adding to his odor of a loser. He had been defeated by Governor Edmund G. Brown, a man who in New York has the reputation of an affable but inconsequential man. Brown's really large achievements in California, his thoroughly grand triumphs in irrigation, education and highways, are unknown beyond the Rockies; so Nixon's defeat by Brown's genuine but unrecognized force marked Nixon as a man who could not carry his own state against a nobody. Moreover, Nixon, with typical bad luck, had announced his permanent departure from California only two days before Rockefeller's remarriage, and thus was a man without a home base. Moreover, in New York Nixon was again unstaffed. A man of modest means, he had come to New York to imitate another Republican loser, Thomas Dewey, a shrewd, controlled operator who had made his mark—and a fortune—as one of New York's highest-priced lawyers in the years since he had left politics. Now Nixon was cool; he was working as a lawyer; he showed no great desire to bestir himself; his own sense of politics told him that if he were to have the nomination in 1964 it could be only after Rockefeller and Goldwater had knocked each other out, when the Party might again turn to him as the healer. He wanted to affront no one at this early stage.

There was George W. Romney, Governor of Michigan. Romney had as yet shown no large appetite for the Presidency, and his first-year record as Governor of the unruly Michigan legislature had ensnared him, so it seemed, in provincial futilities. A feeble attempt had been made to launch Romney as a national figure earlier in the spring of 1963 by several old Nixon men (chiefly Len Hall and Cliff Folger), but it had failed to catch fire.

Then there was Henry Cabot Lodge. Henry Cabot Lodge has, to be sure, all the devotion to public service and the great name to make him a certified member of the Establishment—but with a New England twist. Lodge had the old New England disdain for the New York money men, plus a manner of aristocratic haughtiness that chilled even the oldest New Yorkers. There was no doubt that money could be raised for Lodge in downtown New York (as it was to be later), but not much enthusiasm. He was like medicine—good for you, but hard to take.

The chief thrust for Lodge in early December, 1963, was that he might just possibly have Dwight D. Eisenhower's support—and Dwight D. Eisenhower was still the key. Although wistfully continuing to hope that the mantle could settle on his own brother Milton, the General had, in November, urged Lodge to come home and run for the Presidency. He had spoken to Lodge on the telephone during Lodge's visit to Washington. The General's conscience required that he be sure Lodge, as Am-

bassador to Vietnam, had had nothing to do with the assassination of
the Diem brothers in the November coup ("the King and his brother," as
Eisenhower referred to them); and when Lodge assured him that he indeed had not, but instead had actually offered the asylum of the American Embassy in Saigon to the two murdered men, Eisenhower was
pleased. He then urged Lodge to return permanently from Vietnam
(Eisenhower very much disliked the service in the Kennedy administration of Lodge and Dillon, two of the most eminent members of his own
administration) to stand up and make speeches as a candidate and show
where he stood on the issues of the day. He should run as the "common-
sense" candidate; "common sense" was a term Eisenhower was using
more and more frequently, for he had come to dislike such words as
"moderate," "progressive," and "middle of the road," used as labels by
his kind of Republicans to distinguish themselves from the primitives.

Such conversations with Eisenhower as they leaked to the press
were to baffle and confuse Republicans for months, for when they surfaced, as this conversation did in a front-page story in *The New York
Times,* it would seem always that Eisenhower had finally chosen a candidate. But when checked, Eisenhower would always go back to his original line: he was for a wide-open discussion of issues by *all* possible
candidates, so, in urging as many as he could to run, he was only pursuing his high-minded civic duty. Yet to those who got the word direct
from the General, the distinction between being "urged" to run and the
implicit promise of support if one did run was always obscure—and even
more so to the press.

Thus, there was Lodge, placed square on the front pages by Eisenhower—but from Lodge himself no word. There was much to be said for
Lodge, who could be seen in a role similar to Eisenhower's own in 1952.
In 1952 Eisenhower had been built as the Republican nominee while he
captained the reorganization of Western defense from SHAPE headquarters in Paris; Eisenhower had returned from this great service,
crowned with success, only ten weeks before the Republican convention,
and thus been spared the exertion of the primaries and the inquisition of
a questioning press. Lodge, etched against the sky by the distant war in
Vietnam, similarly had been cast in a heroic role. If Lodge could pull the
Vietnam situation together, he would be a major possibility—but it was
too early to tell, too early to opt for a man whom Kennedy had so resoundingly trounced in Massachusetts eleven years before and who now
served Lyndon Johnson as Kennedy's appointee.

Which left William Warren Scranton, Governor of Pennsylvania.

Of the development of Scranton's character and conscience during
the next year, there will be much to say later in this book. But now, in

December, as he first appeared on the national stage, he was without doubt the most attractive new face on the Republican scene. Lean, handsome, young, polished, endowed with a magnificent wife and four handsome children, he had the quality of a Kennedy—whose friend, indeed, he had been and whose autographed picture (as then-Senator Kennedy) hung in Scranton's home office in Scranton, Pennsylvania. Moreover, in his first year as Governor of Pennsylvania, Scranton had written a startlingly effective record. In the house of mastodons which shelters the Republican legislators of Pennsylvania, the Republican governors for a full century had been gentlemen noted for their florid impotence. Pennsylvania heavy industry ran the state; its lobbyists, when the Republicans controlled the State House, managed its legislation as Boss Croker used to manage Tammany Hall. But Scranton had defeated Philadelphia's Democratic mayor, Richardson Dilworth (uncle of the Rockefeller financial Dilworth of the same name), in a brawling, venomous campaign for the governorship in 1962—and then gone on to amaze all by actually mastering the beasts of his Republican legislature in his own State House. He had reformed the Pennsylvania Civil Service, one of the most spoils-ridden in the nation; doubled education appropriations; passed an unpopular but necessary increase in the sales tax; and begun on a major program of industrial development for the state overdue since the end of the war.

Much more conservative than Rockefeller, Scranton was different from Goldwater too. He had served a year in the State Department as special assistant to John Foster Dulles, and had a finesse in his conversation about foreign affairs that came not only from reading but from an understanding of reality. No breath of scandal touched him. His war record was outstanding. His young men at the State House more than made up in enthusiasm and vigor what they lacked in experience. He was an absolute gentleman, but he had proven himself a major, gut-fighting campaigner.

And he had the finest sponsorship. It was not that Scranton, as a modest millionaire, was a member of the Establishment; he belongs to the squirearchy of American life (like the Knowlands of California) rather than to the Establishment. But his Establishment connections were superb. Not only was his brother-in-law James Linen president of the Time Incorporated publications, but Henry Luce was bound by affection to the Scranton family, which had supported his father's missionary enterprises in China. Scranton's connections with the movers and shakers had been cemented by the proper prep school (Hotchkiss) and the proper college (Yale), and strengthened further by his successful business career in eastern Pennsylvania. He was the favorite of most of the older men who had ever met him, and while his own young staff had begun, without his consent, the most positive and dramatic efforts to call national attention to him, his older friends were proceeding at a much higher level. Chief among these was Tom McCabe, Sr., chairman of the board of Scott Paper Company, a full member of the Philadelphia branch of the Eastern Establishment. McCabe observed once that he knew only two young men who he thought would make great Presidents—Bob McNamara, now Secretary of Defense, and Bill Scranton. Since McNamara was ruled out by his service to Lyndon Johnson, McCabe was now equally pleased to put his energies behind Scranton.

McCabe's luncheons for Scranton were as good a sampling of the names of men who move things in the East as any other, and a typical guest list at a Scranton luncheon in McCabe's office at Scott Paper Company, on November 8th, arranged for display of the young Governor by his elderly patron, ran thus: From New York, William S. Paley, boss of the Columbia Broadcasting System; the New York *Herald Tribune*'s Walter Thayer; Herbert Brownell, Dewey's political chief of staff. From Ohio, George Humphrey, former Secretary of the Treasury under Eisenhower. From New Jersey, William Beverly Murphy, president of Campbell's Soup, plus Bernard Shanley, the state's Republican finance chairman. From Pennsylvania, Thomas Gates, Eisenhower's last Secretary of Defense and president of the Morgan Guaranty Trust Company; plus Walter Annenberg, owner-publisher of the Philadelphia *Inquirer*. From Delaware, a top Du Pont official, plus Harry Haskell, Delaware's Republican National Committeeman. Meade Alcorn of Connecticut, former Republican National Chairman under Eisenhower, was scheduled to come but was weathered-in at a Chicago airport.

Much more important, however, than such quiet exposure as McCabe could make available to the Pennsylvania Governor, or the press exposure that the media began to focus on him soon after the assassination, was the sponsorship of none other than Dwight D. Eisenhower.

For, no more than one week after Eisenhower had catapulted Henry Cabot Lodge into the race with the *New York Times* story of December 8th, Eisenhower did the same thing for Scranton. En route to his winter hibernation at Palm Springs, California, on December 14th Eisenhower halted his special train at the railway yard in Harrisburg for a five-hour dinner with the young Pennsylvania Governor and his wife. To Scranton also the ex-President preached the gospel—get out there and fight, stand up and run, discuss the issues and be a candidate. With this very friendly session (for Mamie Eisenhower and Mary Scranton are extremely fond of each other in the warm relationship that women of different ages can develop), Scranton, too, was faced with the same conundrum as Lodge: Was the old hero simply urging him to run as an exercise in civics or was the old hero implying support? Alas for Scranton, the question was never to be answered, even at the last minute.

Thus, as January began, there was Scranton's name as the subterranean favorite of the entire Eastern Establishment. Yet no one could get a forthright statement from Scranton himself, for Scranton, in his way, was as sibylline as Dwight D. Eisenhower. He would do anything at all for the United States and, in the war, had been willing to die for it. But unless he saw it clearly as his duty, a call to conscience that could not be denied, he would not seek the nomination. A parade of dignitaries and eminent journalists made their way over the miserable rail-and-air connections to Harrisburg, Pennsylvania, to learn what was on Scranton's mind. And they came away confused; the Governor's staff was running him for President as hard as it knew how, yet the Governor insisted he was *not* running for President. Sessions with Scranton, a man of captivating charm, would wind up at his fireplace with Scranton's teasing question, "All right—can you tell me one good reason why I should *want* to be President of the United States?" And there was no answer.

Such a man is difficult for politicians to work with, and thus the weeks idled by into midwinter as observers turned their attention to the man who, on January 3rd, 1964, from his hillside home on the bare crags of Phoenix, Arizona, had announced that he, too, was a candidate for nomination and election as 36th President of the United States. And Barry Goldwater was front and center stage.

Barry Goldwater's announcement came as a surprise to no one.

But who was Barry Goldwater? And what was the Goldwater movement?

For three and a half years, from the tock of the gavel that closed the Republican convention of 1960, politicians and journalists had recognized the "Goldwater movement" as a new force in American politics.

The term "Goldwater movement" was an annoying one.

It would have been more convenient to call it a Goldwater campaign or a Goldwater draft. But "movement" was the proper word. The wordless resentments, angers, frustrations, fears and hopes that were shaping this force were something new and had welled up long before Goldwater himself took his Presidential chances seriously. Always, to the very end of the campaign, two almost independent elements were involved: the movement—and the candidate himself. We must look at the candidate as a man more closely later. But the movement was something deep, a change or a reflection of change in American life that qualified as more than politics—it was history.

One could almost fix the moment of its birth at the Chicago convention of 1960. The Rockefeller-Nixon compact of Fifth Avenue[7] had enraged the primitives of the Party who had written what, for them, was an advanced platform. It had been changed only marginally by the night rendezvous of July 22nd-23rd of Nixon and Rockefeller in Manhattan, when a new draft was imposed on them. But those who were present at the Blackstone Hotel in Chicago in 1960 remember the near violence of the demonstrations against the Rockefeller-Nixon compact—the placards, the shrieks, the emotions, surpassing any of the bitterness the old Taft conservatives had brought to their wars with the Easterners. Nor can anyone forget the smoldering anger of Barry Goldwater, taking the platform in the Hilton Hotel press conference to denounce this Betrayal in Babylon. From Saturday of the compact to Wednesday of the balloting, a procession of men pressed a mission on Goldwater—he must lead a revolt. They promised him 300 votes on the convention floor if he would voice the principles they shared. To them Goldwater said simply, "Get me three hundred names of delegates on paper. Show me." From Saturday to Wednesday they counted; and they counted only 37 names who would openly vote against Nixon on roll call.

Yet the name count was meaningless. For what was there at the Republican convention, however difficult to measure, was a deep fear of what the American Republic was becoming; it was to these Goldwater spoke as he rose from the floor to march to the rostrum and make the nomination unanimous: "Let's grow up, conservatives! If we want to take this Party back, and I think we can some day, let's get to work."

Now, Goldwater's favorite style in politics is exhortation; he is a moralist, not an organizer. He preaches; he does not direct. He arouses emotion—he does not harness it.

Organization was to be the work of other men, for it was almost three years before Goldwater was invited to claim leadership of the Goldwater movement.

What these men could see and could sense was the response and emotional echo to Goldwater's words. They did not think of themselves as primitives. America is changing so rapidly that the principles Americans were taught in school twenty or thirty years ago are challenged at every turn of event or development. What is valid of the old morality and what is not? Across the country, from Maine to California, families and individuals, cherishing the old virtues and seeing them destroyed or ignored or flouted, were in ferment. Across the sky of politics there began to float new names like the John Birch Society, the Minutemen, the

[7] See *The Making of the President—1960*, pp. 196-205.

National Indignation Convention, Freedom-in-Action and other groups. At the extreme of the frustration were madmen and psychopaths disturbed by conspiracy, Negroes, Jews, Catholics, beardies, but toward the center it involved hundreds of thousands of intensely moral people who hated and despised not only adultery but Communism, waste, weakness, government bureaucracy and anarchy. In the cradle of the Rockies the disturbance of conscience could shape itself, unnoticed by the East, in political terms such as those accepted by the legislature of the State of Wyoming, which in 1963 called for supplanting of the Supreme Court of the United States by a "court of the union" made of fifty individual state chief justices; getting the United States out of the United Nations, and the UN out of the U.S.; abolition of all foreign aid; repeal of the Arms Control Act. (It had already, two years earlier, gone on record for repeal of the Federal income tax.) The Negro revolution of 1963 was yet to come, but all over the country there were both men of good conscience and men of evil intent who felt themselves cramped into suffocation by the appetite and need of government for control over their individual destinies. It was a mood entirely different from the mood of the Taft conservatives of the forties and fifties who had wanted, simply, to hold the country still; the new mood of the primitives insisted that the course of affairs be reversed.

This mood was there; but it could be tested only by organization. And though Goldwater himself was not yet willing to test by premature organization the emotions that his incantations aroused, others were. For these were authentic grass-roots American emotions, and by the summer of 1961 the map of politics was sprouting with self-winding political groups which attached their names to Barry Goldwater without any authority from him to do so.

Among these we must single out one group and one individual, in order to pull order out of chaos. All politics, Will Durant has said, is a struggle among the few to control the many. The few must be gathered first, of course, and then they must fight other similar groups. And from the Committees of Correspondence of the American Revolution to the organization of the Kennedy strike in 1960, countless tiny groups have tried to change the course of American history. There is, actually, no other way—Dwight D. Eisenhower to the contrary notwithstanding.

The individual that we pull, then, from the turbulence in the summer of 1961 is F. Clifton White of New York—by no means a member of any Establishment, but a genuine Appleknocker nonetheless. Of old upstate New York stock (his grandfather had been one of the first Civil War volunteers from Cortland County), his mother had actually trekked West as a child of ten in a covered wagon; later she had married and come back to live near Hamilton, New York, where White's father was a farmer and gasoline-station operator. White himself went to Colgate University, majoring in social science; an apprentice schoolteacher when war came, he enlisted and flew twenty-five missions with the Eighth Air Force over Europe in the early days before fighter cover could reach the Ruhr and when his bomb group lost 50 percent of its effectives in its first sixty days of mission. Returning, he became a postgraduate at Colgate—and also chairman of the local chapter of the American Veterans Committee, a champion of public housing for campus veterans, who were being gouged by local landlords. Housing brings one to politics; and in his local housing battles, as a liberal, White learned all about petitions, votes, registrations.

With politics White found his avocation; and as some men become seized by the fascination of a specialty until they become technicians useful only to men of larger vision, White became a technician of politics —one of the finest in America. As a specialist in politics—in petitions, organization of meetings, nominations, convention tactics, floor seating, the buttoning of votes—White moved to lead, and hold the leadership of, the Young Republicans.

The Young Republicans are far, far more important in their party than any junior Democrats on the other side of the divide (young Democrats, apparently, strike for the big leagues immediately, without bothering with junior-league politics). For ten years, from 1950 through 1960, White, as a hobby, exercised a tighter control over the National Federation of Young Republicans and their conventions than anyone had exercised before. But as a professional White discovered his services to be of enormous value, first to the Dewey machine in New York, then to big industry. His own private consulting services to such firms as U. S. Steel, Standard of Indiana, Richardson-Merrell Inc. (Vicks VapoRub) and General Electric had made him an expert in the instruction of aspiring junior executives (both Democratic and Republican) who were assembled by great corporations to be taught how they should participate in public affairs. White could teach them all about county chairmen, convention rules, petitions—the finest minutiae of organizational politics which they, as citizens, must understand. Since executives and employers should at least understand the nerve system of the body politic in which they operate, White became something like a black-belt master lecturing on judo to an audience of nonparticipants.

It was obvious, therefore, to such a lecturer on political judo that something could well be done to cause the emotions of the primitives to flip and flop politically if only the right nerve spots were touched. And in his old friends of the Young Republicans he had, ready-made and avail-

able, a national net. This complicated strand of the Goldwater movement must be followed, however briefly, for this was the group that
seized control of the Republican Party, turned it over to Barry Goldwater, then lost it to Goldwater's inner circle.

White had been used as a technician by the Nixon people in 1960,
and one remembers him then as a pale young man with China-blue eyes,
given to jaunty bow ties, very courteous yet quite tense, in and out of
the offices of the Nixon citizen-committee headquarters in Washington,
never one of the inner circle—a supporting technician. The Nixon campaign upset White, a professional in politics, and caused him to think.
"I'd always been interested in back-room politics. I've probably elected
more presidents of more organizations in back rooms than any other
man in America. But after a while you get to think about the dimensions
of the room. What's going on in the country?" said White later.

White had returned from the Nixon campaign to his own prosperous private business in New York, disturbed. By now he had made the
long journey from his youthful liberalism to a dedicated conservatism,
and something of the alarm that was sounded in Goldwater's phrase
"moral decay" was alive in him. In late summer of 1961 a luncheon
conversation at the Hotel Commodore with some old friends of the
Young Republicans organization provoked him to a new proposition:
Could he exercise his technical skill at a Presidential level? Could or
could not the organization of the National Republican Party be seized
and held as he had seized and held the organization of the Young Republicans years before? Above all, could one reach beyond the convention with a nominee who could rouse the voters themselves in the fall of
1964, three years away?

On October 8, 1961, White gathered a preliminary and absolutely
secret meeting of twenty-two of his national friends at the Avenue Motel,
on South Michigan Avenue in Chicago, to examine this proposal. After a
long afternoon and evening of talk, they decided they would form an ad-
hoc committee to seize the Republican Party for the conservative cause
—and White would report to Goldwater. After some difficulty White saw
Goldwater in November and found the Senator indifferent—but amused.
Unwilling to lend his name officially to their cause, Goldwater was nonetheless unwilling to repudiate this group, which must have seemed like
just another group urging the Presidency on him. A second meeting at
the same motel on December 10th (twenty-seven in attendance, including the Governor of Montana) brought the decision to divide the country
into nine regions for the mobilization of conservatives; to establish an
office that White would run; to raise $60,000 for financing the effort. By
the spring of 1962 White was installed in the Chanin Building of New

York with modest funds, operating out of Suite 3505 and able to travel back and forth across the country contacting each of the nine regions in which regional volunteer directors were already preparing to gather Goldwater delegates for the Convention of 1964.

By April of 1962 White could reassemble his group at a hunting lodge in Minnesota ("I wanted them to spend two days together and not only talk, but drink whiskey together, and get to know each other, and be friends and trust each other"). Not until the next meeting of the cabal— now grown to fifty-five—on December 2nd, 1962, at the Essex Inn Motel in Chicago, did the press finally hear what was under way and announce the formation of a "sinister" Draft Goldwater movement (at this meeting, actually, the session opened with a prayer for God's blessing on its work). White's next meeting with Goldwater was in January. Goldwater, annoyed by the publicity, chilled White, but did not repudiate him outright.

The early months of 1963 were difficult months in which to generalize on politics; Kennedy had lost three Congressmen, but gained four Senators; there was no national mandate of any kind, conservative or liberal; Rockefeller was still the leader in the polls for the Republican nomination. Yet, since nothing stands still, it became more and more obvious that the South was the region of the country in most violent disturbance; and the most impressive and remarkable event of the elections of 1962 was the phenomenal showing that the Republicans had made in the State of Alabama, coming within 6,800 votes of capturing a Senate seat. There was a Southern strategy to be shaped—if the Republican Party did indeed want to court the South.

On February 17th, 1963—again in Chicago, at the O'Hare Inn— the executive committee of the group met once more and decided to go open and national: they would form the National Draft Goldwater Committee. Peter O'Donnell, Texas State Republican Chairman, thirty-six years old, would be chairman; White would be national director; and they would "draft the son of a bitch" whether he wanted to run or not. On July 4th the National Draft Goldwater Committee called a rally for its candidate in the National Guard Armory in Washington, knowing he would not grace it with his presence. By now the White-O'Donnell organization had had almost two years for mobilization. From New York no less than forty-three busloads of Goldwater zealots arrived; from Connecticut, thirteen more; from as far away as Chicago, Indiana and Texas, some 7,000 people chose to give up their holidays to urge that Barry Goldwater be President. Only three gatherings at the armory, it is said, surpassed this initial Draft Goldwater showing: the inaugurals of Kennedy and Eisenhower, and the preachings of Billy Graham.

Goldwater could not be indifferent to this; and, indeed, he was not. Not only was there quite evidently a powerful army of faithful, marshaled and waiting for him to seize the baton, but events were changing the climate of politics too. The Negro revolution that had begun in Birmingham in the spring of 1963 was now spreading like crownfire all across the South; and the North, too, was beginning to see the demonstrations that a year later were to spill over and bloody the big cities. Moreover, Nelson Rockefeller's remarriage in May had erased his political lead—or so said, at least, all the public-opinion polls. Whether he wanted to be or not, Goldwater was by late summer and early fall of 1963 the foremost candidate for the Republican nomination. *Time, Newsweek, Life* all had given him cover displays; the television cameras attended his comings and goings, and, starting in September, his face began to appear with increasing frequency on the evening television shows.

He had not planned it this way. But there it was. Earlier in 1963, when asked whether he was really running for President or not, he had replied, "I'm doing all right just pooping around." It is difficult for anyone, even Goldwater, to say when he decided to run. Essentially, Goldwater thought of himself, and still does, not as a man prepared to or even desiring to run and administer the government of the United States, but as leader of a cause. This cause is precious to him; his loyalty to it is sincere and unblemished. It is the cause of conservative revival and puritan virtues in the United States of America.

Goldwater had begun to talk reflectively with his Arizona cronies about the Presidency as early as 1961—not so much as a strike for power as a strike for control of the Republican Party, and for its purification. He had then set a figure which was to be the benchmark of his inner planning—if he could come within 5 percent of Kennedy in a national contest, then he would have scored a major victory for the conservative cause. A loss by more than 5 percent, he felt, would hurt the conservative cause.

Through 1962 Goldwater toured the country. As Chairman of the Republican Senatorial Campaign Committee—his third round at the job—his senior-level contacts were nationwide, and he could sense the same support for him in the upper hierarchy of the Party in the West and Midwest that Clifton White and Company were discovering at the grass roots.

But the Presidency, as a job, still seemed as remote to him as the day he had discussed it with John F. Kennedy in 1961. Goldwater had gone to visit President Kennedy at the White House on the dismal day of the Bay of Pigs. As Goldwater tells the story, the President's secretary urged him to enter the great Oval Office and wait for the President there. "So I went in and sat on his rocker," recalls Goldwater. "I forget whether I did a test, because we both used the same doctor and she was trying to get me to get one, but I didn't want one. And he came in smoking that little cigar and he looked at me and he said, 'Do you want this job?' I said, 'No, not in my right mind.' So he said, 'I thought I had a good thing going up to this point.' "

It was in the July–August period of 1963, apparently, that decision began to firm in his mind—the Civil Rights Bill was now before Congress; the Negro unrest was growing; Rockefeller seemed out of the race. But Goldwater, as a candidate, proceeded cautiously. He had a sure seat in the Senate coming up for re-election again in 1964. Should he risk that in a long shot at the Presidency?

In August he began to explore the proposition seriously. The Draft Goldwater volunteers were now a national force, operating under White from a fully functioning headquarters at 1025 Connecticut Avenue in Washington, D.C. They could no longer be ignored as just another group —they *were* the combat troops of the Goldwater movement. But they were not Goldwater's troops. There had always been a distance that Goldwater put between himself and White's personality. Goldwater saw himself as a man of purpose and philosophy. He saw White as a technician—loyal but, in Goldwater's eyes, nonetheless limited. To take over the organization without offending White would be difficult. But the organization must be used. To coordinate with the volunteers, while himself remaining independent, Goldwater brought to Washington a polished, gray-haired lawyer from Phoenix—Denison Kitchel, an Ivy Leaguer born and bred who had moved West after graduation from Harvard Law and was now one of Goldwater's closest confidants.

Kitchel arrived in Washington in late August with the title of campaign manager—but campaign manager ostensibly for the Senatorial, not the Presidential, race of 1964.

Actually, Kitchel's duties were no secret to anyone: he was to explore further the Presidential possibilities and help guess whether they could reach the magic 5-percent range against John F. Kennedy.

There was the volunteer organization to be studied and restudied. White had already purged and repurged the volunteer groups of kooks and fanatics; Kitchel now restudied and screened the organization again for extremism; the problem Goldwater might have with his wild-eyed supporters was already apparent. But the prospects looked bright: already the regional volunteers had formal campaign organizations in no less than thirty-two states, and several states had been locked up completely. (In South Carolina the Goldwater volunteers had so buttoned down that state that the *official* State Republican Committee passed an *official* resolution that it was now reconstituted organically and simultaneously as the state's Draft Goldwater Committee.)

There was also the record to be re-examined. And Kitchel undertook, with the assistance of Edward McCabe (later to be named Research Director for Goldwater), the sorting out of the tangle of statements, speeches, judgments, columns and comments that Goldwater had made with such eloquence, intemperance and contradiction over the previous ten years. With a morbid pre-taste of the fun both Rockefeller and the Democrats were later to have extracting nuggets of flamboyant and inflammatory phrases from this lode, Kitchel and McCabe began to catalogue them.

By November of 1963 there was no doubt in the mind of either Kitchel or Goldwater that Goldwater was going to run. For Goldwater, John F. Kennedy was history's perfect opponent—they would debate the issues up and down the country, they would draw the line between the conservative and liberal philosophies. Kennedy would probably win; but Goldwater felt certain of carrying both South and West and, thus, the magic 45-percent mark would be reached or passed.

And then came the assassination. The assassination shocked Goldwater as it shocked every American by its brutality and senselessness. More than that—Goldwater had liked John F. Kennedy, a liking that was reciprocated. When they were fellow Senators, Goldwater had frequently chided the younger man on his voting, saying, "Jack, your father would have spanked you [for that vote]," when they disagreed, which was frequently. Now, after the assassination, he was faced with running against another man, a Southerner, of an entirely different sort. Moreover, his heart was sick within him. Poison-pen letters and hate letters poured across his desk in hundreds as if he, personally, were responsible for the killing of the man he was so fond of. ("Are you happy now?" asked one letter.) And so, for a period of about ten days, Goldwater gave up politics. To his wife he said, "The heck with the Presidential thing," and somberly drew into himself.

Yet a Presidential campaign cannot be dismantled that easily; beyond those of the individual candidate, the ambitions, the energies, the plans of too many other people are engaged. And thus, on December 5th, there gathered in Goldwater's suite at The Westchester apartments in Washington a few senior allies of his campaign. There Kitchel and Goldwater faced Senator Norris Cotton of New Hampshire and ex-Senator William Knowland of California in the presence of a few others. Both Cotton and Knowland needed a decision. If Cotton was to direct the New Hampshire primary for Goldwater in the winter, he had to know now;

and Knowland was well along in organizing the huge apparatus needed for the California primary fight. Goldwater listened to their pleas (including an impassioned peroration by Norris Cotton that no matter what happened Goldwater would never die politically, for the American people were ready for his principles)—and asked more time to think it over. He would get back to them in a week or ten days. A second meeting followed a few days later in Goldwater's Senate chambers—Goldwater and Kitchel, Cotton and Senator Carl Curtis of Nebraska, Dean Burch of Arizona, Clif White of the volunteers, Raymond Moley, an amateur of Presidential politics since his service to Franklin Roosevelt, were all there. Together, once more they reviewed the changed topography of post-assassination politics. Goldwater's mind was slowly changing. This time they were discussing dates and timing—if he were to go, when should he go? Moley urged that Goldwater make the race—but counseled that he stay out of the primaries. Others were of contrary view. Goldwater was perplexed still. It was not to be the campaign he had hoped for against Kennedy; it was to be a struggle with Johnson, and Goldwater disliked Johnson ("the biggest faker in the United States" was the way he would describe the President of the United States at the San Francisco Convention); but, on the other hand, there was his duty to the conservative cause; and the Goldwater volunteers, an entirely new group in politics, might drift away in four years.

Goldwater is unsure himself exactly when he made the decision or how. He was in his den, he recalls, probably sitting at his desk when, according to his own account, he told himself, "You've got to do it." He told his wife first. "Honey," he recalls saying, "what do you think about my running for the Presidency?" And she replied, "Well, if that's what you want to do, you go ahead and do it. I don't particularly want you to, but I'm not going to stand in your way."

When, precisely, this happened is as obscure as are all those obscure moments when a man finds himself on the other side of a decision he has been approaching, step by step, for weeks. By mid-December, Kitchel could call Cotton and tell him secretly that he would have a candidate to run in New Hampshire. By Christmas, Goldwater was home in Phoenix and preparing to announce. He had made, earlier, a serious effort to secure the services of Len Hall, one of the standard East Coast professional campaign managers, to direct his campaign; Hall had turned him down, for his loyalty belonged to Richard M. Nixon and until Nixon was out of it, Hall wanted to make no other commitments. On the evening of New Year's Day Goldwater called a young and dynamic Arizona friend of his, Richard Kleindienst, and gave him an overnight decision to make: Would Kleindienst come to Washington and manage the national campaign that Goldwater was going to announce on January 3rd? Kleindienst had just driven home to Phoenix from the Rose Bowl game; he slept on the decision and, the next morning, enlisted. Forty-eight hours of advance notice was given to press and television nets that the Arizona Senator was going to make an announcement from the terrace of his home on the slope of Camelback Mountain.

On the morning of Friday, January 3rd, 1964, dressed in blue jeans and bedroom slippers, he puttered with his ham radio set. Then he dressed and walked outside and welcomed the nation to Be-Nun-I-Kin, which in Navajo means "House on the Hill," then said that he was openly seeking the Republican Presidential nomination "because of the principles in which I believe and because I'm convinced that millions of Americans share my beliefs in those principles."

What those principles were he did not state at the moment. How many shared them remained to be determined. How he planned to translate them to the voters in the fall of 1964 was obscure; and how his opponents, both Republican and Democratic, would define them in the heat of battle was unknown. But with this the contest was now formally joined.

CHAPTER FOUR

THE PRIMARIES:
DUEL TO THE DEATH

PRESIDENTIAL primaries are always savage. But the Presidential primaries of 1964 were to exceed in savagery and significance any other in modern politics.

For the first time, Americans were required, publicly, to decide their posture on war and peace in an age of nuclear terror.

In 1960 a duel in the Democratic primaries had created two heart-catching national personalities—Kennedy and Humphrey; but one remembers their 1960 duel as a matter of romance and, despite its bitterness, for the attitude of chivalrous respect the duelists showed toward each other. The primaries of 1964 were a Republican duel—and the arena of a far grander and more terrible clash. When the duel was over, the Republican Party was so desperately wounded that its leaders were fitter candidates for political hospitalization than for governmental responsibility.

The map that spread before the duelists of 1964, Rockefeller and Goldwater, was much the same as that offered to Kennedy and Humphrey—sixteen states from coast to coast, plus the District of Columbia, which invited them to enter the lists.[1] Of these, three—New Hampshire, Oregon and California—were chosen for direct clash; and of these three, New Hampshire, with its date set at March 10th, was to be the curtain raiser in the Presidential drama.

---

[1] The following states held primaries in 1964: New Hampshire, Wisconsin, Illinois, New Jersey, Massachusetts, Pennsylvania, Indiana, Ohio, Nebraska, West Virginia, Oregon, Maryland, Florida, California, South Dakota and Texas, as well as the District of Columbia.

New Hampshire was an excellent place for a reporter to begin a study of the campaign and the Republicans. New Hampshire Republicans are as good a cross section of the old-fashioned Republican faith as the nation offers anywhere. Hampshiremen are earnest about their politics; they vote seriously; and in their perplexity, in their vacillating indecision, in the great pulses of changing mood that were to sweep across Granite State Republicans in January and February, one could see foreshadowed all the agony of indecision of millions of Republicans whose conscience, or fear, later in the fall of 1964, would cause them to abandon the party of their fathers.

Myth holds that New Hampshire is full of twangy, skinflint Uncle Ephs or Uncle Calebs who make their living by whittling antiques, milking cows and fleecing tourists. The village greens, the white Christopher Wren steeples, the monuments to the kepied Union veteran who peers out into the distance and into the past, confirm for the stranger who hurries through the green hills to the ski chalets that this bit of green and forested country is a byway, a beautiful anachronism sheltered from the present by an exemption of history.

Nothing, however, could be more wrong. Thriving, throbbing and prosperous, New Hampshire is a state that has won its way back to good health by its own leadership and exertion. The second most highly industrialized state in the entire Union, it is also one of only three whose unemployment is so low that they cannot qualify for Federal aid under the Area Redevelopment Act. Once muscle-bound, as Pennsylvania is still muscle-bound, by two dominant industries (textiles and leather goods), New Hampshire has lived through the technological revolution and, with a proud independence of the freshets of Federal aid that have fertilized such industries in Arizona and California, come out on the winning side in the development of electronic, precision and scientific industries. Its politics, too, are changing as its structure and demography change; and under its effective young Governor, John King, the first Democratic governor in forty years, it is making the slow transition to the Democratic Party that has already swept the southern tier of New England states from their traditional anchorages.

Yet New Hampshire is still basically Republican. Some scholars still debate whether the first meeting of what became the Republican Party took place in Ripon, Wisconsin, in Jackson, Michigan—or in Exeter, New Hampshire. The Union veteran on the village green is more than a monument—he lives in the hearts of Yankees who know their grandfathers fought and died for human rights. Hard-nosed, tight-spending, balanced-budget Yankees most of them, they also display a balancing Yankee virtue—a citizenry with respect for learning, that learns slowly but learns well.

Good political research might have told out-of-staters (as it did tell Nelson Rockefeller) other essentials of New Hampshire politics. It might have revealed, for example, that only three states of the Union (Nebraska, Iowa and Missouri) have a higher percentage of old folks than New Hampshire—and that this state, to which so many older people retire, was no state in which anyone should challenge Social Security.

Good research might have reminded outsiders (as it did remind Nelson Rockefeller) that Hampshiremen are extremely proud of their attachment to the UN—in the spring of 1945, as the United States approached the decision to join the international body, the New Hampshire Legislature insisted that all citizens at their annual town meetings should vote on whether the United States ought to join such an organization or not; and New Hampshire voted twenty to one for the United Nations, the only state in the Union which sought and got citizen approval for this historic step. (New Hampshire children are taught this in their schools, and they are proud of it.)

Good research might, finally, have pointed out to both Goldwater and Rockefeller that New Hampshire Republicans are different from New Hampshire Democrats. The Republicans are overwhelmingly Protestant (Congregationalists are the largest denomination in the state), while the Democrats are overwhelmingly Catholic. The remembered triumphs of Estes Kefauver in New Hampshire in 1952 and 1956 had indeed been won by handshaking and a "Will you he'p me?" approach—but among Democratic French, Italian and Irish working people with an inheritance of underprivilege, a warmth of response to famous names, a delight in contact with the great. New Hampshire Republicans, however, are of Yankee stock—reserved, short-spoken, embarrassed by excessive emotion; they give a courteous handshake to anyone, but reserve judgment until they have thought about things. In 1952 they had turned down the strenuous effort of Robert A. Taft in the Republican primary and voted for Dwight D. Eisenhower, who was still in Paris and who had never visited the state in anyone's memory. One cannot approach the electorate in a Republican primary in upper New England in the same spirit in which one approaches the electorate in a Democratic primary.

These qualities of Hampshiremen were, of course, as little apparent to this correspondent as they were to Barry Goldwater when, toward the end of January, I first accompanied Nelson Rockefeller to the Granite State in the leaden, snow-burdened winter of 1964. But already the state had been crossed and crisscrossed from ski slope in the north to factory town on the Massachusetts border, and the character of the campaign was clear: this was duel to the death, and Nelson Rockefeller was thrusting to kill.

Rockefeller had come this Friday afternoon to open a new headquarters in Nashua. He arrived in his F27 Fairchild, a twin-engine turboprop, about two o'clock, and the usual welcoming group was at the Manchester airport to greet him—his New Hampshire chairman (former Governor Hugh Gregg), his local campaign manager and staff.

His schedule that day called for a reception at a Veterans of Foreign Wars beer hall outside Manchester; a visit to the offices of the Nashua *Telegraph* (which later came out against him); a street tour; the opening of a new store-front headquarters in Nashua (one of ten in New Hampshire) with a speech followed by questions and answers; an address to the Nashua Chamber of Commerce, followed by more questions and answers; then home.

Superficially, it was quite a normal day of primary campaigning: the surprise of people in the streets when approached by a great name (once, making a visit to a workingmen's club in New York, Rockefeller had so startled one of the drinkers that when he said, "I'm Nelson Rockefeller" and extended his large hand, the man had dropped his beer on the floor in surprise); the half-drunk who followed the candidate down the street to solicit a loan of $100 because he knew the candidate could afford it; the curious and the autograph hunters; and a group of children who chanted, "Rocky, Rocky, He's our man, If he can't do it, No one can."

What was different enough to set the tone of the entire primary round was the nature of the questions—they were flat, serious, drilling and, above all, concerned. From beer hall to citizens' reception to Chamber of Commerce, the same questions crowded up at the candidate, questions that were to repeat themselves in the American mind from January to November—and, overwhelmingly, the questions thrust at foreign affairs. War? The Bomb? Castro? Berlin? East-West trade? What about the admittance of Communist China to the UN? (Stay in and fight them there, said Rockefeller, we aren't going to turn the leadership of the UN over to Russia and Red China.) How do you get the Russians out of Cuba? (Re-unify Latin America and the Alliance as the first step.) The Bay of Pigs? (We should have given air cover to the Freedom Fighters.) Over and over again the same questions, with Cuba and the UN uppermost in New Hampshire's mind. (Later in the season and on the West Coast the questioning would give predominance to Vietnam over Cuba.) And the same domestic questions, over and over again: Social Security? (Well, the other candidate wanted to make it voluntary—which would let the young people withdraw and thus wreck the payments the older people had a right to expect.) Negroes and public accommodations? (He was for the public-accommodations section of the Civil Rights Bill, flatly.) How about the Federal Budget? And the income tax? (He was for them both, flatly.)

What was more important than questions and answers, however, was the quality of conversation on the edges of a Rockefeller rally. The people who had come had come not because they loved Rockefeller; [2] they were seeking an alternate to Goldwater. For in less than three weeks since he had first come to New Hampshire, Barry Goldwater had frightened the Hampshiremen, and they were now examining Rockefeller as a way out of their puzzlement. One picked up the echo from the mother with baby in arms who said of Goldwater, "He's making silly statements; he's ambiguous; he's scaring everybody," as well as from the college man, listening to Rockefeller, who shook his head, saying, "We were for Goldwater until two weeks ago—but he's been saying such crazy things"; and (ten days later) from a Concord industrialist, manager of one of the largest printing plants in America, who had begun the campaign as a Goldwater man, but: "If he doesn't mean what he says, then he's just trying to get votes; and if he does mean what he says—then the man is dangerous. So I quit."

On this first trip to New Hampshire in January, only three weeks after the campaign had opened, one already had a sense of Goldwater's doom—but also an illusory sense of Rockefeller's progress. It was progress at tremendous cost and exertion, but, nonetheless, progress. Rockefeller had announced his entry into the New Hampshire primaries early in November at a time when his private polls were showing what the public polls also displayed: that in New Hampshire he could hope for only 20 percent of Republican votes. Rockefeller was campaigning hard all around the country and governing New York at the same time. (He had been in Buffalo on New York State business on Monday of this particular week, spent Tuesday, Wednesday and Thursday in California, returned Thursday night, attended to New York business on Friday morning, and was campaigning in New Hampshire that afternoon.) The schedule was to get worse. But he was allotting no less than twenty-eight days of campaigning to New Hampshire's approximately 100,000 registered Republicans, and he had been gaining: by the end of November he had reduced Goldwater's original 3-to-1 margin to 3 to 2; by January, in

[2] One of Rockefeller's staff members had observed that in the priesthood of government Nelson Rockefeller would end as a Cardinal, not as a Pope—for he could not jump the bridge from the administrative to the spiritual. When Rockefeller responds to questioners, always with good heart, he seems fascinated by the techniques of goodwill rather than by the dream of spirit.

most polls, he was running neck and neck. And he was drawing ahead the way he wanted to—by pinning Goldwater on issues, by forcing the antagonist to utter in public and in print those views and opinions which, for months thereafter, Goldwater was to waste his efforts in explaining away.

The real story in New Hampshire, it was apparent after a short jaunt with Rockefeller, was the Goldwater story—for the man was being destroyed as much by his own candor as by his antagonist's counterattack. And the New Hampshire episode of the Goldwater campaign deserves the most serious attention, for never thereafter was he to recover from it.

For mismanagement, blundering and sheer naïveté, Goldwater's New Hampshire campaign was unique in the campaigns I have seen. He had entered the conservative Granite State with the reputation of the nation's Number One Conservative. He had entered with the backing of what was left of Styles Bridges' once dominant machine, plus the support of New Hampshire's senior Senator, Norris Cotton. He enjoyed the support of the largest newspaper in the state, the paranoid Manchester *Union Leader*. But by January he had shocked New Hampshiremen into realizing that there were several kinds of conservatives—and that he was not their kind.

There is no mystery to the story. It all happened very naturally— but it was probably the most painful lesson ever learned publicly in national politics. No man runs for President of the United States but what, instantly on his announcement, he becomes a public personality on a scale and with an exposure for which no previous experience can prepare him. Goldwater is a man as dedicated as Rockefeller, and as forthright in his answers. But where Rockefeller errs in discussing a complicated world in all its complications, Goldwater errs in a more serious way; for each complication he has a direct and simple answer. His candor is the completely unrestrained candor of old men and little children. A graduate of the school of after-dinner speaking, usually to devoutly conservative audiences, he speaks to the people directly in his presence—and regards as a breach of friendship and confidence the quotation of such remarks when they appear on public record.

He entered on his New Hampshire campaign with an already serious burden of statement on the public record. On October 24th, 1963, in Hartford, Connecticut, he had suggested the use of atomic weaponry by NATO area "commanders" (see Chapter Ten); a week later, in Washington, he had urged that TVA be sold off. Previously he had advocated withdrawal of recognition from the Soviet Union, and abolition of the graduated income tax. But now he was an avowed candidate for the

Presidency—followed everywhere by intrusive newspaper and television reporters jostling one another and the candidate to relay whatever unguarded remark might fall at any moment. And his remarks fell, unguarded, one after the other.

He opened his campaign in New Hampshire—twenty-three days of campaigning were to be spent there to reach the 93,000 Republicans who finally voted—with a fusillade of statements that were to be registered, every one, in the research books of Lyndon Johnson and Nelson Rockefeller and return to haunt him. In his first week he declared that the American missiles were "undependable"; and instantly the vast apparatus of McNamara's Pentagon public-relations staff volleyed back at him. He stood at ease among the newspapermen on January 6th in his headquarters in Concord and, in response to a question, urged that Social Security be made voluntary. (GOLDWATER SETS GOALS: END SOCIAL SECURITY; HIT CASTRO, read the headline of the Concord *Daily Monitor* the next day.)[8] He journeyed down to New York and told the Economic Club there, in a pause from his New Hampshire rounds: "We are told . . . that many people lack skills and cannot find jobs because they did not have an education. . . . The fact is that most people who have no skills have no education for the same reason—low intelligence or ambition." He returned to New Hampshire and insisted that if Red China got into the United Nations, we should get out. When the mad Castro cut off American water at Guantanamo, Goldwater told an audience of 200 people in a rustic square-dance hall that he hoped President Johnson would have the courage to say, "Turn it on or the Marines are going to turn it on for you and keep it on." (Later that day he expanded that statement to say, "Turn the water on or we are going to march out with a detachment of Marines and turn it on.")

Goldwater has, of course, a penchant for the crisp phrase, and, as a conversationalist, it savors his talk with real salt. Some of his finer and more pungent phrases had long been fragrant in recent politics—"Let's lob one into the men's room of the Kremlin," or "Sometimes I think this country would be better off if we could just saw off the Eastern Seaboard and let it float out to sea," or (describing the Eisenhower administration) "a dime-store New Deal." These early phrases, used against him throughout the campaign, had been, however, the legitimate expressions of a private citizen, and, as William Scranton observed, it is entirely unfair to use against a man statements he made before he became a candidate. It is only when he goes on record as a candidate that the public must hold him to account. Yet, now that he was on record, Goldwater had not changed—and every reporter, from the cub following his first

[8] See Chapter Ten for the full statement and its effect.

campaign to the bored veteran covering his tenth Presidential primary, found that covering Goldwater was like shooting fish in a barrel. On any given day, if one listened closely, Goldwater was bound to get off some remark that would make the front page.

One must see the campaign as it was. Goldwater was making his first venture out of his native Arizona into the alien civilization of New Hampshire. Now, whereas the frontier hyperbole is the genuine literary expression of the Old West, the laconic understatement is the idiom of New England. Miserably advised and managed, Goldwater was grotesquely overexposed.[4] Moreover, he had just had an operation for a calcium spur on his right heel; it was so sensitive that he must hobble along in a shoe that had to be cut open on the right side—yet he was scheduled for twenty-three days of campaigning, street tours and handshaking. ("I remember every footstep of that campaign," said Goldwater after his defeat in November.)

In New Hampshire, Goldwater would say exactly what he thought, as he had always thought it—then shake with fury at its quotation in the newspapers, then at its analysis, its re-examination, its full-blown elaboration in the national columns and commentaries. He was offering New Hampshire the gospel of the true faith, hard money and individual rights. Here, if anywhere, it should get sympathetic hearing. And yet, day by day when he read of himself in the papers, it was another Goldwater he was reading about, a wild man seeking to abolish Social Security and go to war with Russia.

I followed him about for several days in the first week in March to observe the phenomenon, and it was quite clear he did not know what he was doing. For example: His strategists in Washington had conceived, for the last ten days of the New Hampshire campaign, a well-reasoned program of speeches on major issues—on morality and violence, on a balanced economy, on foreign policy. The speech scheduled for Keene, New Hampshire (delivered on Wednesday, March 4th), on civil rights, violence and the need for law and order was an excellent speech. But if it was reported anywhere in more than a paragraph, I do not know of it. For Goldwater had preceded and then followed that speech by two au-

---

[4] One of my favorite stories (as well as Goldwater's) is of his visit to a primary school, arranged by a local campaign manager. Goldwater, pressed for time and seeking the Presidency of the United States, found himself in an audience of children so young that their feet dangled from the benches without reaching the ground. Rising to the occasion despite his ire, he entertained the babies with stories of the Arizona Indians, the pioneers and the Grand Canyon—and they loved it. But not one of them could vote for another ten years. He inquired why, *why*, in such a pressing moment they had arranged such an audience; and the local tour manager said, well, after all, if Goldwater was going to be in this district—well, his little kids went to that school and he wanted them to hear Goldwater.

thentic, unpremeditated, beautiful Goldwater performances which explain why reporters, offered such temptations, could never resist them.

In the afternoon before the Keene speech he had appeared at a student gathering at the Keene State College. He had, off the cuff, told the students that government couldn't stop depressions—it only starts them; that we had eleven million out of work as we entered World War II (actual statistic: 5,560,000); that almost all legislation seeks to transfer some local control to Washington; that he had voted against Federal aid to education because he doubted whether the colleges could spend the amount of money they requested; that the recognition of Russia was a dreadful mistake ("We know less about the Soviet Union today than when we recognized her"). One student asked him whether he thought government should take care of poor people. Goldwater asked if the student knew anybody who wasn't being taken care of. The student said yes, and Goldwater flared, "Then why don't you try to help him out, why don't you do what you're supposed to do?"

The morning after the excellently prepared Keene speech on civil rights, Goldwater's cavalcade stopped at the home of Samuel Abbott in Wilton, New Hampshire. The Abbott family, who live on a hill, bestride their little town materially as well as symbolically, with the traditional responsibility of a New England family that hold in private trust the two local textile mills and a manufacturing concern and worry seriously about keeping Wilton's people employed in the face of technological competition. In the Abbott parlor were thirty of the best people in Wilton, with the local divine sedately in the seat of honor—plus fifteen national reporters.

Here, thoroughly at home in the convivial atmosphere of solid, responsible conservatives, Barry Goldwater, once he had sharply told the TV camera crew to turn the lights off his face, was at ease. These were his own kind of people, and, as if at home in front of his own fireplace with a few friends, as if unaware that the national press listened for the national audience, he offered these observations: that he wanted to go to the Republican Convention to see that the good people of the Party wouldn't be kicked around as they were then (in answer to a question of how the nomination was "stolen" from Taft in 1952); advocated the abolition of the electoral college ("the big cities resist us on that"); expressed bitterness about the low tariffs of America ("we have the lowest tariff in the world—about 8 percent"); declared that he advocated "carrying the war to North Vietnam—ten years ago we should have bombed North Vietnam, destroyed the only access they had to South Vietnam, with no risk to our lives" (the previous day he had mused aloud to the students how ten years ago we might have dropped a low-yield atom bomb on North Vietnam to defoliate the trees); pointed out that we were a trillion dollars in debt—$20,000 per family; declared that the great depression had started in Austria because Austria did precisely what the American government was doing today; observed that he didn't know a state in the Union with more discrimination (against Negroes) than New York; defended his position on sending in the Marines at Guantanamo by saying that "a dictator was able to push the United States around and get away with it; sooner or later we must stop this, our embassies being burned up, our flag being torn down and scoffed at"; and ended his forty-five-minute visit by saying he'd been glad to be with them—"it gave me a chance to find out what's on your mind and you to find out whether I have one."

Flanked by two such provocative impromptu news sessions, Goldwater's major speech, so carefully worked over and drafted in Washington, received only minimal press attention. And it raised, as early as winter, the problem of the press: How could one be fair to Goldwater—by quoting what he said or by explaining what he thought? To quote him directly was manifestly unfair, but if he insisted on speaking thus in public, how could one resist quoting him? Later, when Goldwater would hang himself with some quick rejoinder, the reporters who had grown fond of him would laboriously quiz him again and again until they could find a few safe quotations that reflected what they thought he really thought. They were protective of Goldwater in a way that those zealots who denounced the Eastern press could never imagine. This change of attitude was a long time coming, however; and in the New Hampshire episode it was difficult for anyone to believe that any Presidential candidate could be so innocent of the importance and attention of the national press in a major campaign. Goldwater's irresistible candor, translated by the reporters into print and then fed back out to the electorate by Rockefeller's research staff, was, ultimately, to work his destruction. But in the beginning it was self-destruction, and reporters played the game by the standard rules. "He continues," wrote Loye Miller of *Time* Magazine in an unpublished summary of Goldwater's first January week of campaigning in New Hampshire, "to roll out rather remarkable statements which sooner or later seem bound to get him into trouble. . . . Now that Barry Goldwater is a formally declared candidate for the Presidency, he can expect intensive scrutiny of every remark from the rest of the nation. And this week he got his facts so muddled on at least a couple of the leading issues of the day that it will be surprising if Nelson Rockefeller doesn't pick up the inaccuracies and imprecision and use it to portray Barry as an intellectual lightweight."

It was Rockefeller, of course, who destroyed Goldwater—not out of malice but, at this stage, out of the need of his own campaign to score through on New Hampshire conservatives and win them to his own Presidential cause. If Rockefeller could expose Goldwater, then unpin the natural conservative loyalties of the upper New Englanders, he might bring them to his own standard. By the end of January he had, by every poll, succeeded in accomplishing the unpinning. Yet by January every sampler of public opinion had become aware of the quaking instability of Republican loyalties, that questing and groping for a *Republican* way out which was to continue through spring and into the fall elections themselves. For Rockefeller was not only a New York liberal (and a New Hampshire liberal is several degrees to the right of a New York conservative) but a big spender; and, however much Rockefeller explained the huge New York State $2.83 billion budget to the Granite State men (who brook neither general sales tax nor income tax in their state), he could not explain away the fact that one hell of a lot of money was being spent in New York. Moreover, if Goldwater could not escape the burden of his candor, Nelson Rockefeller could not escape the burden of his own honor. As a woman at one of his rallies expressed herself, "Oh, he sounds all right—it's just his family life, if you know what I mean." It was not the divorce so much as the remarriage—"It's not right for a man to throw away his old wife and take up with a younger woman," or "She looks nice enough when you meet her, but I still can't understand—about those children, I mean."

I followed Rockefeller in the last week of campaigning, and in the town of Hollis, New Hampshire, on the Saturday before primary day, I preceded him to a midmorning rally. The shiny new gymnasium of Hollis High School was the meeting place, and by eleven o'clock in the morning a large band of intent citizens had gathered on the glistening floor to hear the candidate. This time he arrived with Mrs. Rockefeller, and since she was heavy with child, someone hastily brought a little folding chair upon which she might sit. The TV crew quickly rearranged its lights, and their glare flooded her. Her husband stood beside her; she sat—the only seated person in the hall—as if in a witness chair; the townspeople gathered in a semicircle somewhat apart from the candidate and the seated woman, as if they were a jury. And then, as he spoke—excellently that day on the subject of leadership and responsibility in a world at war—I watched the audience. Whether or not they were listening to him I cannot say, but their eyes were all staring unnervingly at the handsome woman with bowed head and curled legs who sat in the spot of light. They might have been a gathering of Puritans come to examine the accused. They had made up their minds. Hollis was to vote: for Henry Cabot Lodge, 238; for Rockefeller, 120; for Goldwater, 85.

What would have happened had Henry Cabot Lodge's name *not* been entered in the New Hampshire race, no one can guess. It is this reporter's opinion that Nelson Rockefeller would have won by a flat majority, gone on to a larger majority in Oregon and then probably carried California to defeat Goldwater conclusively.

One should have space to tell the Lodge campaign in full; for the Lodge campaign had little, if anything, to do with Henry Cabot Lodge in far-off Vietnam—who, a few days before he left for Southeast Asia in August of 1963, had personally assured Nelson Rockefeller that he, Cabot Lodge, was supporting and would support Nelson Rockefeller as Republican nominee.

The Lodge campaign as seen in the field was a madcap adventure, the gayest, the happiest, the most lighthearted enterprise of the entire year 1964. Its entrepreneurs were four of the most attractive amateurs ever to trigger a political upheaval. There was, to begin with, an elfin New England businessman, Paul Grindle, forty-three, a man with a delicate, satiric touch of mischief. A Harvard man who had left the university in his freshman year, a newspaper reporter later, then a New York public-relations man, he had finally come home to Boston, after having married an equestrian artist of the Barnum-Bailey enterprise, to develop, by the most novel direct-mail techniques, a highly successful scientific-instrument firm.

His companion was an exceptionally wise young lawyer, deceptively innocent in appearance, who had grown bored with law practice in Boston—David Goldberg, thirty-four. Their staff was two of Boston's finest young ladies, both immaculately groomed, healthily handsome, fresh with first bloom—Sally Saltonstall and Caroline Williams, both twenty-three. Wherever this group of four appeared in the long season—and they traveled as a do-it-yourself package, from New England to Oregon to Los Angeles to San Francisco—their presence made a party.

Their operations, even in retrospect, seem so simple and childlike that to compare them with the operations of genuine professionals like Lawrence O'Brien and Clif White is impossible. They had campaigned together as a group for George Lodge for Senator (against Teddy Kennedy) in Massachusetts in 1962; maintained their friendship; scraped together $300 to open a Lodge-for-President headquarters on State Street in Boston in late December, 1963; and then been shut down by the Massachusetts law which forbids such self-winding headquarters without a candidate's approval. On Friday, January 10th, they drove to Concord, New Hampshire, to sample the atmosphere. Amateurs all, as Grindle said later, they thought it would be "fun" just to try. They stopped that very afternoon, hired an empty store for $400 (from January 10th to March

10th), put a deposit down for a telephone, borrowed—for free—some furniture and folding chairs from New Hampshire Republican State Headquarters and returned to Boston.

The following Monday they hired a truck in Boston, put on overalls, dismantled their plywood partitions in the banned State Street headquarters, packed their literature, petitions, stationery in cartons, drove through the snow to Concord again, hired a signpainter (for $162) to paint a huge sign saying LODGE FOR PRESIDENT over their headquarters across from the State Capitol. And they were in business.

Their first step, using Grindle's professional knowledge of direct-mail technique, was to take an old statewide list of registered Republicans and send out 96,000 letters inviting the recipients to write Ambassador Henry Cabot Lodge, Saigon, c/o Concord Headquarters, pledging support.

In politics it is essential that a group give the impression of a force in being, for everywhere, in thousands of American homes, are individual citizens who believe themselves helpless because they are alone. It is only when they find they are not alone that they become a political force. The response to the first direct-mail appeal was unbelievable—no less than 8,600 pledges of support bouncing back by return mail. Rockefeller had dislodged New Hampshire's conservatives from Goldwater; they had paused briefly to consider the New York Governor; now, here was a new consideration. And as the four young people sensed the response, they began in February to build toward a climax. The first mail had brought names; names could be organized by town and county into volunteer groups; a second mailing ("Pick Up Two More Votes") brought an even greater response. Multiply the almost 9,000 original pledges by three —and one gets 26,000 votes.

Out of an expected total vote of 100,000, this was enormous—if it could be delivered. Lodge was not on the ballot, and his name would have to be written in. Out went a third mailing that included a clear sample ballot with fine red markings showing exactly how to write in the name of a candidate on the ballot. Grindle went off to New York to gather up old film clips of Lodge with Eisenhower, left over from the 1960 campaign; and, for $750, was able to produce a television show that refreshed every memory of Lodge's career, from Lodge as a Lieutenant Colonel in the U.S. Tank Corps, to Lodge of the UN shouting down Russians.

Meanwhile, in an odd and completely unplanned syncopation, the Lodge campaign bounced about in the press. At each mailing, at each new mobilization of volunteers, there would be a burst of news stories by correspondents who had already written themselves out on

Goldwater and Rockefeller. Each new effort would be measured by the public-opinion analysts, who would, all through February, find a rising tide for Lodge. As this would make a new story, the four pranksters ("What fun this is," said Grindle one day in the back room of his store-front headquarters. "Whatever happens, this proves politics can be fun.") would whack the story up another degree with some new statement or device. And as correspondents would come to visit them in Concord (as much to chat with their handsome young ladies as to escape the weight of Western civilization being debated by Goldwater and Rockefeller), new stories would appear. The story grew and grew, swelled and swelled. But was the Lodge boom real? Could a write-in candidate, 10,-000 miles away, managed by amateurs, beat two such highly organized campaigns so profoundly serious in meaning and intent?

On Tuesday, March 10th, New Hampshire enjoyed an old-fashioned New England blizzard: up to fourteen inches of snow from the Canadian to the Massachusetts border—snow crusting the kepis of the Union veterans, snow blocking Governor John King's new state highways, snow slushing the streets of Manchester, snow over mill and factory and ski slope and farm. New Hampshire's polls closed at seven P.M. Portsmouth, New Hampshire, the only town in the state that uses voting machines, is the first sizable center to report. By 7:18, Walter Cronkite announced over CBS that Henry Cabot Lodge had won New Hampshire. Half an hour later the predictions were precise, only a fraction off the final count: Henry Cabot Lodge, 33,000; Barry Goldwater, 20,700; Nelson Rockefeller, 19,500; Richard M. Nixon (also a write-in), 15,600.

Shortly after eleven P.M. the national television shows displayed on camera the lean, pleasant image of George Lodge, the candidate's son, gravely thanking the people of New Hampshire for this vote of confidence in his father—while, mad to the end, the merry crew was grappling just off camera with a giant French-Canadian drunk who was trying to strangle David Goldberg for some fancied slight. Two state policemen finally joined the free-for-all and saved Goldberg so that that good young man could go on, with Grindle, Williams and Saltonstall, to become a continuing factor in the politics of the next few months.

"Kickoff in a driving rain on a muddy field to no decision" was the way one observer described the New Hampshire primary. It was Lodge now who commanded the headlines. Nixon was silent. And the next move was up to Goldwater and Rockefeller.

Goldwater had been as seriously wounded by the New Hampshire results as Rockefeller; he had started higher and fallen farther. ("We have it made," he had said as he left New Hampshire for the last time, guessing he might pick up 40 percent of the vote.) He had relied on local advisers as inexperienced in Presidential politics as he was himself, and they had misled him. Now, exercising authority for the first time over his own campaign, Goldwater decided that he would skip the next primary round—Oregon—entirely and concentrate all effort in the final round, California. In deep discouragement, he toyed briefly with the idea of quitting the campaign entirely. His advisers dissuaded him—the Clif White operation was moving silently through convention after convention in the Western states, rounding up delegates; he still could make it, they insisted. But, whatever the delegate count, Goldwater knew he must have one win at the open polls to prove to the Party that his support could reach beyond hard-core conservatives and zealots to ordinary people who would vote at the polls in November. Thus, for Goldwater, all effort was now to be concentrated on California—let Lodge have the victory in Oregon which seemed certain.

For Rockefeller the decision was harder. Yet of all the qualities that Nelson Rockefeller possesses which earn him the respect, sometimes grudging, of other men, the chief is courage. And it was Rockefeller courage that was to dominate the Oregon primary.

For Rockefeller, the name of the game was now Impact. From New Hampshire on, there was no longer any realistic chance of his becoming the Republican nominee. But to veto the choice of Goldwater, he must prove before the convention assembled that Republican voters would not have Goldwater on any terms—West, no more than East. Rockefeller received the New Hampshire returns at his Fifth Avenue apartment—and at five o'clock the next morning was up, aching from defeat, and off at seven A.M. in a family six-passenger Saberliner jet to the West Coast to show his Oregon and California troops that he was still fighting and meant to fight to the end.

It was as if the Dodo in *Alice's Adventures in Wonderland* had planned the elections of 1964—all should win, all should have a prize. Lodge should have New Hampshire, Rockefeller should have Oregon, Goldwater should have California—and Lyndon Johnson should have the country.

Yet it was not quite that clear when Rockefeller arrived in Oregon to campaign as a hopeless loser. Lodge's picture was now on the magazine covers across the country; Lodge led every poll from coast to coast. From 19 percent in the polls in January he had become the leading choice of the Republican rank and file in April. In the aftermath of the New Hampshire victory, Oregon Republicans shifted as the nation's Republicans shifted, and in the first Harris samplings of Oregon showed thus: for Lodge, 46 percent; for Nixon, 17 percent; for Goldwater, 14 percent; for Rockefeller, 13 percent.

Perhaps it was the very hopelessness of the outlook that now charged Rockefeller's effort with a warmth, a yearning, an intensity of emotion that transmitted better than his normally jolly yet precise speeches. Only twice was he to stir emotion in his 1964 campaign—during the Oregon campaign and then in his climactic appearance at the San Francisco Convention.

He was tired when he arrived in Oregon for his last week of campaigning. The boils that had begun to fester on his neck in New Hampshire hurt him still; an osteopath, Dr. Kenneth Riland, a long-time friend, accompanied him on his private plane to rub the weariness away between flights. He was grayer, thinner, even hungry-looking. Back in New York a liquor scandal, involving men he had trusted and personally appointed, had embrangled him with his own legislature; he was engaged in a homefront war with recalcitrant New York State legislators that contrasted badly with Scranton's spanking victories in the Pennsylvania legislature. Sensitive for his wife, trying to protect her, Rockefeller realized that the birth of their first child would fall approximately on the weekend before the California primary—and knew that his own staff, as well as hostile politicians, were callously debating this matter of privacy and betting whether a baby would or would not help him in the contest if the birth fell before the Tuesday of voting in California. Over weekends he would fly to New York to be with his wife, then flog himself back without rest to the Oregon campaign. From mid-March to voting day in California on June 2nd, it was on the West Coast he must prove himself—as much for her as for what he believed in.

Real emotion can always get through in direct contact with people. And Oregon is a small enough state to permit such direct contact—with only 1,826,000 people (1963 estimate), Oregon is far closer in texture to tiny New Hampshire (with 627,000) than to its giant neighbor, California (17,590,000). Nor was Rockefeller overexposed, as in New Hampshire; and he had the state to himself.

Sitting on the bus with the newsmen who followed him and now began to respect his courage, he would be depressed and confide to one or another, "I don't care any longer if I win—but I feel strongly about these things." If he could not stop Goldwater here, then he saw the Party and the country in desperate trouble.

Now, at this juncture, he developed the term "mainstream" that was to irritate Goldwater so badly and later be borrowed by the Democrats to hammer the Arizonan. For if Rockefeller had been boxing in New Hampshire, now he was slugging. A composite of his many crossroads, factory and supermarket speeches might run thus. If, say, he were at a sunny supermarket plaza in mid-morning with the parking lot full of handsome young Western mothers, their babies in arms and toddlers chasing around, he would begin with children: Where were these children going to school when they grew up? What opportunities would they have? They needed schools, and he was *for* aid to schools, Federal aid to schools (score one against Goldwater). And let's not forget the older people too ("the people who built this country and made it great"); some people would wreck Social Security by making it voluntary, but he was for Social Security *and* Medicare (scores two and three against Goldwater). But what was most important? That these young people and old people alike look forward to a world of peace. The United States needed a *responsible* man for President, not one who might set off war unwittingly (score four against Goldwater). If, say, he were addressing a group of workers at one of the new electronic or technical plants in Oregon, he would begin with complimenting them on their techniques, what they were doing, explaining that they were living in a world of change, that America (and their work) led these changes—and that America must move with these changes; it couldn't stand still; the Republican Party must move with the times. Whether at factory, supermarket, Elks' meeting or farmers' gathering, he would close by driving it home: These people here, right in front of him, were going to decide in which direction the Republican Party moved—and it must move with the "mainstream" of American life.

As for himself? A little girl in the sun at the Raleigh Hills Plaza in a Portland suburb asked him why he wanted to be President with all those problems. And he replied, "This young lady asked me why I wanted to run for President. Well, that's a good question. I'll tell you. I decided to run for the Presidency because I really care about America, about freedom—I think every citizen has a responsibility to do his or her best if we are going to help keep a free society strong. I've been in appointive jobs such as Assistant Secretary of State, working with the President, but I came to the conclusion that if you really wanted to have a say in the future of this country, the only way to do it is to get into elective office because that's where the decisions are made, and I want to have a voice where decisions are made."

Underlying the personal drive of the candidate were two other elements.

There was, first, a campaign drive captained by one of New York's best political technicians, Robert Price, the permanent campaign manager of Congressman John Lindsay, to whom his loyalties belong. Price had been released to the Rockefeller campaign for this emergency. Arriving in Oregon, he set up the same kind of direct-mail campaign that was normally effective in Lindsay's Manhattan district, but on a statewide scale overmatching Grindle's effort in New Hampshire. With unlimited funds and all the personnel he needed, Price in three weeks developed out of his mailing lists a volunteer organization that began to flourish from the Columbia River to the California border. To this mail campaign was added a carefully pinpointed newspaper advertising campaign, plus a use of TV that came to a crescendo in a series of Rockefeller appearances, paid and unpaid, that totally dominated the Oregon home screens for the forty-eight hours before voting. And capping the entire technique, as the candidate spoke of "responsibility" and "mainstream," was the emotional phrase: "He Cares Enough to Come." The other candidates were absent—but Rockefeller was there. HE CARES ENOUGH TO COME—THE GATEWAY SHOPPING CENTER, 3:30 P.M. WEDNESDAY, MAY 13TH would read a street sign, or ROCKEFELLER CARES ENOUGH TO COME—UNIVERSITY OF PORTLAND, 8:00 P.M.

Buttressing both candidate and campaign was yet another factor— the local bitterness between Oregon Republicans. For Eastern Republicans, "extremists" are hyperthyroid citizens who inhabit a fuzzy two thirds of America that begins across the Alleghenies where the anti-fluoride, anti-UN, anti-income-tax, anti-Supreme Court groups pullulate. But beyond the Alleghenies there are as many, proportionately, of liberal and moderate Republicans as there are anti-fluoride or anti-Supreme Court Republicans in the East; and in the West the moderate Republicans had already been humiliated by the effort of extremists to seize control of the Party machinery wherever a coup might be staged. Such a coup had been staged in Multnomah County (Portland): the Republican Party machinery of that largest county in the state had been snatched from the hands of its moderate business and professional leaders by right-wing zealots only the year before the primary. To those defeated in the coup, Rockefeller's name was a flag—a man who would stand up and fight extremism anywhere in the country. Their organization on his behalf was animated by a zeal, a conviction and a skill that Rockefeller's local allies in New Hampshire had never been able to generate—and the momentum carried them on a few weeks later, after their primary victory, to recapture Multnomah County's organization from the right-wingers.

One could sense the indecision as sharply in Oregon as one had felt it in New Hampshire. Only this time it worked the other way. From a 46-percent lead in early April the Lodge margin began to fade—to 40 percent in the first week in May, to a final 35 percent (though still in the lead) a few days before the primary. It is difficult to tell in a primary where a shift in voting patterns begins—for a primary involves no party loyalties, and there are no stones to anchor the voters' moorings. Louis Harris, analyzing later his own misjudgment of the swift shift, declares that in Oregon the Lodge vote began to crumble first in the suburbs—at the professional, doctor, lawyer, junior-executive level. And from the opinion-making level in the suburbs this detachment from Lodge spread with gathering speed down to other groups, urged on by Price's media campaign and Rockefeller's vigorous presence. Yet whether it was shifting from Lodge to Rockefeller or from Lodge to Nixon (on whose behalf an amateurish campaign was belatedly organized), no one could tell when the voting began on the clear and sunny Friday of May 15th.

Oregon's polls close at eight, and votes are largely counted on paper; but by 8:03 early returns were already filtering through NBC's calculators clearly enough for that network to announce (the first nationally) that Nelson Rockefeller was the winner of the Oregon primary. The final scores read: Rockefeller—93,000; Lodge—78,000; Goldwater—50,000; Nixon—48,000. And it was on to California for the final phase of the duel in the sun.

No state could have been selected for more dramatic effect to climax the Republican primary duel of 1964 than California—for the history of California Republicans had preceded or mirrored for three quarters of a century every stroke and counterstroke of the civil war which rages in the Republican Party nationally. Students generally remember the date 1912—when Theodore Roosevelt split his Bull Moosers from the Party and led 350 delegates out of the Republican Convention to "stand at Armageddon and . . . battle for the Lord"—as the most dramatic marker of the irreparable breach. But for a full decade before that, California Republicans had already been at war with each other in the gold and tawny slopes and valleys of the magnificent state for control of the machinery at their capital, Sacramento.

The hero of that early war was a waspish, stocky, hot-tempered, self-opinionated little man called Hiram Johnson, without mention of whom, even today, no one can understand California politics. Johnson's enemies were the Neanderthals of the turn-of-the-century California Republican leadership who had sold out their state, lock, stock and barrel, to the lobby of the Southern Pacific Railroad. In leading the moralists and ordinary Republicans of his time in a crusade against the Southern Pacific Railroad and its legislative puppets, Johnson created a Progressive Republican Party that not only purged the state of evil but went on to design one of the finest state governments, best school systems and cleanest systems of politics in the nation. His revolt—coded in laws very similar to the progressive revolution in Wisconsin[5]—wiped out political patronage as known elsewhere in the Union, forbade conventions to nominate candidates, fumigated every crevice and cranny of conventional party politics. California has never since had an effective "boss" or an Eastern-style "machine." In California, up until the last few years, candidates have had to contend with each other on the basis of personalities, excitement, public relations, media control or volunteer groups to get out the vote that elsewhere is delivered by permanent party organization.

This makes for unstable politics. But this instability is further exaggerated by a unique and dominant local condition: California is a state of newcomers, the last stop of the westering impulse of American movement. In the single generation from 1930 to 1964, California has more than tripled its population (from 5,677,000 in 1930 to about 18 million today). This growth has resulted in one of the most triumphant demonstrations of American democratic achievement—no other political community in the world has undergone so violent a population explosion in so short a period of time and simultaneously provided great schools for its children, great roads for its cars, good houses, shelter, hospitals and universities for all who need them while, throughout the upheaval, always preserving the dignities of law and order in freedom.

It is remarkable that there is any stability at all to this vast community of quivering, inrushing, ever-changing system of human atoms. (Sixty percent of California's population, at the last census, was born out of state—as compared to New York State's 33 percent—and among voters over twenty-one the disproportion is even higher.) Some have come west to the sun to sit in the afternoon and doze as they wait for the second stroke and the end. Others who passed through in the war as GIs have returned to raise children who grow six feet tall—without overshoes, mufflers, overcoats or sniffles. Others have come lured by new technological industries and the horizons of an expanding economy. But all, alike, have left some condition back home that vexed or exasperated them—the congestion of the Eastern cities, the clash of race hatreds, the cramping of regulations, alien tongues, faiths and faces that have overwhelmed their native neighborhoods. And it is only natural that their new community should be racked by political fevers unknown elsewhere, for the emotions of these restless newcomers are rocked and tormented and confused by a questing for the certainties and safe moorings that they lost in their homelands.

These emotions produce the "kooks" for which California is famous—the kooks of the Republican far right and the kooks of the Democratic far left, the little old ladies in tennis shoes as well as the beardies

---

[5] See *The Making of the President—1960*, pp. 80-82.

in similar tennis shoes. These emotions produce the comic and flamboyant surface ripples on California politics—the nature faddists, the evangelical sects, the pension-plan extremists, the EPIC planners of the thirties, the vegetarians, the prohibitionists and one-Godders. But not since Upton Sinclair's EPIC campaign of 1934 have these emotions played so great a role in the basic politics of the state as they did in the Republican primary of 1964 when Goldwater and Rockefeller squared off against each other. For now the emotions were able to lock on to the great split first opened in the Republican Party by Hiram Johnson so long before.

That split had been plastered over for sixteen years by two Republican Governors—Earl Warren and Goodwin J. Knight—who shared a simple strategy. Their strategy held that the Republicans, to win, must offer clean, good, effective government and, as the minority party in a state where Democratic registration outnumbered Republican registration by almost three to two, must appeal to moderates and men of goodwill, whatever their party label. It was an aggressive, active Republicanism they offered. It had to be: to govern a state in such turbulent growth, government needed constantly to be *doing* something. Yet the aggressive Republican philosophy irritated, during all the years of its dominance, the whole school of primitive Republicans who feel that government's role is to keep out of people's private lives. And when, in 1958, clumsy and stupid party bungling forced Governor Goodwin Knight to give up his chair and let Senator William Knowland run for it, the Republican Party collapsed again into its twin wings of progressive and primitive, hating each other with fine, undiminished hatred.

It was this ancient split among California Republicans, freshly renewed and bitterly fought since 1958, that had already set the stage for the last primary when Goldwater and Rockefeller and the trailing corps of reporters, all thoroughly exhausted, arrived in Los Angeles for the final round of the duel.

Already Los Angeles, which receives a heavier dose of political billboarding in every campaign than any city outside of Boston, was wearing battle colors. VOTE FOR THE RESPONSIBLE REPUBLICAN, read the Rockefeller signs, and the Goldwater banners in blue and gold read, WE WANT A CHOICE, NOT AN ECHO. $AU_3H_2O=1964$, read the bumpers on Goldwater cars; $AU_3H_2O=H_2S$ countered the university car bumpers. Huge Goldwater posters read: YOU KNOW WHERE HE STANDS—VOTE FOR THE MAN YOU CAN TRUST. KEEP AMERICA IN THE MAINSTREAM, replied the more modest white Rockefeller posters. (Rockefeller's advisers felt that too many overlarge signs, of Goldwater size, would advertise and flaunt his personal wealth.)

The Goldwater campaign was easier to describe than the Rockefeller campaign. At 4848 Wilshire Boulevard, on the corner of Tremayne, above a green lawn and some yellowing pine trees, rose an adobe building with red-tile roof, of earliest California-Moorish style. Its balcony, its windows, its doors all ornamented with blue-and-gold Goldwater banners, it hummed with activity. At any hour of the afternoon the good ladies were to be seen arriving and departing, carrying out to the family cars cartons of posters and pamphlets for distribution. The switchboard was always so jammed that it was sometimes easier to visit than to telephone ("It's all these ladies," the switchboard operator told me, "they just keep our lines tied up calling in with advice for Barry and telling him what the Rockefeller people are doing"). This was ostensibly the command post. From these headquarters, organization and mechanics were directed by eminently respectable Southern California conservatives who reported to former Senator William Knowland, chairman for the state, in Oakland. Goldwater had chosen Knowland for just this quality of ponderous respectability.

Less than a mile down Wilshire rose the Los Altos Apartments. And here, in the fourth-floor corner suite once occupied by Greta Garbo, the national command of the Goldwater team had arrived in force for its on-the-spot strategy meetings. Here were gathered all those who made up the candidate's inner group. Denison Kitchel, his chief of staff and Karl Hess, the new speech writer; Clif White and his field operators; Richard Kleindienst and Dean Burch of the Arizona group. Here were plotted and prepared the tactics, the blows, the volley and counter-volley of public statement designed to answer Rockefeller in headlines and on TV.

And, lastly, with no definable office anywhere, there was a third force with a purpose of its own. This was the force dominated by a pale, thin and tense young man named Robert Gaston, undisputed and autocratic boss of the Young Republicans of California. Gaston, thirty-four (a transplanted New Yorker, like several of the right-wing leaders in the West), had made the Young Republicans of California into the most severely disciplined body of youngsters the state had ever seen; he had masterminded the rowdy control by right-wingers of the National Young Republicans' Convention in San Francisco in June the year before—the episode that triggered off Nelson Rockefeller's blast against "extremists." Intelligent, eloquent, purposeful, Gaston is without doubt the most engaging of the dedicated right-wingers in American politics. To him even the conservatism of William Knowland is suspect. ("That Knowland," said Gaston, "talks conservative when he's back here, but when he was in Washington he voted for foreign aid and the UN.") In the great rally of the fall of 1963, when Gaston's Young Republicans had packed the Dodgers stadium for one of the most frenzied Goldwater receptions ever staged, Gaston had excluded Knowland from the speaker's dais. Now that Knowland had been officially named chairman of the Goldwater effort in California, he had retaliated by wiping Gaston's name from the slate of Goldwater delegates to the convention.

This, however, had not diminished Gaston's ardor to do battle and destroy Nelson Rockefeller and the liberal Republicans. In Gaston's mind, Goldwater was only a beginning and moderate phase of the conservative revolution that would come later—and Gaston controlled troops. Since early spring his troops and operatives had been in action at every level. The official Goldwater headquarters at Wilshire and Tremayne had, for example, divided Los Angeles County, which holds 36 percent of all California votes, into six areas in which 10,000 volunteer doorbell-ringers were out canvassing registered Republican voters. Of the six area commanders, five also reported privately to Gaston as Young Republican chieftains; of all the volunteers, Gaston estimated he controlled 8,000. Headquarters, throughout, were infiltrated by Gaston operatives up to the senior levels of campaign planning. In some counties, like Orange, San Bernardino and Santa Barbara, the volunteer operation was entirely his. With his own funds (and with at least one major last-weekend mailing) Gaston conducted his own campaign for Goldwater as if he were an independent ally of the Arizona Senator. And beyond Gaston, of course, were the real kooks: the Birchers, zealots, fanatics and sectarian crusaders whom the Goldwater senior staff was desperately engaged in stripping out and excluding from formal campaigning. Said one Goldwater leader, "We've got superpatriots running through the woods like a collection of firebugs, and I keep running after them, like Smokey Bear, putting out fires. We just don't need any more enemies."

What volunteers can do has never been better demonstrated than by the combined Goldwater-Gaston forces. March 4th had been the opening of filing in California, and 13,000 signatures were needed to qualify a candidate for statewide ballot. Thirty days were allowed for signature collection, and the first candidate to collect the required number would occupy first place on the ballot (estimated by some politicians to be worth 5 percent of the vote alone). Goldwater volunteers had set up an "Operation Q" (for Quiet) for the March 4th opening, as meticulously drilled and planned as an expeditionary force waiting for D-Day. By early morning they were ringing doorbells all over California; by noon they had collected 36,000 signatures; within two days, 85,000. Rockefeller men, by contrast, had to pay for signature collecting and required weeks to get 44,000, of which 22 percent were found invalid. (In Orange

County, Robert Gaston's troops were so pleased by their quick success that when they found out that Rockefeller's organizers were paying their workers fifty cents per signature, they mobilized and collected 7,000 more signatures for Rockefeller—and turned the money over to the Goldwater campaign!)

The view of the California campaign from the Rockefeller encampment was entirely different, and for the political tourist the best guide was George Hinman, the New York National Committeeman. Hinman had arrived in late fall to study the political geography of the huge state, and by January he had made a bungalow in the tropical garden of the Los Angeles Ambassador Hotel his home and headquarters.

The first observation that came to Hinman's analytical mind as he tried to describe the problems he faced in invading the society of the Golden State, so different from rural upstate New York, was that California life was "unstructured." To the grave view of this Binghamton, New York, lawyer, California was a vast, whirling puzzle. Who were the natural leaders of the new communities of California? Where could Rockefeller find native allies through whom he might reach the mass of Republican voters?

Hinman's approach to the puzzle was through the delegate problem. California was to have 86 delegates and 86 alternates at the Republican Convention; and California Republicans would be offered, in the primary, a Goldwater slate and a Rockefeller slate, winner take all. Whom should Rockefeller choose to stand for him? How should he get them? For three months of quiet and patient missionary work, Hinman labored on the problem to come up, finally, with a little political masterpiece. In his delegate and alternate slate the New York lawyer had made a public diagram of what he considered to be the best in California civic leadership. At the top of the slate was Senator Thomas H. Kuchel, the most distinguished of California Republican liberals; then came former Governor Goodwin J. Knight; then San Francisco's ex-mayor, George Christopher. A Chandler of the dominant Los Angeles Chandler family, and a Sutro of San Francisco's Sutro banking family, added the oldest names of California's pioneering history. A Firestone (of Firestone Tire), Theodore S. Petersen (former chief of Standard Oil of California), Jack L. Warner (of Warner Brothers Pictures), Justin W. Dart (of Rexall Drugs) spoke for the best industrial leadership. A distinguished artist (Millard O. Sheets), a distinguished scientist (Dr. Joseph Kaplan), several architects and doctors brought the slate the luster of arts and sciences, James P. Mitchell, Eisenhower's former Secretary of Labor, industrialist Charles Ducommun, plus several score of the best civic and citizen leaders in all of California's Congressional districts, filled out the roster of 86 delegates and 86 alternates. All in all, the Rockefeller delegates shaped up as the most glistening slate offered any state by either party in 1964, an honor roll of achievement. The slate could be read, either cynically or naïvely, as an Easterner's view of what the California Establishment was. But courage, too, of a rare order was involved: from January of 1964, when Hinman began his recruiting effort, to March 3rd, when delegate names were announced, Rockefeller had never scored higher in a California poll than 31 percent. These delegates had been mobilized to support the New Yorker as much by their real fear of the California right-putsch in state politics as by Hinman's persuasion. They were willing to stand up and be counted.

These were leadership names, however—not doorbell-ringers, not troops. Somehow, in some way, the voters, too, must be energized. If they were to be energized, volunteers must be found among the "moderates"—and, as one California politician put it, "moderates are the people who don't go to the polls on election day." In the dark days of midwinter and early spring, as Rockefeller bounced about at the bottom of the polling lists, the problem of energizing the grass roots had to be approached in a way quite different from the approach of Goldwater leadership, which relied on crusading volunteers. To this end, Hinman had to seek professional help—and went to the firm of Spencer-Roberts, a team highly experienced in the California specialty of professional campaign management. Its list of triumphs had included, in 1960, the election of Congressman John Roussellot, the John Birch Society chieftain on the West Coast, and thus it knew, out of past partnership, exactly what the opposition troops could this time deliver. With absolutely zero troops of their own to start with, Spencer-Roberts had to use money to recruit troops: by opening 40 to 50 Rockefeller headquarters across the state (6 in Los Angeles County alone) to show the flag and banner where moderate volunteers might rally; by publishing a newsletter whose circulation rose from 4,500 to 25,000 in a few months; by direct-mail techniques; by billboards, advertisements, television and the drenching of radio, which in automobile-happy California has a dimension of its own.

The purpose of political campaigns is to gain attention first; then, with attention caught, win voters to identify with the candidate's personality or goals. And nothing attracts attention more than victory. Until the Oregon victory, despite all the Hinman and Spencer-Roberts efforts, Rockefeller had failed to score through on California's attention. But with his Oregon victory he overnight scored through on an audience that his unlimited resources had hitherto been unable to reach. A woman at an airport rally expressed it in simplest terms to me: "Oh—I guess I made up my mind after the Oregon election. I started to listen to him on television and then I read what he said."

And with the attention of California finally fixed on Rockefeller, there came again one of those giant emotional swings that marked the rending effect of the Republican civil war all through 1964. It was useless trying to measure the swing by talking to individual voters, for in the final stages of any campaign, voter polling gives only feedback of what the candidates and their propaganda machines have drummed out. This reporter's notebooks were crammed, in the final weeks of the campaign, with such random quotations as:

§ "He's come out against Social Security and all that stuff—and he'd get us into the war, too."

§ "He's not for the working person at all."

§ "To be honest with you, my problem is Rockefeller's home life."

§ "My problem is with Goldwater's rash statements."

§ "I'm still looking for another. Lodge would be my choice, so I guess I'd vote Rockefeller."

§ "I just don't like the way Goldwater talks—he's against everything."

§ "He can't buy an election out here with all his money—I'm for Goldwater."

Such notes are of use only when scientifically sampled and added up in digits. Which is what drove the pollsters to despair, for the digits would not stay stable. And the volatility that had plagued the pollsters all through spring now reached a peak. As the blows of the two rivals grew more punishing, as the Republicans in California realized that in their state no other names on the ballot gave them an escape hatch from choice, they would shift almost day by day in their loyalties.

In the last three-week period, intensive polling by the Louis Harris samplers found:

§ *In the last poll before Oregon*   GOLDWATER, 48; ROCKEFELLER, 39; undecided, 13

§ *Five days* AFTER *Oregon*   GOLDWATER, 36; ROCKEFELLER, 47; undecided, 17

§ *Ten days after Oregon*   GOLDWATER, 39; ROCKEFELLER, 48; undecided, 13

§ *Friday before voting*   GOLDWATER, 40; ROCKEFELLER, 49; undecided, 11

§ *Sunday (after the birth of the Rockefeller baby)*   GOLDWATER, 40; ROCKEFELLER, 42; undecided, 18

§ *Monday (after the Goldwater*   GOLDWATER, 44; ROCKEFELLER, 44;
    press barrage)   undecided, 12

They were hammering, now, in these last ten days.

Rockefeller had come into California from Oregon, flush and vibrant and buoyed by his Oregon victory. But now he was barnstorming the state from end to end, from Disneyland in the south to the High Sierras, shaking hands at receptions (organized by Spencer-Roberts) of as many as 7,000 or 8,000 people in a few hours, speaking from early-morning take-off to late-evening reception.

Politics gives a reporter the opportunity to see the change in a man's life at fixed two-year intervals, so that one observes him grow, mature, then age, as if in slow motion. This was the fourth campaign in which this reporter had followed Rockefeller, and he had changed from the ebullient amateur of 1958, almost boyishly thin and athletic, to a much heavier, graver man, his square jaw now fixed as if cut in stone, his body thicker, his hair changing from a light brown to a brown that was graying. He walked now more stiffly, stiff at the haunches, weary with many weights. He had lost twenty pounds this spring. He had changed from the citizen enthusiast of 1958 to the brusque administrator of a great state; and in the process he had learned all the chicaneries, the uncertainties, the depth of other men's passions and ambitions. Outwardly, on the platform, he displayed his usual jolliness of personality and delivered the standard, unyielding, uncompromising credo of the liberal Republican. His surge in the polls cheered his troops on; so did the unqualified support of the Lodge volunteers, led again by Grindle and Goldberg, who had moved into California after their Oregon defeat; so did a last-week statement by Eisenhower, almost—but not quite—excluding Goldwater from his list of acceptable Republicans. But to counterbalance this was the knowledge that the support Rockefeller had expected in California from other liberal Republicans—from Romney, from Scranton, from Nixon—had disappeared. If Rockefeller was indeed to win, as the polls now seemed to indicate, then the other Easterners apparently felt it unnecessary to add to his margin by their support. Their attitude could be read simply: Let Rockefeller dispose of Goldwater in California, then we will finish off Rockefeller at the convention. And, above all, there was the condition of Mrs. Rockefeller, expecting their first child momentarily. ("I have a show opening on both sides of the continent the same weekend," he said.) His advisers urged him *not* to fly back to New York every weekend to be with her—it drew attention to the marital situation they wished to have ignored. And

Rockefeller told them he would rather not be President if it meant leaving Happy to face her hour alone.

Moreover, now in the final desperate days, the Spencer-Roberts people found themselves entrapped by the simple problem of how to spend money to good effect. They had fired out a mailing to all of California's two million registered Republicans (estimated cost of the single mailing: $120,000), entitled simply "Who Do You Want in the Room with the H Bomb?" A collection of authentic Goldwater quotes, savagely presented, it had triggered a back reaction, as much for its cost and size as for its shrill tone. They had prepared an expensive half-hour television film called simply *The Extremists,* which was an incandescent visual display of the fate of a number of Californians persecuted by the witch-hunters of the California southland. But Rockefeller's principles insisted it be canceled, as "McCarthyism-in-reverse"; he distinguished between Goldwater as a man and the extremists of the Goldwater movement, and he felt the film was unfair. Moreover, the Spencer-Roberts people had established what are called, in California politics, "telephone mills"— banks of telephone operators calling up name after name on registration lists to locate or persuade votes for Rockefeller. In the final weeks of the campaign some 1,200 paid telephone workers were operating out of such mills across the state. But the effort to recruit them had had to be a commercial, not a volunteer, one—and Rockefeller's principles required open non-discriminatory hiring. In the largest telephone mill (600 callers) in the Los Angeles area, the operators hired were, substantially, Negro girls (who were overwhelmingly Democratic voters themselves). And, screened as carefully as they might be, the Rockefeller voices over the telephone were voices with a Negro accent, a Southern inflection—in a community and electorate mild or hostile to Negroes and civil rights. Further, with the Goldwater leaders hammering at the alliance of the labor-union "COPE conspiracy" with the Rockefeller fortune, the Rockefeller campaign managers grew gun-shy of any last-minute drenching of press, radio and television with a planned, paid, terminal barrage such as Price had orchestrated in Oregon.

And then, finally, on Saturday afternoon, May 30th, Nelson A. Rockefeller, Jr., was born—with the results apparent in the voter fluctuation in the Harris Poll next day. (Rockefeller had been lucky enough to be home for the arrival—he had telephoned Happy from the San Francisco airport at 4:30 on Friday, May 29th, climbed aboard his private plane immediately and, with the pilot pouring on the coal, had arrived in New York before she was taken to the hospital.)

Goldwater, too, had learned. He was no longer, as in New Hampshire, exposing himself in parlors or coffee gatherings where scores of reporters could snatch a conversational hyperbole from a question-and-answer period and make a headline of it. Rarely available now to the press, he appeared on the platform at great gatherings, concentrating in Southern California, barnstorming not on issues but on fundamentals like patriotism, decency, liberty and law and order. Freed from the restraint of the sophisticated East, where to speak of flag, God and mother is to paint oneself as cornball, Goldwater could talk from the heart. Here in Southern California, where the demonology of the primitives reaches its crispest black and white, the Goldwater volunteers were battling for the Lord against the twin evils of Eastern labor-union conspiracy and Eastern liberal money. If he could not prove in southern California that conservatives were a majority of Republicans, he could not prove it anywhere; and he sermonized over his last-week rallies like Cromwell praying over the Roundheads before taking them into battle.

Any of these meetings could be taken as a sample. But on May 30th, Memorial Day, as Goldwater barnstormed through southern California, he arrived shortly before noon at the high school of the small town of Riverside, where 1,200 adoring supporters waited for him. The meeting opened with prayer; then came the trooping of the colors; then the pledge of allegiance to the flag; and then Goldwater took up the liberal theory that Communism is growing softer. The full flavor of his message is only faintly recaptured in these rough reportorial notes of a typical passage. Since it was Memorial Day and he was discussing Communism, Goldwater's patriotism was throbbing. He remembered, as one of his first memories, watching his mother sew two new stars on the flag when his state and New Mexico had become part of the Union; he wanted that flag to fly proudly everywhere; men had died for that flag; and now "they" were telling us that the Communists were growing softer, not to worry.

> We are gathered [so my notes read] here in memory of those boys who gave their lives for us . . . and wherever they are in heaven they are trying to warn us, don't buy that phony sort of nonsense. . . .
>
> Some wobbly thinkers think that laws will stop you from hating, laws will make you generous. But when I read about street crimes, about hatred covered with blood, I ask what's happening to the land of the free, what's happened to the principles these men died for? Did they die in vain? Sometimes I think they did.
>
> . . . and the fault is not only in government, but in us, too. Ask yourself before going to bed tonight: Did I live

today with hate? Did I steal, cheat, hate, take shortcuts? If you answer yes, you haven't been a good American. Now that may sound corny—but I find nothing wrong with what my grandfather and grandmother could have told me. How often have you asked yourself—What would I pay to be free? Would you be willing to give up material things to remain free? Will you work hard on election day or are you going to stay home, watch TV, drink a few cans of beer and play with the kids? If you say yes, then you haven't been a good American.

. . . I deplore those far-out partisans of principles that are trying to tear the American people apart, trying to tear the home apart, trying to assume they can do such things for your children as care for their education and health better than you can. When you say to some bureaucrat in Washington—"You take care of the kids' education"; when they say to you, "Don't worry about Mom and Pop, don't lay aside any money, enjoy yourself, the Federal Government will take care of Mom and Pop"—then *this* is the ultimate destruction of the American family. When this happens, Communism will have won.

Thus Goldwater. And beneath him the crusaders of Gaston's legions were organizing for election day. In Los Angeles County alone, 9,500 volunteers ready to go from door to door on election day to make sure that every indicated Goldwater voter marched to the polls to vote for the moral society; 10,000 more in southern California; in northern California, 5,000. Just as lavishly financed as Rockefeller, the Goldwater strategists held off their press barrage to the last weekend, and then, on Sunday, in the Los Angeles *Times* alone, one of the most expensive advertising media in the nation, laid down four and a half full pages of display, exhortation and message in the weekend edition—with parallel efforts on TV and radio. (The Rockefellers bought only one page in the *Times*, with a restraint advised by their campaign managers because they were tagged as the millionaire's campaign.) And in the late afternoon of Monday, before the voting, churches opened (six in Los Angeles alone) for prayer meetings for the Goldwater volunteers, that the Lord might bless them in their effort of the morrow.

The most vivid analysis—and the most correct—this correspondent heard of the prospect of the primary came from the Los Angeles Republican County Chairman, Julius A. Leetham, a crew-cut young lawyer who had held himself scrupulously neutral in the campaign. I reproduce it at some length, as he talked out his thinking over lunch in a dimly lit, highly fashionable restaurant, because it is the sound of a sensitive politician thinking deeply about his community, and understanding it.

The old West [said "Jud" Leetham] was different. They banded together for defense, and for entertainment, and for economic reasons. The new West is different.

There's a transitory quality about this electorate. Our turnover of homes in this country is twenty percent a year. Most of the time it's a very odd person who's lived in the same house for very long; people who've lived in the same place for ten years are really rare. They move fast—and you get this shyness. Fundamentally, Californians aren't in contact with their neighbors. So the impact media have a flamboyant effect. All the psychological factors have conditioned Californians to a Rockefeller win. But what you don't see is the effect of these bumbling, disoriented, inept amateurs who are out stomping around from door to door asking for votes for Goldwater. They make people feel that someone is interested in *them*.

Nobody's got a bigger objection to the mass techniques of media than the guy who came out of the Army, who got brainwashed by indoctrination, who stood in line for chow. That's the guy who came here after the war. With his wife. So this Los Angeles *Times* editorial has created a hell of a furor. [The Los Angeles *Times,* the dominant paper in southern California and now, at last, one of the best papers in the country, had just come out with an editorial endorsing Rockefeller over Goldwater.] They resent the Los Angeles *Times*.

This common run-of-the-mill guy—he's beginning to resent being called an extremist. And all of these people running around doing all the wrong things, and if they say the right thing it's only by accident—they're making contact; they're reaching voters; they're touching them. Mathematically, they could compensate for Rockefeller's support in the media.

Out here, people are lost. In California they read about their neighbors only in the papers. They have no one to talk to. And the doorbell-ringer has an importance far beyond his normal pictorial quality.

This is the psychological factor. These guys who win elections by surprise—they've reached out and touched. What's the reason for the success of Disneyland and bowling alleys? Loneliness. The family unit, the neighborhood unit, is of decreasing importance, so they seek a commercial warmth. The political decisions of the next generation are going to be founded on just these psychological factors.

When you talk about Los Angeles, you're talking about the drifter. You can't find out what moves him in his economic status, or even in his hope of the future. You've got to find it in the approach by someone or something that suggests an interest in *him*—as a person.

What we're talking about is the Listener-at-the-Door. This Goldwater doorbell-ringer—she's probably very shy. And she waits a minute before she rings the bell, and she pushes it, and the other girl comes out. So the woman at the doorstep fidgets around and she says, "Where are you from?" The other gal says, "I'm from Bangor, Maine" or "Fall River, Massachusetts." So the first gal waits a second and she says, "Isn't that interesting—I'm from Finlay, Ohio."

What's the next question? The next question is "How long have you been here?" And normally the answer is something between six months and eighteen months.

Well—the whole interchange hasn't taken longer than fifteen seconds. But two lonely women have met and touched. It's an experience in the girls' lives. She can tell her husband about it when he comes home from work. Nobody else has been around to see her all day long except maybe the guy from the Fuller Brush Company or the Alta-Dena Dairy. Neither one of the two women knows which is the best man, Goldwater or Rockefeller. The question is—which of the two they'll remember when they go to vote. The Los Angeles *Times* editorial? Or the other girl who came to the door? The Rockefeller group is class. The Goldwater group is the most messed-up, cluttered-up organization I've ever seen. But I think Mr. Goldwater is going to win at the doorstep.

Mr. Leetham's instinct was unerring.

California Republicans voted on Tuesday, June 2nd, 1964. No electorate had had a clearer choice given it. How great was the tug of the past and all they had been taught in Sunday schools from Bangor, Maine, to Finlay, Ohio? How much could they be moved by the leaders whom Rockefeller had persuaded to explain the world as he saw it?

Voting starts at seven in the morning in California and closes at seven in the south, at eight in the north. No greater competitive effort had been made to collect and analyze a vote than was made in the State of California for its Republican primary. For 2,172,456 votes involved, an estimated 50,000 poll counters were deployed by the three television networks. Within twenty minutes the computers (this time the CBS-IBM computers) had done their work. By 7:20—while voters of the Bay Area around San Francisco were still trudging to the polls—the Columbia Broadcasting System, through Lou Harris, announced that Barry Goldwater would carry California Republicans by approximately 53 percent. (Final count: 51.6 percent or 1,120,403 to 1,052,053.) Nelson Rockefeller was to carry northern California (by a margin of 106,000 in the Bay Area counties around San Francisco), but the restless people of southern California, seeking older and more stable attitudes, were to reverse this decisively (by a margin of 207,000 votes in the two counties of Los Angeles and Orange alone).

And with that, the Republican voters had been consulted on their choice for the last time. Power, in America, comes from voters—their quiet, non-violent expression of taste or conviction. Once the voters have voted, however, power has been gathered in a package that can be manipulated. Goldwater had won only one contested primary—and that by a scanty margin in a state unique in feel and longings. It was power, though. Could it be stopped? Could other power systems in the Republican Party frustrate this narrow statement of the minority party of California?

The answers to all these questions are quite clear in retrospect. But they were not at all clear in early June of 1964, for over the previous eight weeks, while the duelists fought in the open-primary arena, other Republican leaders had been feverishly active in the brokerage of hidden power that might control the nomination.

CHAPTER FIVE

THE DANCE OF THE ELEPHANTS

Q. "Where do elephants go when they die?"
A. "To Cleveland."
(Overheard after the Governors' Conference at Cleveland,
June 10th, 1964)

POWER in America rises in its most primitive form from the citizens voting their will at secret polling places.

Nothing can frustrate the voter in America if his statement is clear enough. But Republican voters had spoken three times in three contested primaries in the spring of 1964, only with utmost obscurity—once for Lodge, once for Rockefeller, once, finally, in California, for Barry Goldwater. And when the voters' statement is unclear, it becomes the time of the power broker.

Thus, then, in the aftermath of Goldwater's California victory it was the time of the power broker—or, rather, one suddenly became aware that the power brokers had been quietly operating for months behind the public scenery of American politics, hidden from national attention. The nation had marked the week beginning July 13th as the climax of the Republican struggle, to be staged openly at the national convention in San Francisco, where one or another of the aspirants must win the 655 delegate votes needed for the nomination. And, had Goldwater lost to Rockefeller in California, the calendar of Republican politics in 1964 would indeed have read like the Democratic calendar of 1960—an open crescendo rising this time from the New Hampshire primary to the Oregon primary to the California primary to climax in the Republican Convention at San Francisco.

Now, suddenly, with a great burst of illumination, one realized that this public calendar had become an anachronism. Another calendar, precisely marked and meticulously kept, had to be seen, in retrospect, as the governing calendar of 1964.

It is well to examine the background of this calendar in some detail, for it prepared the *coup d'état* which the nation would witness a few weeks later in San Francisco.

This was Clifton White's calendar—and his geography of American politics, his deployment of pawns, knights and delegates.

To understand the White operation, it is important to grasp that the 1,300 to 3,000 delegates who appear at the two great national conventions—the most romantic forums of American decision—are chosen not by national laws but by state laws. There are fifty such independent systems of law on delegate choice, and as they change from year to year, only a political technician with staff can keep in mind all their intricacies, dates, sharp and pointed legal distinctions, and dominant personalities. Only fifteen American states invite their people to vote openly in primaries for some or all of the state's national delegates. (For the Republican Convention in 1964, some 540 delegates were so chosen.) Some states leave the choosing of delegates to state conventions; others to local and district conventions; others to party committees; others to a procedure combining all of these elements. In almost all states, however, the key unit for selection of delegates is the Congressional district; in some states one can lose a statewide primary and still pick up a majority of delegates; and organization at the grass roots, at precinct or county caucus, is the fundamental precondition for conquest of the Congressional-district delegates.[1] Hundreds of delegates are available for the harvest in such caucus-and-convention states in every Presidential year; and the major candidates of both parties always open hospitality suites at such state conventions, to lure, persuade or convince the grass-roots delegates that they should send on to the great national convention national delegates who will vote "right."

[1] The bookkeeping of delegates in a Presidential contest is so confused that any simplification is a distortion. By and large, though, the largest states—California, New York, Pennsylvania, Illinois, Ohio, Florida, Massachusetts, New Jersey—choose all or most of their delegates, whether statewide or by Congressional district, in open primary, governed and controlled by the normal laws of public voting. Seven smaller states also have open, legally controlled, public primaries. But most of the smaller states permit the selection of party delegates by party machinery in a strange system of precinct caucuses that rise to district caucuses or conventions, that come to a surface summit in state conventions. The laws governing these tiny precinct gatherings of citizens are very fuzzy; and these grass-roots meetings, so lavishly praised in textbooks on democratic practice, are precisely the kind that can be dominated by the energetic, the motivated, the dedicated. It requires deep commitment for an ordinary person to seek out the apartment house, the parlor, the school hall where a local precinct may gather its Republicans or Democrats on some obscure and sometimes unpublicized evening to choose delegates to a higher district caucus or convention; and even deeper commitment for him to devote attention to the behavior of those delegates at the district caucus, or follow their choices up to the state conventions.

But the harvest of such delegates is most successfully reaped by those who sow early.

And long, long before Nelson Rockefeller or Grindle & Goldberg had begun to tabulate delegate count, before William Warren Scranton was even elected as Governor of Pennsylvania, while John F. Kennedy still governed and Lyndon Johnson was reflecting on retirement, F. Clifton White and his volunteers had, as far back as 1961, marked out each state with its dates, its procedures and its delegate count. Their planning was more careful, more detailed, and earlier begun than that of any previous Presidential campaign; it was a masterpiece of politics.

The chart that hung in Clifton White's shabby office at 1025 Connecticut Avenue in Washington in January of 1964 was, in form, a calendar of the current year with specific dates marked in red. But it did not start with March 10th, the date of the open New Hampshire primary; it started in early February in North Carolina. There, district conventions would be choosing delegates before the state convention met on February 28th and 29th; and days before the state convention met to ratify district choices, these districts would already have chosen 22 national delegates. The calendar noted that Georgia's Fifth Congressional District (which includes the city of Atlanta, hitherto a bastion of liberal Republicans) would meet on February 23rd, the day after Washington's Birthday, to choose 2 fresh delegates (both, as it turned out, for Goldwater)—and on February 29th Oklahoma would choose its delegates at a state convention (all 22 for Goldwater)—and so on through the spring season while the nation would be watching the more dramatic jousting at the open tourneys in New Hampshire, Oregon and California.

Presumably, similar calendars hung at headquarters of other aspirants. But White's calendar had roots—roots reaching as far back as the first meeting of his adventurers in October, 1961, at the Avenue Motel in Chicago. For one cannot grasp what White and his volunteers had been doing simply by reading a calendar which says: 25 of 26 delegates from North Carolina on February 29th, and all 22 delegates from Oklahoma the same day; 16 for Goldwater in South Carolina on March 21st; or a committed total to Goldwater of 147 votes by mid-April with another 300, already known by name, available for harvest. The sowing of the harvest had been going on for two years.

One must translate arithmetic into people and passions, and thus—to choose just one state—by April, 1964, when the Republicans of Washington State gathered in precinct caucuses, the dominant Republican figure in that state was, already, an unrecognized political newcomer named Luke Williams. It is impossible, on meeting Luke Williams, with his brown, thinning hair, his round face, his earnest eyes, his two-tone plastic eyeglasses, to think of him as a dominant figure. Seeing him in a coffee shop or a living room, one would think of him as a high-school principal, a country doctor or an affable mortician. Luke Williams, now an energetic forty-one, was as unlikely a figure in the demonology that the East painted of Goldwater figures as, say, Gabriel Hauge, the thoughtful chief of the Manufacturers Hanover Trust, is unlikely in the demonology that the Goldwater primitives describe as Eastern bankers.

Luke Williams had enlisted in the Navy at the age of eighteen, and had learned to be a sign painter in the service. When he came out of the Navy at twenty-two, sign painting was his trade. Like a good citizen, he took out a card in the painters' union (which he still holds), and in 1946 he and his brother began painting signs in Spokane, Washington. But, as the American success story requires, they were strivers; at night they tinkered on an invention for illuminated signs that would flash both time and temperature in huge lights readable miles away. They sold their first time-temperature sign in Seattle in 1952; six more the next year outside Washington; twelve more the next year (and by now their time-temperature signs flash over every city in the country). Hard work, exuberant effort and a good product led Luke Williams on the road to success. Flying, traveling, opening branch offices around the country, he made the business grow national. And by 1959, when he could finally come back to Spokane to be a family man again (he has a daughter eight and a son fifteen), Luke Williams discovered the vacuum that comes in the life of every man who arrives at success early. In 1960, for the first time, he interested himself in politics, for he was bothered by a government which seemed to cramp him, control the labor he used, supervise wages, hours and depreciation so that the larger he grew, the more government seemed his enemy. He participated in several statewide campaigns—and lost. Nor did his own fellow Republicans seem to be aware of the danger of government; power in Washington State lies in the city of Seattle; and Seattle, so dependent on the Federal Government for its livelihood (the Boeing Airplane Plant is the largest but not the only government client in Seattle—the Bremerton Navy Yard, the Army's Fort Lawton, the Air Force's McChord Field all join in underbracing its economy), was dominated by Republicans who were doing business with the Federal Government. "Doing business with the enemy," Luke Williams called it.

Williams' philosophy is the philosophy of an honest man who has made it on his own, by wit, ingenuity and hard work; and the government that hampers him as his efforts flourish seems a hostile one. (In Luke's philosophy, as he explained to me, "police protection, fire protection, and sanitation and education at a local level—these are the basic things government should do. But we've gone dangerously beyond these basic things, and the government needs pruning back—that's my concept of it.")

Luke thinks that if "we really had a President who would expound this philosophy of government, instead of more and more regulation," the Republic would be set in safe hands. Goldwater has been one of his heroes for years. So when Clifton White came into Seattle in June of 1963 to talk about drafting Goldwater for President, Luke Williams was ready to listen—and then to work.

Efficient and affluent, with a Seattle apartment of his own (to which he could fly in his own plane from Spokane, 225 miles away), knowledgeable in directing a staff, Luke—an outstandingly effective businessman—could grasp what Clif was talking about. Seattle, which dominates King County, has 1,800 precincts; but the Republican Party had precinct leaders in only 300 or 400 of these precincts. If you went down into the streets; if you gave the conservatives you found there a vent for their emotions; if you registered them on the rolls as you found them, the vacant precincts would have Republican precinct chairmen. The Republican organization in Seattle, headed by Mort Frayn, who was deputy national chairman of Rockefeller's national campaign, was glad to have precinct captains in blank areas; but "once we got a precinct captain named," said Luke, "we knew where he'd be when it came to voting." Only 2,500 precincts in the State of Washington (out of 5,500) had Republican precinct organization when Luke began his work—of which perhaps 50 percent were led by Republican moderates and 30 percent by natural conservatives. But when all 5,500 precincts in Washington were filled with Republican activists by the end of 1963, 65 percent were people controlled by Williams—and by Clif White. The people whom Williams organized were people who felt, as he did, that the Republic was in danger—people willing to go up and down the street ringing doorbells to find their kin-in-spirit, and register and motivate kinsmen when they found them.

Precinct caucuses are held in Washington every two years—in homes, in apartments, in libraries, some attended by a few hundred people, some by only three or four. A good guess is that only some 60,000 Washington Republicans turn out for their precinct caucuses. Of these, 15,000 get elected to county conventions; of these, 877 go on to the state convention, which chooses national delegates. It does not take too much money to dominate the system, and the Goldwater Republicans in Washington did the entire job on $35,000—"not a nickel from out of state," says Luke Williams. One old doctor died and left 5 percent of his estate to the cause; one little old lady endorsed her Social Security check over to the cause (she felt Goldwater was the only man who would protect her from the Negroes). In February the precinct groups caucused, and the Goldwater people were all there; the delegates they elected met in county committees in March and April; by June 11th to 13th, when the counties had sent their delegates to the state convention to choose 24 national delegates, the Goldwater people, under Luke Williams, controlled the state convention lock stock and barrel. At least 70 percent of the delegates were his; not even Edith Williams, a granddaughter of Theodore Roosevelt, could be elected a delegate to the national convention; Mort Frayn, former Republican State Chairman, was turned down by the convention as delegate from his congressional district; Washington belonged to Goldwater and would have, says Luke Williams, whether he had won or lost the California primary the week before, simply because the people who had been at work so long felt, as Luke did, that "Barry Goldwater will help this country."

All over the country, in the spring months, such precinct, county and state conventions gathered without national notice. It would be superfluous to catalogue them all, yet it is useful to break the over-all Goldwater enterprise down into its major regional elements.

In the Clifton White plan there were the Westerners, the Southerners, and the old Taft conservatives of the Midwest.

One can take the Luke Williams operation in the Northwest as a good example of Goldwater leadership in the Mountain and Pacific states—homespun people dedicated to saving an America that seemed to be repudiating its own past.

Then there was the South—and the South presented an entirely different problem.

So many kinds of revolution are simmering in the South in our time that it is impossible to paste on this vast and fermenting geographical area any of the labels so easily used by Northerners. Generally, the simplest way of saying it is that the South—white and black—is in full revolt against its past. In some states, captained by such men as Terry Sanford of North Carolina or Carl Sanders of Georgia, the revolt against the past leads to perspectives of the happiest and most civilized kinds. In other states, like Mississippi and Alabama, the revolt expresses itself through civil-rights and Negro leaders who have converted large areas into a terrain of guerilla war. In yet other states the revolt is headed by citizens' councils and state's-righters of the most desperate nature. And in all the states the revolt is motored by the growth of industries and the swelling of cities which hold a new middle class—white and black—who find the red-neck, demoniac leaders of the old Democratic Party repulsive to both intelligence and taste.

The Republican revolt in all these states is an expression of this general condition—and is inspired by a deep and abiding contempt on the part of the educated white middle class of the new cities for the one-party courthouse politicians who have chained the White South to the past by promising, in recompense, only to shackle the Negroes in an endless segregated servitude. This Southern revolt has been going on now for years—some date it from 1928, when the Protestant South so heavily voted against Catholic Alfred Emanuel Smith. Others date it more accurately from the election of 1952, when Eisenhower broke, and broke permanently, the grip of the Democratic Party on Southern electoral votes. But the creation of a respectable, broad-based party has been a major objective of the Republican Party for the past decade, ever since Meade Alcorn, the best National Chairman the Party has enjoyed in modern times, mounted Operation Dixie in 1957. And the capture and distortion of this Republican rising in the South by the Goldwater forces is one of the dominant political facts in the history of 1964.

The young men who directed this seizure across the South are almost all of a kind—men between thirty and forty years old, city people, well-bred, moderate segregationists, efficient, more at ease at suburban cocktail parties than when whiskey-belting in courthouse chambers. One could start with almost any of them—Peter O'Donnell of Texas, Wirt Yerger of Mississippi, Drake Edens of South Carolina, or John Grenier of Alabama.

It is best to start with Grenier, for his career in politics in the South has been the shortest, swiftest and most striking. In the summer of 1960, Grenier, twenty-nine years old, a lithe, handsome, youthful lawyer of Birmingham, became chairman of one of Clifton White's Young Republican clubs: paid membership in Birmingham at that time—twelve.

Five days later he discovered he was supposed to organize a rally for Nixon's campaign visit to Birmingham—and discovered simultaneously that, for all intents and purposes, no Republican Party at all existed in the State of Alabama. Yet twelve people, effectively organized, can achieve wonders if they feel deeply and act as they feel—even if they must start, as these twelve did, with a pouchful of dimes, calling up unrecognized names out of the telephone book from drugstore telephone booths. The Birmingham rally was more than a success; it was a demonstration of what ripening emotions were waiting a call. "You start with twelve people in phone booths," reflected Grenier during the campaign, "and you locate two or three people in each precinct, and you go on from there. I went to forty-four precinct meetings in forty days that year, and we called some of our meetings in drugstores. For three years I traveled the state five nights a week, talking to two or three people in each county to get one leader and then train him.''

In 1962 Grenier organized the statewide campaign of James Martin, who ran within 7,000 votes of Senator Lister Hill. By early 1964 Alabama was, from a Republican point of view, perhaps the best-organized state of the Union. Thirty thousand of Grenier's volunteers, organized in 64 of the state's 67 counties, had been prepared, recruited, trained and motivated to make that state—as it almost certainly will be in the near future—the first Republican-governed state of the South. But by then Grenier was already the Southern regional captain of White's national network (he had joined the inner group at the secret meeting at the Essex Inn Motel in December of 1962) and was spreading his gospel and his organizational procedures all through neighboring states of the Confederacy, from Virginia to Texas.

What Grenier found everywhere in the South was leadership like his own—local leadership of young businessmen and their wives, men who had outgrown the primitive system of courthouse politics which, in return for segregation, corrupted and cramped all local progress. What he could offer in Goldwater was a flag, a national symbol, or, as Grenier put it: "Goldwater was horsepower. We needed voter sentiment in depth. We had the local leadership without the horsepower, but we couldn't get the voters to switch parties until we provided them with the catalyst."

By late May of 1964, when the Southern Republican state conventions were well on their way, it was superfluous for any observer to wait for the final and climactic Republican state convention in Texas. Already, for all those who could read politics, there were 300 Southern delegates as sure for Goldwater as sure can be in politics—out of the 655 necessary to nominate.

There remained the third region of operation: the Midwest—and the Midwest was critical. New England had had to be written off. The Northeast was spotty—New Jersey bright, but Maryland hopeless; New York was Rockefeller's; Pennsylvania, Scranton's. In the Midwest, however, another kind of Republican Party reigned—solid men, neither homespun nor racist, Hoover conservatives of the classic mold, traditionally hostile to the Establishment Republicans of the Atlantic Seaboard, their hearts still bound to the memory of Robert A. Taft. Here matters were more delicate—requiring not only an effort at the grass roots but organizational negotiation. Ohio, Wisconsin, Iowa, Michigan (if Romney were indeed out of it), Indiana—all these were states led by Republicans of the old order, whose emotions might warm to Goldwater as much as any homespun fundamentalist's, but who knew that in politics the primary objective is victory. In the broad Midwest, White's chief deputy was Wayne Hood, a former Wisconsin Republican State Chairman who had quit politics in 1953 after the liberal Eisenhower victory, but now, in 1963, had returned to enlist for the genuine conservative, Goldwater. All sorts of tentative alliances might be negotiated in the Midwest, but could be made firm only if Goldwater could show in California that his appeal ran deep enough to carry an open voting contest as well as a formal convention. The West and the South belonged to Goldwater whether or not he won in California; the Midwest would fall firmly in line only if Goldwater succeeded in the Golden State. Which, on June 2nd, 1964, he did.

Thus the pattern as the dramatic first week in June came to a close and ushered in the even more dramatic second week. California had been won by Goldwater on Tuesday by the scant margin of 68,350 votes, or 51.6 percent. Headlines of *The New York Times* that Sunday reflected the tattoo of his progress: ALABAMA NAMES TWENTY. FOURTEEN FROM COLORADO. TWENTY-TWO OF TWENTY-FOUR IN WASHINGTON. FOUR OF EIGHT IN HAWAII. FOUR VIRGINIA DELEGATES. By the weekend of June 7th, 532 delegates were publicly or legally committed to the Arizona Senator; and the Texas convention was to select and instruct 56 more Goldwater delegates nine days later. In White's rolls the nomination was all but locked up, his alliances secretly indicating 200 more ready to break in the next few days.

Could anything stop Goldwater?

What happened next was farce, cast from a script which required all actors to style themselves "men of principle."

No word is more hallowed in the Republican Party than "principles." It is difficult to get any senior Republican leader to define the "principles" that guide his Party, beyond the often repeated defensive statement, "I believe in the preservation of the two-party system." ("Principles" is as vague yet as important a word among Republicans as the word "humanitarian" among Democrats.)

Somehow, however, all those with a foretaste of what a Goldwater nomination would do to the Republican Party in November must have known, even before the California primary, the principles for which Goldwater stood. They knew these principles could not possibly mesh with the principles of what Nelson Rockefeller called the "mainstream" of the Party. But none—except for the Lodge leaders—had seen fit to give Rockefeller, in the California contest, the slightest support for the "principles" which now, suddenly, they found endangered by the imminent Goldwater coup. Like the Kerensky government, they were unaware of revolution until the Red guards were already ringing the Winter Palace.

The principles that moved the great Republican chieftains to their behavior in the first week of June are difficult enough to define, but the personalities involved seemed determined to confuse them yet further. In the surprising victory of Goldwater in California, the only liberal Republican candidate to challenge him in the field had been bested. If Rockefeller's principles were to prevail, the last chance for them would be at Cleveland, where on June 7th a Governors' Conference was to convene. The Republican governors would gather there, along with the Democrats, and might find an opportunity to coalesce around a fresh candidate of principle for a last-ditch stand.

Yet if any one of them thought or seriously prepared to seize the opportunity which the Cleveland conference was about to offer, it escaped this correspondent. Instead, the week before and the week after the California primary of June 2nd exposed all of them in the worst possible light.

Rockefeller is, indeed, an exception. He had conducted himself with courage and honor through the long year despite the fact that occasionally he reminded one of Bishop Odo of Bayeux, who would lustily shatter an enemy's skull on the battlefield, convinced that it was God who helped him wield the mace. Rockefeller had fought to exhaustion. He knew now he could not be President. Riding home to Tarrytown from New York a few days after his California defeat, he had leaned wearily back against the headrest of his chauffeured car and rambled briefly about the worsening situation in Vietnam, then called himself back to the reality of his personal defeat and added, "Sometimes with a defeat there's a gain—there are things you don't have to worry about any more, like do you have to take the country into war over Vietnam." And so, finally, with sardonic wit, he let himself enjoy the spectacle of all the latecomers who now, suddenly and belatedly, discovered that Goldwater was about to lead their party into new avenues of history.

The Lodge volunteers had also behaved with honor. From a gay and lighthearted group of pranksters who had merrily disturbed the politics of the spring in New Hampshire, they had slowly, in the spring months, come to realize that they had danced into the most serious business of the world—the Presidency of the United States. They, too, now supported Rockefeller and his principles.[2] But their support for Rocke-

---

[2] One must say farewell to the Lodge volunteers in a footnote, whose brevity they will understand. Their campaign in Oregon died chiefly because of Rockefeller's exertion; but also because as the war in Vietnam worsened, the silhouette of their candidate in Saigon fuzzed in the general mess; Lodge was obviously no triumphant hero to be brought back like Eisenhower from SHAPE in Paris as from a

feller's principles, after the Oregon primary, was like a crushed Christmas ornament on the day after New Year's.

All other men of principle among the Republicans had carefully withdrawn themselves from the California fight before June 2nd. "It's understandable," said Rockefeller mischievously four days before the primary in San Francisco. "All of them are available for the nomination and hoping that lightning will strike. What other position could they take?" Sure that Rockefeller's last-minute burst in California, after the Oregon triumph, would sweep the Golden State, none had felt it necessary to stump for, or even announce his support of, the principles they now found were in danger. Indeed, for all of them, no more tantalizing or appetizing political prospect could have been imagined than the probable outcome of the California struggle: Rockefeller would finish off Goldwater at the polls in California; thus all Goldwater's accumulated delegate strength would come to nought. Yet Goldwater, with 500-plus delegates pledged to him, would still retain a veto power over the nominee at the Convention. So after Rockefeller had finished Goldwater at the polls, Goldwater would finish Rockefeller at the Convention. And then other men would gather in the rooms of San Francisco or New York to decide which unity candidate would carry the fight against Lyndon Johnson for a united party.

Thus the unwritten script, adhered to, whether consciously or unconsciously, by all other senior Republicans in the party.

Dwight D. Eisenhower, for example, a week before the California primary, had been persuaded to issue a statement to the New York *Herald Tribune* defining the ideal candidate in terms of such rare (for him) precision as to be almost a perfect description of Nelson Rockefeller. This had been, momentarily, a punishing blow to the Goldwater campaigners in California—until other men persuaded Dwight D. Eisen-

---

pinnacle of achievement. When the Lodge campaign ended in Oregon, Grindle & Goldberg flew to California the next morning to mobilize all their strength behind Rockefeller. Nothing ever transpired in the firm of Grindle, Goldberg, Saltonstall & Williams that did not bubble with hilarity. The first week in California they found delivered to them a package of $10,000 in hot money—four packets of cash in $10 bills which they had to return to the donors, if they could find them (in Denver, finally), lest their patrician New England chief blow a fuse. The final evening of their effort was spent at Los Angeles International Airport, where an overpowering crowd of Goldwater supporters had gathered to boo, jeer and heckle Rockefeller's last appearance. Grindle & Goldberg stood on the edge of the crowd, earnestly debating whether they should plow into the hecklers and convert the demonstration into a full-blooded riot so as to show on the attendant television cameras, and thus to California and the nation, how "extremists" behaved when anger was engaged. They made the wise American decision not to convert the demonstration into a riot, although tactically they might have gained enough votes for victory by provoking bloodshed.

hower to redefine the meaning of his statement all over again on June 1st and declare, in fact, that he was by no means urging anyone to vote for Rockefeller or against Goldwater—he was simply speaking on civics as a citizen.

Richard M. Nixon had also followed the script, but his principle was the principle of neutrality; no other script was open to him. For Nixon, who in the years of his winter has mellowed far more than his public statements indicate, the nomination could never be won by challenging for it. It could be his only by deadlock—if other Republican leaders, remembering his rugged services to the party in the past and the closeness of his 1960 race, would turn to him as harmony candidate. But there had been much breakage in his solitary campaign of 1960; he had further won the embittered enmity of the hard-rock Republican right for his California campaign in 1962, when he had, with courage, flatly repudiated the John Birch Society. But his name was still a political force; a weekend campaign for him in Nebraska had easily drawn 31.5 percent of the voters, as write-ins, in a primary where Goldwater, on the ballot, could get only 49.5 percent. In the public-opinion polls Nixon's name constantly turned up ahead of Goldwater's or Rockefeller's as the choice of the rank-and-file Republicans. Yet with no staff, no money, and tactically trapped by lack of a base, he must wait.

The waiting game is always intricate. In California, Nixon's name still had clout; thrown into the scales of the primary fight there, it might have swung the contest either way; if he had actively campaigned in California, he could certainly have carried the state for the candidate of his choice. Yet if he campaigned for Rockefeller, he would have the everlasting enmity of the Goldwater delegates at the Convention; and if he campaigned for Goldwater, he would be denying all the principles that he had championed as Vice-President under Eisenhower. The solution was obvious—too obvious for word or phrase to conceal. Nixon would be friends with everybody; he would telephone Goldwater in California to discuss Convention arrangements, assuring Goldwater of his friendship and neutrality (but he would whisk his mother to a visit in New York to keep her from appearing on a platform at a Goldwater rally). He would also assure Nelson Rockefeller of his friendship through intermediaries, yet refuse to put this friendship into public print, where it might help in the California struggle. There was no other solution but this for Nixon; yet the result, too, was predictably obvious. Neither Goldwater nor Rockefeller trusted him in the slightest; both recognized that Nixon's only hope of nomination lay in their mutual destruction.

Only once in the campaign did Nixon approach the problem of his nomination positively. On Memorial Day, Saturday, May 30th, 1964 he assembled his old followers and lancebearers (Fred Seaton, Robert Finch, L. Richard Guylay, half a dozen others) in a suite at the Waldorf-Astoria Hotel in New York to consider the possibility of his nomination. The underlying assumption, of course, was that Rockefeller would win in California. What should Nixon do thereafter? Should he establish immediately a campaign organization to seize the nomination for himself? Should he open a Washington office, let Nixon clubs be organized around the country? Or should he be quiet and let the Republican Party boil over in fratricidal strife at the Convention, then wait until friends offered him to the Convention as a peacemaker? His panel discussed the matter for several hours, and Nixon listened silently while they examined the merits of both courses and disagreed. It is certain that some decision from Nixon would have been forthcoming on the Wednesday after the California primary, had the assumption of a Rockefeller victory been correct. But it was not. Goldwater won in California. By then, news of the Waldorf gathering was known everywhere. It had leaked, actually, the day before they gathered—and now Nixon was suspect to both wings of his party.

Next after Nixon, on the list of the perplexed was George Romney of Michigan. Mr. Romney is as much a newcomer to politics as John Grenier of Alabama. A vigorous, masculine man of serious religious conviction, a successful automobile manufacturer, an evangelist whose civic attitude is closer to Dwight D. Eisenhower's than anyone else's in his Party, he can endow his ideas with a stirring fervor. A man who insists on his principles as much as Rockefeller or Goldwater, Romney perplexes less-principled observers, for no one is ever exactly sure what his principles will tell him to do.

He had for months tried, in all sincerity, to find out from Goldwater what the Arizonan really stood for. Goldwater had avoided any direct meeting with a finesse and persistence that slowly enraged Romney. Yet when Goldwater and Rockefeller clashed in California, Romney had not only maintained a rigid neutrality but fired off to Senator William Knowland, Goldwater's California campaign chairman, a public telegram which declared that he was not "supporting nor opposing any candidate" in the California primary. Goldwater read the telegram aloud to refute the claim that Romney stood with Rockefeller "in the mainstream." Romney approached the dance of the elephants at Cleveland as the governor of a great state where, gradually, he was making practical a new and enlightened constitution (that he had largely designed)—but with no open record of visceral commitment to either side in the civil war in the Republican Party.

Thus—Eisenhower, Nixon, Romney. And beyond them was William Warren Scranton. Scranton possesses some of the finest qualities any man in public life can offer: first, a sense of duty; beyond that, intelligence, administrative competence, a knowledge of the great foreign world beyond the borders, a love of country and a deep sense of roots. But what Scranton lacks is appetite—appetite for power, appetite for command. He fights best and fights well in hard-knuckled fashion only when duty summons him; but he would rather not brawl.

Yet he had become, despite himself, the center of the Republican struggle. Acceptable to Goldwater (who at this time yearned to have him as a Vice-Presidential running mate); respected by all who like a clean, sharp administrative record; acceptable also to Rockefeller, who, though he thought Scranton personally soft, respected him as a governor, Scranton had been the favorite underground candidate of the majority of the Eastern Establishment for over a year. Scranton, however, had played it cool. Like Nixon, he had preferred to stay out of the fight in California rather than give offense to Goldwater (whom he liked personally but with whom he disagreed violently on politics) or to Rockefeller (whom he did not know well, but with whom he agreed fundamentally on politics). He had flatly repudiated an effort of Pennsylvania Senator Hugh Scott to make an alliance with the Rockefeller forces before the California primary. Scranton was, indeed, about to play host to a quiet gathering of Eastern politicians, scheduled for the Thursday after the California primary, to review the expected Rockefeller victory. Unlike Nixon, Scranton had wisely settled on holding his meeting *after* that primary.

The scheduled meeting at Harrisburg never took place. Goldwater won on Tuesday. And Scranton was left alone with his wife, his conscience—and Dwight D. Eisenhower.

Whether Scranton initiated a meeting with Dwight Eisenhower at this point or Eisenhower initiated a meeting with Scranton is totally unclear, even at this date. And will never be clearer. Confusion had settled on the entire Party after the California primary. Scranton had appeared on television after the primary and, when asked whether he would accept the Vice-Presidential nomination under Goldwater, had said yes. It was a sincere remark, understandable only to those who knew the modesty of the Pennsylvania Governor, who felt it arrogant for anyone to refuse the Vice-Presidency of the United States. Yet, as seen on the screen of television, it had a craven quality—a bowing before the inevitable. And then, Saturday, two days later, fresh news claimed political attention: Scranton had motored down to Gettysburg, Pennsylvania, had talked with Dwight Eisenhower for an hour and a quarter—and had emerged, apparently, as Eisenhower's candidate for the Republican nomination for

President of the United States. Eisenhower, for Republicans, is like the Holy Ark that the Israelites carried into battle against the Philistines. Somewhere deep in the mystery of Eisenhower lies that which most Republicans think their Party is all about. By Saturday nightfall after the Gettysburg morning meeting the nation knew that Scranton was about to run—and with Eisenhower's support.

The best report of the meeting that this correspondent has been able to get is that Eisenhower telephoned Scranton on Friday evening and asked if he, the old lion, might come to Harrisburg and see the Governor. Scranton replied with nice delicacy that, no, he himself would drive to Gettysburg to see the former President. It was a rainy, drizzling day. Eisenhower drove to his office in town, and Scranton disappeared inside its glass door for an hour and a quarter of conversation while Mary Scranton went to chat with Barbara Eisenhower, the General's daughter-in-law. The two men spent a good deal of their time talking about Goldwater; Eisenhower was appalled at the prospect of Goldwater's nomination, and they discussed the effect it would have both locally and nationally. When they were through, it was Scranton's impression that Eisenhower had urged him, once more, to make an open fight of it; he had not received an open endorsement, but he felt he had the assurance that if he, Scranton, opened the battle with Goldwater at Cleveland, he, Scranton, would then receive the support of the General as an open ally.

By nightfall, all across the country, the word was out that Eisenhower was about to move, finally. And that Scranton was his candidate. Scranton was to appear on a television program called *Face the Nation* the next day, Sunday, from the Governors' Conference at Cleveland, and Eisenhower had told him, "I'll be watching." Late as it was, there appeared to be still time for riposte to the challenge from the West.

The yearly Conference of the Governors of the States of the Union is an annual political carnival underlain with serious purpose. It is important that the governors meet annually—not so much for what they do publicly as for the discussions that go on between them, and between their staffs, when all the details of state administration (of mental health, of aid to schools, of highway financing, of safety patrols, of air pollution, of sales tax, of death penalty) become the informed chatter of governmental housekeeping. This vital swapping of information, ideas and gossip helps move American government forward. Yet it is always overlain by the more dramatic surface news that the governors make as they volley back and forth on great national issues at press conferences, when they grow heady with the attention of national press and television. In a Presidential year, particularly, the Governors' Conference is overburdened with dramatic political clash and maneuvering—and never more so than the Governors' Conference of 1964 at Cleveland.

One must see this conference in the context of the Republican dilemma of 1964.

When Republicans propose to govern the American union of states, their philosophy insists that the maximum effort of government be made at the level closest to the people—at the statehouses. But their philosophy insists also that the Federal Government confine itself to foreign affairs, national defense and the condition of the dollar. Thus, by definition, two arms of Republican purpose are set up, apparently at war with each other. The Congressional arm of the Republican Party, reflecting its "no-no" wing, sits permanently in Washington as a watchman's society, scrutinizing, denouncing and restraining every effort of the Federal Government to act on tomorrow's problems; while the governors' arm of the Republican Party meets but rarely, and when it meets it reflects the thinking of men who must act, govern and grapple with those problems. There are many cleavages in the Republican Party—between East and West, between fundamentalist and interpretationist, between Calvinist and Baptist, between small family businessmen and large administrative business. But of all the cleavages, the most important is that between the Republican governors, who know that government is necessary, and Congressional Republicans, who think that strong government is bad.

If there is any connection between philosophy and strategy, between principle and tactic, it should lead the Republicans to act from strength, to present their philosophy to the nation from the broadest base and broadest coalition of local government and local triumphs it can find. Goldwater was the negative, or Congressional, face of the Republican Party. But in 1964, on the governors' level, the Republicans were enjoying a vintage year. Smylie of Idaho and Hatfield of Oregon, Chafee of Rhode Island and Love of Colorado, Scranton of Pennsylvania, Romney of Michigan, Rockefeller of New York were among the finest governors of their generation. Vigorous, virile, energetic men, they could offer the nation a team of glowing faces. Had they been able to combine, they would have had—and if ever in the future such Republican governors are able to make of themselves a combination, they will have— the key to the only viable victorious Republican strategy on any principle apart from racism.

Nor were they in a minority at Cleveland.

Of the fifty governors in the Union who gathered in Cleveland in 1964, only sixteen were Republicans. But of these sixteen, only three (Babcock of Montana, Fannin of Arizona, and Bellmon of Oklahoma) were Goldwater men, while the other thirteen were men who knew that in their states a Goldwater nomination would mean catastrophe for the Party. All of them, suddenly, as they translated the Goldwater image into the idiom and problems of their own states, could see him fresh and real as a choice Democratic target: here was a man who would sell off TVA, cripple the Rural Electrification Administration, cut down on matching Federal grants for welfare, put Social Security on a voluntary basis, abolish farm subsidies, strip labor of its power. However perilously narcotic such Federal actions may seem to men of stern fiber, the practical governors could foresee the instant pangs and anguish of great masses of voters suddenly cut off cold from things they needed or wanted—voters who would vote against any man who would cut them off. Moreover—the small handful of Republican governors knew they lived under a special, untimely menace: the spring 1964 decision of the Supreme Court requiring each state to redistrict its own legislatures to correct for rural imbalance. The state legislatures elected in the fall of 1964 would redraw the political map of their states; and if ever the Republican Party needed a candidate of broadest appeal to help them, it needed such a candidate now.

Thus they arrived, the Republican governors, over the weekend of June 6th and 7th, each bearing his own worries and alarms, for the encampment at the huge Sheraton-Cleveland Hotel—and for the next three days, down its green-carpeted corridors, through its suites and golden banquet rooms, governors with white rosettes, their aides with blue rosettes, the newsmen with red rosettes, TV men with their creeping jungles of cable, tumbled over each other in a historic exercise in futility.

Some of the Republican leaders were for unity now, at any cost. Hatfield of Oregon said that the game was over, that they all should have supported Rockefeller in California (though he had not supported Rockefeller in his own state primary in Oregon), and for better or worse they must now close ranks behind the obvious national candidate. From Michigan, where he was journeying that weekend, Richard Nixon checked in with similar views at a press conference—while declaring that the race was still open, he said he would have no part of a Stop Goldwater movement, and urged the apparent candidate to join the rest of the Party closer to the center of the mainstream.

But others were in a more combative mood, and the first to erupt was George Romney. For months, with the dogged tenacity that had made him a great businessman, Romney had been pursuing Goldwater for a private meeting where Goldwater would explain his views; with agility equal to Romney's persistence, Goldwater had avoided the meeting. Romney felt he had a commitment (given by Clif White) for a showdown with Goldwater immediately after the California primary; but he could not reach the Arizona candidate. Romney normally refuses to discuss politics on Sunday, but he now felt the crisis of the Party was so deep that he must breach his practice. Romney is an early riser, and starting shortly after dawn on Sunday, opening day of the conference, he dictated a blast against Goldwater which had all the fervor of a missionary statement. With this draft he went downstairs to a preliminary breakfast of the Republican governors in the Chieftain Room and sounded off with an eloquence that stirred the depressed moderate governors to a turmoil of conscience. Scranton, to be sure, was reported en route from Pennsylvania and scheduled, as all knew, to appear a few hours later on *Face the Nation* to unveil the candidacy heralded the day before. But here, in their presence, was a firebrand of the moderate cause and liberal principles, already blazing away. Rockefeller, whose rancor at Romney's repudiation in California a week before was still fresh, now generously applauded and urged the Michigan Governor on. Should such a blazing draft statement be issued publicly to the press, however, tearing the Party apart? This the governors discussed.

It was at this point, about half past ten in the morning, that now arrived to join them William Warren Scranton. He had just flown in in a Pennsylvania National Guard Constellation in whose seventy-five seats had sat three aides and his wife, Mary, tensely considering the unrolling of the day when he should announce to the nation his open candidacy. Motoring directly from the airport to the hotel, he had arrived in his ninth-floor suite to be told that General Eisenhower was trying to reach him.

He returned the call.

The General wanted to make sure that Bill hadn't misunderstood the Gettysburg conversation of the previous morning. It was all over the papers that he, Eisenhower, was supporting Bill for the Presidency—but the General wanted Bill to know that he could not be part of any "cabal" to stop Barry. Bill was on his own.

Jolted but silent, Scranton had no time for reflection; he was due downstairs at the breakfast in the Chieftain Room, where, suddenly, his discouragement was further deepened. For days now, since the primary in California, the name of the game had been Who Will Bell the Cat? Scranton had decided to bell the cat himself if no other would—expecting all the moderates to fall in line. Now, as he entered the Chieftain Room, he found the governors divided and debating whether Romney should or should not issue his statement, and whether it was or was not too late for anyone to stop Goldwater.

Scranton arrived at the breakfast late and left early—he was due at the studio to appear before the nation at 12:30. Pocketing a prepared statement of his candidacy, he now sat down, flustered, for one of the most miserable ordeals and poorest performances a major politician has ever lived through. His arms folded, his eyes downcast, nervously fidgeting, not knowing his own mind, he stumbled through half an hour of awkward question-and-answer, taking refuge in "principle." He felt strongly about the traditional principles of the Republican Party, he said, but he did not feel strongly about stopping Goldwater. He was "available" for the nomination, yet he would not fight for it. "I don't plan to go out to try and defeat Senator Goldwater. I have no such intention. I do think it is important . . . that the party keep to its sound footing," he responded in answers comprehensible only to Doctor Dolittle's Pushmi-Pullyu.

The nation and all the governors at the conference had been waiting for Scranton's appearance as the high act of the day. It had been preceded, immediately before air time, by a George Romney press conference in which he had, after all, read his blazing crusader's statement to the press: "If his [Goldwater's] views deviate as indicated from the heritage of our Party, I will do everything within my power to keep him from becoming the Party's Presidential candidate" (while, as an obbligato in the next room, as if ordered to play mood music, a band practiced "Stars and Stripes Forever").

Then, after the TV show, Nelson Rockefeller followed with a press conference, sardonic and sarcastic:

Q. "Governor Scranton said today he is available for the nomination. Would you support Governor Scranton?" *A.* "Governor Scranton said that he was waiting to see where Senator Goldwater stood. And after listening to his press conference, I think I've got to wait to see where he stands."

And again: *Q.* "Governor, do you feel that Governor Scranton is displaying . . . responsibility of leadership?" *A.* "Did you see him on television?"

Or again: *A.* (from Rockefeller) "The statement is that Senator Goldwater has been in public life a long time, and that he has made many speeches, written many articles, had a column, voted on the floor of the Senate . . . and why is there a mood now to try to find out where Senator Goldwater stands? I share the point of view of the questioner."

Q. "It does seem puzzling that people at the last minute want to find out now." *A.* "We should live so long."

The New York Governor finished his press conference in high spirits and, like a runner doing an extra lap after the race is over, went on to the Negro district of Cleveland to stump for civil rights.

One of Goldwater's staff was asked for a comment on the Scranton television appearance and replied in one word: "Admirable."

It was now, apparently, all over for Scranton. Yet by late afternoon he had begun to recover from his numbness and confusion. The first trickle of reaction to his appearance had grown into a freshet, and he began to recognize the picture he had given of himself—a puppet on a string, put into the race by Eisenhower, pulled out by Eisenhower, dancing to music other than his own. Yet the Republican Party was his, his since his grandfather's day, his by inheritance from mother and father. If a Stop Goldwater candidate were to be found, it must be done here and now. Thus, Scranton took the lead in a series of fermenting conferences and frantic emergency gatherings. Finally, in the suite of Ohio's Governor James Rhodes (offered by Rhodes as the "anti-Goldwater tent") Scranton met with the Ohio and New York governors, and by midnight they had decided: Romney must be the man. Romney had by now flown back to Michigan—but a late-night call reached him in Lansing and informed him that the other governors would fall into line as lancebearers in the crusade he had so eloquently preached that morning—if he would but lead.

It would be ridiculous to follow in detail the events of the next forty-eight hours as the Republican governors squirmed. Some drafted unity-and-harmony statements to paper over the Party rift. Others demanded that Goldwater explain himself to the nation. While still others saddled up on the instant and began to deploy their forces to prepare the way for George Romney's crusade. Trailed everywhere they went by television cameras and newspapermen, knowing that wherever they paused or gathered some news medium would leap on them to make permanent any chance remarks in a situation that changed hourly, they made a circus of their doings. "What fun," said a Democratic governor, "I've never been to a Republican convention before."

Yet what they were doing was not a circus. One must draw away to see what was happening. Together the thirteen anti-Goldwater governors governed some 58 million Americans in the name of the Republican Party. The great cities, communities and states they tended were of world importance; their resources of talent and wisdom could shake the nation. Yet all of them, together, could find no way of shaping an alliance that might confront the legions whom Clif White, John Grenier and the Luke Williamses had mobilized under the Goldwater flag. Staff work broke down; details snared mighty administrators. Out of the bedlam of the mad forty-eight hours one can point to a single episode that was almost symbolic. Over and over again, since Sunday, the Republican governors had demanded that Goldwater come before them and explain his views; press and radio echoed their demand. But no one thought to extend the invitation to Goldwater directly, to set time or place or request the courtesy of private response. At the Monday-evening banquet at which Goldwater was present, he recalls, "I sat there just waiting for them to ask me—I wanted to talk to them. I don't know whether it would have done any good, but I was waiting for the chance." No one asked.

It was Nixon who, on Tuesday, June 9th, in the last twenty-four hours of the proceedings, added the crowning touch of confusion to the conference. And one must see Nixon as a victim of circumstance rather than a major actor in the pandemonium.[3] He had not wanted to come to Cleveland. Instinct had told him, correctly, that the gathering would be a free-for-all from which any participant would walk away wounded and bleeding. He wanted to maintain his posture of neutrality until the San Francisco Convention and then let the Party turn to him if it would. Distance makes men larger. But against his initial judgment he had been persuaded by Governor Rhodes to come for at least one appearance. Eisenhower was coming, Goldwater was coming—why not Nixon? Nixon was to be exposing himself in his non-campaign in Michigan the very weekend the conference began. He was to be in Muskegon on Monday evening. Rhodes would send the Ohio Governor's plane to Muskegon to bring Nixon to Cleveland if only he would appear on Tuesday.

[3] For years Mr. Nixon has been plagued by a malevolent bad luck. In 1960 he had, quite innocently, roused the emotions of Dallas the night before the citizens of that violent town roughed up Lyndon and Lady Bird Johnson—and been blamed for it. He had been running even with Governor Brown of California in the contest of 1962 in his home state—when the Cuba missile crisis erased him from the front pages just as he was beginning to bring his campaign to peak. He had, in one of what Nelson Rockefeller calls his "peevish" outbursts, denounced the press and declared he would never give them another press conference or opportunity to "kick Nixon around" after his California defeat. Bad luck had chosen him as its child even during the Eisenhower administration—he had been spat upon during his South American trip; and in the opening weeks of his 1960 campaign a chance blow on his knee by an automobile door had provoked so serious an infection that he had been hospitalized for a critical ten days at the beginning of his campaign against Kennedy. He had announced that he was moving to New York, thus retiring from politics, only two days before Nelson Rockefeller's marriage to Happy Murphy, which, too late, opened all Presidential politics to Nixon again. Thus, he arrived in New York as a refugee from the California scene, trapped in a state where Nelson Rockefeller, who disdained him, controlled every crevice and cranny of the Republican Party which might possibly have offered a new base. To top it all, he moved into the same apartment block that sheltered Nelson Rockefeller and his new wife, as well as his first wife, and must live in property controlled by the Rockefellers. Napoleon, when asked how he chose his generals, had replied, "I choose lucky ones." The sense of bad luck that Nixon carried impressed other Republicans, as it would have impressed Napoleon. Of all these curses of fortune, however, none was more mischievous than that which brought him to Cleveland against his better instinct.

Thus, Monday was a strange day for Nixon—he breakfasted with Romney in Michigan and discussed the general Republican situation in terms on which the two men's memories completely disagree. In the afternoon Nixon went on to his public appearances in upper Michigan—and, as usual, with no staff but one aide. So he was already exhausted when late Monday evening Governor Rhodes's plane arrived in Muskegon to take him to Cleveland.

If one lingers on the events of the next day, Tuesday, it is not so much for amusement as for an illustration of how human great men become under stress.

Mr. Nixon arrived in Cleveland about midnight—still a possible Republican nominee. On the way in from the airport, members of Governor Rhodes's staff briefed him. According to them, a counterforce was already in being: Romney, Rhodes, Rockefeller and Scranton were all getting together to stop Goldwater. On arrival in his suite, Mr. Nixon was informed that Governor Romney himself would be up for breakfast at seven in the morning. Meanwhile, however, Mr. Nixon had to learn further from Rhodes's staff what was going on, and it was not until two o'clock that the tired man got to bed.

Bright and early, Mr. Romney, very fresh, arrived—to drink his morning milk at breakfast while Mr. Nixon concentrated over black coffee. Mr. Romney was perplexed: principle demanded that he honor his pledge to the people of Michigan to serve out his governorship and *not* run for President in 1964. But here were the pledges from Rhodes, Rockefeller and Scranton and their offer of full support. It was a hasty breakfast, for there was to be yet another closed breakfast downstairs at which all the Republican governors would privately discuss, finally, the campaign against Lyndon Johnson.

That breakfast was a thoroughly busy political one—while the main speakers talked, private conversation whispered between table neighbors. Governor Scranton, sitting next to Governor Romney, told him of his continuing work in cementing the coalition—of his efforts to contact Herb Brownell, Ray Bliss, Len Hall to run the mechanics of the campaign, of his own and Rockefeller's willingness to turn over all their staffs to the cause. At some point Mr. Romney managed to get Mr. Rockefeller aside and suggest that all of them get together—Rockefeller, Rhodes, Scranton, Nixon, himself—for a quick powwow. Mr. Rockefeller declared that he would not sit in the same room with Nixon to talk about a matter like this. He did not trust Nixon. Hastily, it was arranged that the four, minus Nixon, would meet in Romney's suite after the breakfast.

Meanwhile, formal discussion of the campaign was going on. When it was Mr. Nixon's turn to speak, he tried to cheer the gloomy governors by a fighting talk about their prospects—of how they must hammer LBJ as a man, and get under his skin. Mr. Nixon was offering them his own old gloves-off style: riddle Johnson with scandal and tag him with Bobby Baker. Nelson Rockefeller, whose researchers had thoroughly investigated the Bobby Baker scandal, dryly offered the comment that if they really did open the Bobby Baker case, they'd find as many Republican Senators involved as Democratic Senators; and that the files of the Department of Justice were crammed to bursting with smear material on Republicans as vivid as any they might develop against Johnson.

When they rose to leave the room, the press was jostling and crowding, waiting for news. To escape to privacy, Mr. Romney tugged Mr. Nixon by the sleeve to the men's room. There he hastily explained Scranton's offer at the breakfast to organize the governors to support him, to find Brownell or Hall or someone else to manage his campaign. Was this a draft or not, asked Romney. Mr. Romney's horizons were being enlarged now by the hour—he had been a novice in politics two years before; now he was being asked to run for President.

Mr. Nixon agreed to meet later in the morning with Mr. Romney and review the situation once more; meanwhile he offered to leave the men's room and proceed to tell the press that support was growing among the governors for one of their own to make the run against Goldwater and keep the nomination open. Thereafter—chaos.

The press has a disaffection for Nixon, but he is always good copy; he provides exciting stories, as he observed himself after his California defeat with his bitter farewell: "You won't have Nixon to kick around any more." By now the press had decided to kick Mr. Nixon around—but thoroughly. Wherever he went that day, he was trailed by reporters seeking a news break. After his men's-room session with Mr. Romney, Mr. Nixon was caught by a press conference in which he declared there was, apparently, "a new force in being to stop extremism." Mr. Nixon was merely trying to keep the Republican nomination from being locked up. But the press would have it otherwise. His breakfasts with Romney, in Michigan the day before and again this morning, were used to make it appear that he was using Romney as a stalking horse to create the deadlock out of which his own nomination would come. His weekend in Michigan was described by one as a fishing expedition in which he had caught George Romney and cut him up like a piece of bait.

Actually, events were entirely otherwise. Mr. Nixon wanted to be as far as possible from this scene of confusion. If there was any hope of salvage for his position as the great middle-ground compromise candidate, he must get out of here. On his way upstairs from this press conference he asked his solitary aide, Sherman Unger, to locate the earliest possible flight and get him out. But he was not to be let off that easily.

If at this point the reader becomes confused, it is because confusion was total. In George Romney's suite was now going on a meeting of Romney, Rockefeller and Scranton to discuss the organization of the campaign. Rhodes was to have been present at this meeting. Instead, Rhodes was on his way upstairs to the Nixon suite to discuss the Romney campaign with the champion of 1960. There he asked Mr. Nixon for advice, and Mr. Nixon replied that it should be a "positive" campaign, not simply a Stop Goldwater campaign.

Next into the Nixon suite came George Romney himself, fresh from the meeting in his room from which Rockefeller had excluded Nixon. By now several-score reporters and TV cameramen were pressing themselves into the corridor outside, sweating, steaming, fouling all access. After fighting his way through this crowd into Nixon's suite, Mr. Romney explained what had just gone on—he had told the upstairs people that he would not argue for his own candidacy but would crusade for his own principles. Could he honorably get off the hook of his pledge to the people of Michigan not to run for the Presidency? If a majority of the Republican governors insisted he run, was this a genuine draft? Was this the occasion for a crusade for moderate princples?

At this point, accounts diverge. It is a trap of history to believe that eyewitnesses remember accurately what they have lived through. Mr. Romney and Mr. Nixon remember events differently. According to Mr. Romney's memory: Nixon urged him to become an open candidate. Romney insisted that only principle should guide him. Nixon urged that time was running out, that Romney must walk out of the door and tell the jostling press that Nixon had urged him to run, so had other governors, and he was going back to Michigan to think quietly on whether he should or should not become a candidate.

The Nixon acccount is different: that he urged Romney to go out and barnstorm the country for moderate principles, but explained that he, Nixon, had to remain neutral. When Romney, by this account, asked for help, Nixon replied that he could not and would not change his posture of neutrality.

An eyewitness, present as the two men talked, reports that Mr. Romney, coming up from his conference with Scranton and Rockefeller, was on the edge of decision. Mr. Nixon, familiar with the enormous dimensions and complexity of a national campaign, asked Mr. Romney, still a novice in national politics, who was going to manage the campaign. Mr. Romney rattled off names like Ray Bliss, Herb Brownell, Len Hall, all of whom he thought might be available. Mr. Nixon, an old veteran examining a combat replacement, observed that it would be difficult. Mr. Romney sat down to pencil out on the back of an envelope a statement whereby he might acknowledge his pledge to the people of Michigan *not* to run for the Presidency, yet simultaneously announce the beginning of a national crusade for moderate principles in the Republican Party. Mr. Nixon began to pack his bags. He had been there too long.

Thus, Mr. Romney was alone when he left Mr. Nixon's suite and faced the press in the corridor to declare that Mr. Nixon had urged him to run for the Presidency, so had other governors, and he was now about to make up his mind. (For five or six hours, thus, Mr. Romney was the leading Republican candidate in the headlines. Not until late afternoon, after hearing from his advisers in Michigan that the Detroit newspapers would beat his brains out if he broke his pledge, did his short-lived candidacy die. By then, too, he had spoken to Brownell in New York, who could not undertake to manage his campaign, for Brownell's wife was ill; and to Len Hall, who advised against the whole idea.)

Mr. Nixon, left alone in his suite, unaware that he was being named public foster-father of the candidacy of George Romney outside his door, was now in haste to get out. Except that he must receive one more visitor, Governor Scranton, who further complicated his understanding of events. Scranton said that he was willing to mobilize the Governors *if* Romney first took the lead, whereas Romney had told him that Scranton was organizing the governors to *draft* him and only then would he have to take the lead. Mr. Nixon made no comment at this point, and was off to catch the plane for Baltimore while the press wires fed the confusion to Washington. There, Barry Goldwater's easily-crisping soul was at once seared by the news. To Goldwater the various dispatches made it seem obvious that Nixon was trying to deadlock the nomination against him, using Romney as the pawn. "Nixon," said Barry Goldwater, "sounds more and more like Harold Stassen every day." Nixon's posture of neutrality and party-healer was now shattered. The next morning, with the flak of the Goldwater batteries still trailing him, Nixon flew off to London, wondering how he had been mousetrapped. He had become the target for all factions of his Party, and his Presidential hopes for 1964 were over.

By Tuesday evening, June 9th, the gloom that rested over the Republican governors at the conference was complete. They had come on Sunday, expecting Scranton to blow the trumpet. But that trumpet had blown an uncertain note. All day Monday and through Tuesday morning Romney had been their leading candidate. But by Tuesday evening Romney, too, was out—principle had held him to finish his work in Michigan and honor his pledge to its people.

Romney stayed in Cleveland for the Governors' Ball, as did Scranton and his wife. The Democrats were gay. When asked whether the Democrats, too, could make any news of this conference, their National Chairman, John Bailey, replied laconically, "I hope not." The Democratic governors danced merrily, as befits men who have been entertained for free. The Republican governors also danced under the gold lights of the Grand Ballroom, from tables set with gold napkins, but danced because they must, lest it delight the enemy to see their discomfiture. The band played "Everything's Coming Up Roses," and Mary and Bill Scranton danced gracefully, refusing to show their hurts.

* * *

Two deeply humiliated men returned to Pennsylvania from the Cleveland conference.

One was Dwight D. Eisenhower. He had not realized how his good-willed, scrupulous, but contrary influences on Scranton would be described in public print. On his way back from Cleveland by train, a savage newspaper cartoon of himself as a puffy, baby-faced old soldier leaning over his desk and crying in indecision first alerted him to the impression he had made over the previous weekend. He was already upset when he returned to Gettysburg. On Wednesday, June 10th, a deputation of thirty Pennsylvania state legislators came to plead that he lead an anti-Goldwater move even at this late date because the Goldwater candidacy imperiled their chances at the grass roots of his home state. His brother Milton Eisenhower and historian Malcolm Moos called on him Thursday morning, June 11th, and confirmed that, in his well-meaning meddling, he had muddled. After Milton left, Moos lingered for another hour of conversation with Eisenhower and offered the thought that he, the ex-President, would probably be held responsible for Goldwater's nomination and that, even now, he should urge Scranton into the race.

William Warren Scranton had also been humiliated.

He had danced at the Tuesday-evening ball, but his grace had covered a deep and growing bitterness. He had known immediately after his Sunday *Face the Nation* appearance how weak and vacillating a personality he had displayed. His aides, with a frankness permitted them because of their adoration, had told him immediately after the broadcast that it had been a "complete and utter bomb." Scranton was almost a joke; one of his fellow governors had referred to him as "Gutless Bill." The press called him "the toothless tiger." He had tried to organize, overnight, a campaign for Romney, but on Tuesday afternoon Romney had quit. He had come to distrust Rhodes completely (during the hectic effort to draft Romney, Rhodes had said to Scranton: "Romney's ready to go—

let's get him out front there and then we'll decide what to do"). He was all alone.

Scranton is not without ego; he is a man of enormous personal pride, apparent both in his manner and his bearing. He could either live with the humiliation the rest of his life—or act. His public posture was galling. Even his wife, who all through the spring had urged him to stay out of the race, was embattled.

All this, however, might have washed away on Wednesday morning as they flew moodily back to Harrisburg on his National Guard plane, talking to no one, had not another major event now burst on American politics. This event—far more important than any governors' gathering anywhere at any time—was one of those majestic legislative colons by which the American system introduces the reality of revolution and change.

As the governors had conferred and then danced in Cleveland on Tuesday, Lyndon Johnson and Hubert Humphrey were moving to throttle off the three-month filibuster over the Civil Rights Bill which John F. Kennedy had proposed to the nation a year before. A cloture vote was to be taken on Wednesday so that the Senate might peacefully decide on a revolution which otherwise would be decided in blood on the streets. It was time to stand and be counted.

On the day when Governor Scranton of Pennsylvania returned to Pennsylvania—Wednesday, June 10th—Barry Goldwater stood in Washington to be counted. And he voted against cloture, in effect voting against the Civil Rights Bill (which he was to do formally nine days later); and also, in effect, declaring that the apparent Republican nominee for the Presidency was unalterably opposed to intervention by the Federal Government to secure the human liberties and civil rights of all citizens, black or white, in any state where such fundamental rights might have been denied them by previous Constitutional interpretations of states' rights.

Scranton's own personal humiliation was now joined to a deeper motivating force: his sense of duty. Was the Republican Party, the creation of Abraham Lincoln and his own forebears, to become the party of segregationists?

It was about noon on Thursday, June 11th, a day after his return to his capital, that the decision jelled. Whether before or after a last telephone conversation with Eisenhower is uncertain—for Eisenhower had called on Wednesday, too, with another interpretation of their misunderstanding. On Thursday, however, at his office in Harrisburg, the Pennsylvania Governor was told that Eisenhower was once again on the phone. He answered the call. This time the former President raised a fresh mat-

ter: Eisenhower had been the first modern President of the United States
who had caused any bill on civil rights to pass an American Congress,
and he was now upset by Goldwater's vote the day before against the
new Civil Rights Bill. He wanted to know what Scranton thought of
Goldwater's vote. Scranton replied that the vote had made him sick—
that it, more than anything else, had made him want to run for the Presidency. Eisenhower responded that he was certainly glad to hear that.
Eisenhower was furious himself. He hoped Scranton would say what he
thought out loud, publicly. Eisenhower said that he was sending Malcolm Moos up from Gettysburg to Harrisburg to explain his own feelings.
Scranton said he would be glad to see Moos.

As for himself, however, Scranton had already made up his mind
by that Thursday morning. More important things than delegates, nominations or public images were involved. For William Scranton, the Republican Party takes second place in his loyalties only to the United
States itself; it is his personal and family heritage as much as his
church and faith. What was at issue was the bearing and conduct of his
Republican Party in the presence of revolution. He had seen no reason,
no *duty* to become President before; but there was now, clearly, a duty
he owed to his Party. Thus, he would run. His wife, Mary, had sought
him out in his chambers that Thursday at lunch—she, privately, after
long inner reluctance, had come to the conclusion that Bill now *had* to
run. When she arrived at his chambers she found he had been trying
to reach her as she drove in—he had wanted to tell her he had come to
the same conclusion.

It is impossible to consider the short-lived Scranton campaign as a
serious exercise in politics; but as an exercise in gallantry it may have
saved the soul of the Republican Party. The entire Scranton campaign
lasted less than five weeks, from June 11th of 1964 to July 15th; it swept
across the country, from coast to coast, with a breathless recklessness; it
cost the Governor of Pennsylvania and his friends over $800,000; it
was hopeless to begin with—a miscast horse opera in which the young
hero saddled up his white horse to overtake the kidnapper and headed
him off at the pass only in time to witness the final act of ravishing.
And yet—over the long range the Scranton effort, coupled with the
Rockefeller effort, preserved the base from which the Republican Party
might, if it would, some day rebuild into a major force.

Wednesday night, after the Cleveland conference, the Pennsylvania Governor had wordlessly brooded in the Governor's mansion at
Indiantown Gap, alone with his wife. By Thursday morning Scranton had
made his decision. By Thursday evening he had summoned his staff of

young men to the mansion and, with little fanfare, no forewarning and less preparation, flung them into action.

One can see their exertions in the next forty-eight hours and next three weeks as almost heroic: instant arrangement of an invitation to the convention of the Maryland Republicans the next day to proclaim the candidacy; conscription of the private plane of John Hay Whitney of the New York *Herald Tribune* for the flight; immediate assignment of speech writers to detail, of schedule makers to tour plans; appointment of a standby government of the state to direct its affairs for the next month; immediate liaison with the Rockefeller and Lodge forces; dispatch of a task force to San Francisco to make battle and communications arrangements, to find floor space and command space at the Cow Palace, scene of the Convention.

But politics gives little credit for heroics.

There was, to begin with, the iron count of Convention delegates— and on the weekend of the Scranton announcement the Associated Press already counted 604 sure delegates for Goldwater. To shake this Convention, Scranton would have had to mobilize every uncommitted delegate in the Republican Party—and then shake away perhaps 200 of the delegates already committed or leaning to Goldwater. To do this would have required a mastery of the mechanisms of politics and national relations equal to that of the teams which had thrust Willkie in 1940, and Eisenhower in 1952, down the throat of hostile Republican Conventions —plus the time and the alliances and the connections those teams had enjoyed. It was an operation requiring a master plan and a master staff, and Scranton had neither. Instead the Scranton campaign relied on hope and faith and a series of "ifs":

If the Rockefeller and Lodge forces would now turn over their entire apparatus, research, troops, organization and delegates—

If the Midwestern favorite sons—in Wisconsin, Michigan, Ohio and elsewhere—would hold firm against the Goldwater bandwagon—

If some of the convention-state delegates already chosen could be lured, persuaded, or pressured away from Goldwater—

If the old Eastern Establishment still had muscle and the will to operate in this emergency—

If Eisenhower would, finally, be provoked by Goldwater's vote on civil rights to descend into the arena and fight openly for Scranton—

If, finally and above all, the conscience of the Party could be aroused and public opinion be mobilized so vividly as to shake the convictions of the San Francisco-bound delegates—

If all these things worked, then perhaps there might be the faintest hope of a miracle.

But there were too many ifs. Scranton's personal staff of young men were as promising, virile, enthusiastic and imaginative a crew as any that took to the field in 1964. But they were his personal crew. In 1962 they had braced their young candidate for a stunning, violent victory over Philadelphia's Richardson Dilworth; but they had operated then from a base of organization prepared for them by an alliance of all the Pennsylvania county Republican machines. Scranton had been combat commander of an army in being. In national politics, no such mobilized army existed—and his staff lacked totally the manpower and the experience for the national adventure. Their energies were devoted to their candidate; there was nothing to spare for organization. The Rockefeller machinery was turned over to Scranton, lock, stock and barrel, within two days of his announcement of candidacy. Yet it was ten days before the ablest of the Rockefeller delegate managers, young Robert Douglass, was brought to Harrisburg to explain to his Pennsylvania contemporaries just how enormously intricate, and how completely different from a state contest, a national contest is; and not until Convention Eve itself was the actual management of delegates turned over to the Rockefeller regional directors.

From Saigon, Henry Cabot Lodge announced his resignation on June 23rd and flew back to add what he could to the Scranton campaign; Lodge proceeded directly to visit Eisenhower and ask for his help. But Eisenhower would not move publicly. Nelson Rockefeller, battling as valiantly for the Pennsylvania Governor as he had for himself, telephoned the former President to plead for support. The former President said he could not come out publicly, he had to preserve his influence. Acidly, Rockefeller inquired: For what?

Nor could the Eastern Establishment succor Scranton. Personally, Scranton telephoned former Presidential candidate Thomas E. Dewey in New York: Would Dewey help? Dewey would try, and so the old President-Maker gathered a sample of top Eastern Establishment leaders for lunch at The Recess club. Along with himself, he drew two members of his old Presidential team, Herbert Brownell and Lucius Clay. Three Wall Street lawyers of the loftiest Wall Street firms gathered with him: David W. Peck of the august firm of Sullivan & Cromwell (John Foster Dulles' old firm), Frederick M. Eaton (of Shearman & Sterling), Judge Bruce Bromley (of Cravath, Swaine and Moore); two Pennsylvanians— Thomas Gates (Eisenhower's last Secretary of Defense) and Robert Dechert (speaking for Tom McCabe, Sr., of the Scott Paper Company)— and two Rockefeller men—John A. Wells and Robert Douglass. The question before them was whether Dewey could exercise his old magic. Could this summit group of Wall Street lawyers exert (as they had twenty-four years before for Willkie and twelve years before for Eisenhower) that influence they once had held over scores of regional law firms of their association to influence regional delegates to the Republican Convention in favor of Scranton. A list of delegates committed to Goldwater who might just possibly be swayed was prepared; and the eminent leaders did indeed try. "But it was," said someone familiar with the operation, "just incredible. We called all the old names; but they weren't there any longer, or they weren't in politics any longer. It was as if the Goldwater people had rewired the switchboard of the Party and the numbers we had were all dead." Within two weeks Dewey had decided it was hopeless and, thus deciding, declined to attend the coronation of Goldwater—the first Republican Convention that Thomas E. Dewey skipped in twenty-eight years of politics.

One could most easily see what was happening by following Scranton on the road. His headlines through June had swollen day by day as he flew from state to state; his television appearances had begun to draw a wider and wider response—30,000 letters after one appearance alone. Thus, on June 30th he arrived at the caucus of the Illinois delegates at the O'Hare Inn near Chicago (one of Clif White's rendezvous in the early 1963 planning) to bite hard on the nail. Of 58 Illinois delegates already chosen and leaning to Goldwater, how many could he shake? The Republican Party in Illinois had long been balkanized; it no longer recognized a general master; so one could negotiate with these delegates only as individuals, sway them only by charm and an appeal to conscience.

The O'Hare Inn on the eve of the caucus was a study in contrasts. Hastily a few Scranton enthusiasts were hand-painting signs in a gloomy, empty room for the morrow. The names of the delegates were unknown to them; the hour of their chief's arrival the next day was uncertain; they did not know where he was now. One floor down and in a farther wing the Goldwater forces had opened a hospitality room—brilliantly lit, crowded with delegates drinking from the flowing bowl and munching canapés, the room pulsed with energy. The delegates were almost unaware that a Scranton headquarters existed at their hotel at all. And as for the caucus tomorrow? Their hearts were Barry's. They were bound to him not by law but by something stronger. "I used to read the newspapers," said a middle-aged manufacturer going to his first national convention as a delegate, "and every time I read a headline I got mad. Then when I went to the hospital I had time to think, and I decided it was no sense getting mad; we had to do something about it; we started to organize for Goldwater in my district in 1962."

Scranton arrived at Chicago early in the morning, drove to the O'Hare Inn—to discover that he was being picketed by two groups,

Goldwater pickets and civil-rights extremists. The Goldwater pickets were to follow him around the country at every stop with parades, placards, demonstrations in the first manifestation of the kind of counter-picketing that was to plague Goldwater himself so persistently in the later campaign against Johnson. But this time they were joined by the civil-rights pickets, who, by some surrealistic logic, had decided that in Chicago they would demonstrate against Scranton, who was *for* the Civil Rights Bill, rather than Goldwater, who was *against* it.

The delegates gave Scranton his hearing; gave Goldwater a hearing; gave Harold Stassen a hearing too—and then took their vote. For Goldwater, 48; abstentions, 8; passes, 2 (from two Negro delegates whose hearts still held for Rockefeller); for Scranton—none.

It was this way wherever Scranton went. He flew to Salt Lake City from Chicago. At the airport friendly signs greeted him (SMITH GIRLS FOR SCRANTON, read one). But at a caucus with the delegates he learned the truth. All Utah's delegates had been taken over by the Young Republican organization as early as the fall before. He spoke well, and pleased them indeed. But they were locked. He flew to Seattle and spoke eloquently—but Luke Williams had preceded him there by over a year; 22 of Washington's 24 delegates were unshakable.

One could not but have sympathy for the man. Urbane, polished, unwearying, he could still quip. Of the Nielsen polls being circulated, he remarked, "Yes—I know the Nielsens are split. There are two A. C. Nielsens—the father is for Goldwater, the son is for me." Visiting North Dakota's State Fair, he went up in a balloon and, when asked how the balloon ran, replied, "It runs on hot air, just like my campaign."

Yet while he quipped, he pleaded—and punched.

And it is only in the long range that the Scranton campaign made any sense at all. For somehow, as he said, "even if we lose—we have to create a rallying point in our Party for people of our generation." For the Party of the Republicans was as much his heritage as the mansion at Marworth which his parents had built, or the town of Scranton which his great-grandfather had founded. On the walls of his library hung a U.S. Army commission to a Scranton forebear signed by Thomas Jefferson with a later promotion by the hand of James Madison; and a commission to Joseph Hand Scranton signed by Abraham Lincoln. His family, like other Pennsylvania industrialists, had helped fuel and finance the Republican Party of Pennsylvania for generations; just as they had financed China missionaries and good works of all kinds. It was *his* party; and if, to save it, he had to punish Goldwater, an old friend, and destroy, in 1964, the value of its nomination, then so it had to be.

Trailing in his wake the same headlines that Rockefeller had first stimulated, Scranton painted Goldwater, from East Coast to West Coast, as a warmonger ("Please, I implore you," he shouted from the canyons of downtown Chicago, "send to the White House a man who thinks deeply, who is not impulsive"), as a man who would destroy Social Security ("This man has no part of the heritage of a true conservative like Bob Taft. Would Bob Taft destroy Social Security?") and as a candidate doomed to disastrous failure.

All this, and then the crushing blow: Could the party of Abraham Lincoln now become the party of the racists? Could the Republican Party present to the nation a candidate who, denying the will of the majority of his own party in Congress, had voted *against* the Civil Rights Bill? Scranton's own Republicanism, he said in Seattle, was to "take the best there is of the past, and then apply those principles to the problems of today," and in Eugene, Oregon: "I can conceive of nothing more terrible for the United States than to have a violent disruption over these matters, racial matters; this is not in our tradition, not in our custom, not in our manner."

Within three weeks all the pragmatic and mechanical calculations of Scranton's campaign had come to naught. It was a battle of Young Lochinvar against Field Marshal Ludendorff. There remained, as the Convention approached, only the hope of miracle—that somehow the mass of the Republican Party might feel moved to press its delegates at San Francisco to reconsider. Did they indeed wish to have the new law of the land enforced by a man who had voted against that law? Like an undertone, all through the campaign of 1964, ran the mocking question of history: What is the relation of liberty to law, of freedom to government? Scranton had posed the question fairly to his Party in the brief campaign; if there was an echo deep enough, he might somehow prevail; and if there was no echo in June of 1964, Scranton could nonetheless content himself with having posed his question clearly enough to public conscience to earn a place in history.

The great Civil Rights Bill had now, as he barnstormed, been passed by the Senate on June 19th and been signed as law by President Johnson on July 2nd. And one cannot go on to the climax of the Republican struggle of 1964 at San Francisco without first examining this law which so divided Republicans among themselves and from their leading candidate.

A revolution was taking place.

And it is at this tragic and tormenting revolution we should now look. For the politics of America are made as much in the streets of its cities and the hearts of its citizens as in their forums of formal legislation—or national conventions of delegates.

Black and white Americans had begun to take their bitterness into the streets and country lanes, into the piney-woods ambuscades and slum-alley butchering places. This revolution was to overhang the entire election of 1964—and will intrude in all the elections of the foreseeable future.

alley bordering places. This re .... was to overhang the entire election of 1964—and will intrude in all the elections of the foreseeable

CHAPTER SIX

FREEDOM NOW—
THE NEGRO REVOLUTION

IT was almost absurd for a visitor to the United States Senate in the late winter or early spring of 1964 to think of himself as witness to a ceremony of revolution. What he could see from the press or visitors' gallery of the Senate was so unutterably commonplace.

At noon each day, after the invocation of prayer, the swinging doors that lead to the Senate floor from the rear cloakroom and side entrances would slowly open, then shut, and some twenty or thirty Senators would unhurriedly enter, wander down the red carpeting of the wide aisles to their desks, shuffle their papers, then cock an ear, half interested, to what was being said. It was as if, rather than pausing to listen to words, they were sampling sound.

The open floor of the United States Senate is a singularly unromantic place, and its proceedings are, on the surface at least, the dreariest to be heard in any of the great lawmaking bodies in the world. Its sound is normally a drone, lacking the flash of the old French Assembly or the deep masculine roar of the House of Commons. The drone comes in every accent of America—the rasp of the Midwest, the drawl of the South, the sing-song of the big-city states. Yet in the spring of 1964 one listened more closely to the drone than ever before. The Senators below also listened: it was as if the *tone* rather than the mumbled words held their ear. If the sound were normal, they would rise from their desks after a few minutes, amble about the hall to chat with their neighbors, then casually filter out the back doors to lounge and talk in the more comfortable black leather chairs and couches of the cloakroom, screened from public sight.

By half an hour after the noon opening, normally, there would remain on the floor only four out of the one hundred Senators of the United States—two Southerners and two Northerners, each pair watching the other, as they were to watch each other for seventy-five full days while the business of the nation was suspended in the longest filibuster in American history. An Ervin and an Ellender for the South against a Javits and a McGee; a Russell and a Long against a McIntyre and a Keating; a Talmadge and an Eastland against a Case and a Kennedy, assigned by their leaders to patrol the floor, two against two—apparently evenly matched.

Yet this was no even match. The Humphreys, Kuchels, Javitses and Clarks, just as certainly as the Russells, the Longs, the Stennises, the Thurmonds, knew who would win: a passage of history was coming to a close with relentless and inexorable predetermination.

For eighty years, ever since the end of Reconstruction, a coalition of Southern Senators had dominated the Senate of the United States, letting it act only if the Senate chose to act outside of, and indifferent to, the condition of humiliation of the Negro American in the South. Year after year, by threat or use of their unlimited freedom of debate, they had blackmailed Senate and country not to intrude in the relation between black and white Americans in the states of the old Confederacy. Their numbers, their seniority, the outstanding ability of several of them, and, above all, their skill in filibuster and parliamentary process had made the United States Senate their club, their home, their institution. Here they had reconquered by wile and defended by cunning what Lee had surrendered at Appomattox. Now it was being taken away from them.

We shall examine later the mechanical process and the strategy by which the Northern and Western Senators, under the leadership of Hubert Humphrey and Thomas Kuchel, were to impose the will of the nation on the Southern irreconcilables.

But it was indeed the will of the nation that was being expressed in the drone on the Senate floor; and the will of the nation, in the winter and spring of 1964, was to test fate. This country had been built on the thought that all men should be equal in opportunity and that all must obey the laws that guard these opportunities. Unequal as men may be in talent, temper and ability, each man must be allowed to go as far as his talent, temper and abilities would carry him; and ever since the Pilgrims had signed the Mayflower Compact in 1620, this country had pledged itself to abide by the laws of the majority. The Federal laws of the United States had ignored the need of Negroes for equal opportunity and the dignity that comes with it; the new law was designed to provide this opportunity, even at the cost of wrenching the Constitution out of the thinking of the fathers. What was being debated was whether the Constitution should be so wrenched—whether the nation should tempt fate with this legislation to preserve the spirit of opportunity in the American Constitution; or whether it should cling to the cool words so precisely tooled by its authors and let the issue of opportunity be settled by blood in the streets.

The Senate was to choose—every man knew this from the beginning—to gamble that the reason embodied in this legislation was wiser than a test by violence. But one cannot understand the mood either of the Senate or of the country unless one goes beyond the drone of debate on the red-carpeted floor to the reality of the revolution that had exploded a year before, in 1963, in the states of slavery.

There are any number of Negro communities in the United States—and though they differ among themselves as much as Bedford-Stuyvesant in New York does from Conant Gardens in Detroit, though their leaders vary from authentic saints to potential killers, they are locked together not only by the color of their skin but by their consciousness of humiliation. This humiliation of the black man lies at the roots of the revolution, and it colors with uncontrollable emotion every backward glance at, or forward analysis of, the tragedy of the Negro in America. The black man, looking back on his trackless history in this country, can feel only indignation. But the white man must see the story as the history of sin: of men willing to use other men as animals, and the women of such men as vessels of carnal pleasure. Where guilt for the past lies is now too far away in time to be relevant—for the guilt of slavery is apparently shared as much by African chieftains who sold men of their own kind as beasts as by white traders who bought the black bodies for gain and profit.

Of the history of the black man in Africa almost nothing is known, and of his history in the United States little more. Yet through this history in America one can see shaping over the centuries only one black institution Negroes could call their own: a genuine but independent branch of the Christian church. Deprived of both personality and identity, stripped of tongue and culture, blacks were civilized by the lash—and, as a guilty afterthought, by the planting among them of some sense of the Christian message through preachers and prayers of their own. Now, the Christian message is one of mercy; but its bearer, Christ, was a passionate revolutionary, his words an inflammatory denial of both Roman and Judaic law. What is amazing is that it took so long—a full century after the Civil War—for the Christian message to come to flame in the Negro community of America.

One must start with this Christian message and the Christian leadership of the Negro revolt in the South, for one cannot understand the practical politics of 1964 unless one makes the distinction between this Southern Christian leadership and the leadership of the newer, more perplexing Negro communities of the Northern cities.

The politics of the Southern Negro had only glancingly touched the campaign of 1960.[1] Historians of the Negro revolt trace its beginnings, of course, much further back than 1960—to the Montgomery, Alabama, bus strike of 1955-1956, the Autherine Lucy case of 1956 or, if one wishes, to Nat Turner's slave revolt in 1831. Yet even in 1960 the struggle of the Southern Negro had seemed peripheral to national politics, intruding on the Kennedy-Nixon campaign only by Kennedy's masterful reaction to the arrest of Martin Luther King, Jr.

But the smolder was already there. Early in 1960, on February 1st, four brave Negro students had sat themselves down at a lunch counter in Greensboro, North Carolina, and demanded, as their human right, that they be allowed to sit and eat the food they wanted to buy. In the spring of 1961 a handful of white youngsters, kindled by the contagion of a righteous cause, had joined a number of Negro youngsters and ridden through the South on "Freedom Buses," demanding that they share alike in public facilities which, theoretically, all were entitled to share. By 1962 the map of the South, as viewed from the offices of the Department of Justice, was beflagged and pocked with little pins where here, there or elsewhere, Negroes and their white friends were demanding that they be treated as human beings. And emissaries of the Department of Justice were traveling the Southland like emergency squads, putting out fires, conciliating, pleading, coaxing local communities and their leaders to be good. For at this time, and all through 1963, there existed no base in law, either Federal or local, from which a Southern Negro might demand that a court protect his right to sit and eat in any public place, to use the facilities of the roadside wherever the general public was so permitted, to enjoy not the courtesies but the necessaries without which travel, work, learning or dignity is impossible in our day and time.

However devoted the Department of Justice and its Attorney General, Robert F. Kennedy, might be to the cause of human rights—and no more dedicated Attorney General has sat in that office in modern times —the Department was crippled for lack of such a law and authority to act; and it was powerless to demand such a law because nothing yet had burdened or seared the general American conscience with the nature of Negro humiliation in the South.

It was this that Martin Luther King, Jr., the Police Department of Birmingham, and American television were to provide in 1963: the visual

[1] See *The Making of the President, 1960*, pp. 321-323 *et seq.*

demonstration of sin, vivid enough to rouse the conscience of the entire nation. "I saw no way of dealing with things," reflected King a year later, "without bringing the indignation to the attention of the nation. There *had* to be a method."

Of medium height (five feet seven inches), close cropped of hair, full and ruby-red of underlip, a man with almond-shaped eyes and a thin mustache, Martin Luther King, Jr., is, at thirty-seven, one of the more handsome Negro Americans among the exceptionally handsome new Negro leaders of America. And to his neatness of personal habit he adds a neatness of thought and a clarity of expression that make him (along with Roy Wilkins of the NAACP) one of the two most impressive revolutionary leaders in America.

For King, 1963 was the year to move. Even now he cannot give any one reason for this choice of time.[2] He offers several: because it was a century from the Emancipation Proclamation and Negroes were still held in servile condition; because it was almost a decade from the Supreme Court's 1954 decision on school desegregation and the glacial pace of desegregation had been tragically disappointing; because all over Africa in the previous decade black men had reached self-expression under their own leadership; because, finally, the movement he led as president of the Southern Christian Leadership Conference had found, as he calls it, "its undergirding philosophy" of nonviolence. The Negro churchmen of the South had been developing their technique of passive Christian protest for almost a decade—from the Montgomery bus strike of 1955-1956 to the Albany, Georgia, protest of 1961-1962. By the fall of 1962 the Southern Christian Leadership Conference, founded in 1957, had expanded to embrace twenty-five affiliates; of these, one of the strongest was the Alabama Christian Movement for Human Rights, headed by the

---

[2] Perhaps his best expression of mood came in his later famous "Letter from a Birmingham Jail," which was addressed to the white clergymen of Alabama: "My friends, I must say to you that we have not made a single gain in civil rights without determined legal and nonviolent pressure. History is the long and tragic story of the fact that privileged groups seldom give up their unjust posture; but as Reinhold Niebuhr has reminded us, groups are more immoral than individuals.

"We know through painful experience that freedom is never voluntarily given by the oppressor; it must be demanded by the oppressed. Frankly, I have never yet engaged in a direct action movement that was 'well-timed' according to the timetable of those who have not suffered unduly from the disease of segregation. For years now I have heard the word 'Wait.' It rings in the ear of every Negro with a piercing familiarity. This 'Wait' has always meant 'Never.' It has been a tranquilizing thalidomide, relieving the emotional stress for a moment only to give birth to an ill-formed infant of frustration. We must come to see with the distinguished jurist of yesterday that 'justice too long delayed is justice denied.' We have waited for more than three hundred and forty years for our constitutional and God-given rights."

Reverend Fred Shuttlesworth of Birmingham; and the movement, by 1963, says King, was "ready for full maturity"—ready to take on Birmingham.

Birmingham is described as one of the two meanest cities in the South—the other being Jackson, Mississippi. But while Jackson is a clean, handsome little town of 140,000 citizens, Birmingham is an ugly industrial metropolitan area of 635,000 people with a tradition of violence and killing that goes back to the turn of the century, when convict labor worked its mines and the unions won recognition by the law of tooth and fang. Grimy and smoky now from the fallout of its steel mills, Birmingham has experienced violence almost continuously to this day. In the six years preceding the Martin Luther King invasion it had known fifty cross-burnings, eighteen racial bombings. Symbolically, its leading citizen was Eugene "Bull" Connor, Commissioner of Public Safety for twenty-three years and chief of one of the most brutal racist police forces of the South. Its municipal administration lay in the hands of men so backward that the white citizens of Birmingham themselves, in the spring of 1963, were preparing to turn them out in the election of April.[8] In all downtown Birmingham there was no general restaurant where a Negro might go in and sit down to eat, no public restroom where he could go to relieve himself, no theater where he might freely enter to see a movie of his choosing. Nor was anyone in Birmingham particularly disturbed by this. (Once in Jackson, Mississippi, I asked a particularly intelligent state official whether it didn't gnaw at him or bother him in any way at all that Negroes could find no place in the downtown heart of his city where they might sit and eat like human beings. He replied: "Not one goddamned bit.")

In any great political clash or movement, men must be bound together by a "doctrine." The doctrine King offered for the assault on Birmingham was his famous doctrine of "nonviolence," derived partly from his reading of Gandhi's experience with the British Raj in India, but distilled in large part from Negro experience in America. Violence would lead to killing and bloodshed—in which, more likely than not, Negroes would be massacred and the white conscience hardened. Nonviolence, in the various forms of passive resistance, was more applicable to a system which based itself on laws supposedly drawn from conscience. King distinguished between "moral" laws and "unjust" laws. An "unjust law" was one that flouted morality directly—as do segregationist laws. "In no

---

[8] A good part of the emotional steam in John Grenier's captaincy of the Alabama Republicans (see Chapter Five, pages 135-137) lay in the disgust of the white middle-class Alabamians with just such men as Connor and the city administration.

sense," King had written, "do I advocate evading or defying the law as the rabid segregationists would do. This would lead to anarchy. One who breaks an unjust law must do it openly, lovingly. . . . I submit that an individual who breaks a law that conscience tells him is unjust, and willingly accepts the penalty by staying in jail to arouse the conscience of the community over its injustice, is in reality expressing the very highest respect for law."

All through 1962 the Southern Christian Leadership Conference leaders debated whether this doctrine was sufficient to armor them for the assault on Birmingham. "I contended," recalls King, "that Birmingham was pivotal. That we had to go there. It is the largest segregated city in the United States. If we could break through the barriers in Birmingham—if Birmingham went—all the South would go the same way."

All through the fall of 1962, as their decision hardened, the Negro Christian leaders discussed the assault on Birmingham. They had learned from their 1962 demonstration in Albany, Georgia, that a demonstration, to be successful, must have a specified target—it must not diffuse itself. They had decided after Albany that 1963 was too early for political protests in the Deep South to be useful—for the non-voting Southern Negro has little leverage on politicians elected to suppress him. It was better, they concluded, to pinpoint specific targets—areas of humiliation or exclusion, lunch counters, buses, libraries. In Birmingham, whose population is 35 percent Negro, the purchasing power of the Negroes is tremendous; a well-aimed economic strike could all but strangle the major merchants of the town—nonviolently. Thus they began to prepare for "Operation C"—C for Confrontation.

The demands finally settled on by Negro leaders were innocently primitive: that the downtown stores which thrived on Negro business let their Negro customers eat at the same lunch counters as whites and use the same water fountains and restrooms; that the stores employ a few Negroes on the sales floor as a token of equal opportunity; and that a biracial committee be set up to study steps for desegregating the schools. It was months before the white citizens of Birmingham learned of these demands; no white newspaper in Birmingham would publish them until long after the city had become, as King hoped it would be, " a fuse which would speed all across the country."

Full plans were drawn up after Thanksgiving, 1962. In December the Alabama Christian Movement leaders met with King at his Atlanta headquarters in the Masonic Lodge building at 334 Auburn Avenue and decided to launch their protest just before Easter—one of the two main shopping seasons of the year. An election of a new city administration was approaching, in which Bull Connor was running for mayor; thus the demonstrations would begin immediately after the elections, so that the demonstrations would not convert white wrath into votes for Connor. In January, King's aides flew from Atlanta for reconnaissance. In February they began the recruiting of volunteers: 300 Negroes of Birmingham, each committed to nonviolence and pledged to be willing to spend at least ten days in jail. All of these, moreover, would be drilled in nonviolence—how not to resist, how to cover head, eyes and vital organs while being beaten, how to go limp and let the body become a sack of flesh to be hauled away.

Elections were to be held in Birmingham on Tuesday, April 2nd—twelve days before Easter Sunday, including ten critical shopping days for downtown merchants stocked with holiday inventory. On Election Day, late in the afternoon as the polls were closing, King himself flew into Birmingham and proceeded directly to the First Baptist Church in suburban Ensley, where his brother, A. D. King, was minister. There some 75 of the volunteers were waiting. King lectured them on the philosophy of nonviolence and the situations of combat they would encounter; and the meeting closed with the singing of "We Shall Overcome."

Wednesday, April 3rd, was a bright and sunny day, and the first 40 volunteers, in groups of 8 each, proceeded to their assignments—five downtown stores: Woolworth's, Kress's, Green's and two others. Five of each group of 8 would walk into each store to sit down at the lunch counters; 3 of each group would peacefully picket outside the store in protest. With 300 volunteers ready to go, King's platoons of 40 men each would last almost ten days if, as expected, each day 40 more would be arrested. But who would follow?

All 40 were arrested the first day; the second 40 the following day; and by the third day the downtown stores had begun closing their lunch counters entirely. On Saturday King escalated—a march on City Hall from the Sixteenth Street Baptist Church, joined by new converts: and 125 arrested. On Palm Sunday following, another march—and another 100 arrested. By now King and Shuttlesworth had fused the Negroes of Birmingham, originally split and divided as to whether to back the challenge, into a communion. This was more than a protest—this was "Freedom Now."

On Good Friday, April 12th, 1963, King himself made his witness in a march that moved no more than eight blocks before he was arrested and thrown in jail; and by Easter Sunday his arrest had made the Negro revolution central to national and world attention.

One must see the weeks of April and May of 1963 in all the complexity of changing American life.

National television had discovered the drama in Birmingham; and though news of these events was pushed back to a few paragraphs on the inside pages of the local Birmingham newspapers, it was not only on front pages everywhere else in the country—it was the prime spectacle of the evening television news. TV does nothing better than spectacle—and the spectacle it showed the nation now, day after day, entered every Negro home and most of the white homes too. Fifty or sixty people were being arrested in Birmingham every day, but they were mainly adults, not students. Now television, showing the drama of heroism, began to stir the Negro youngsters of Birmingham, too; and as the students entered the streets, the Birmingham police and Bull Connor [4] were suddenly faced with a problem for which no police manual has a solution. On Thursday, May 2nd, 1,000 students marched and were arrested. The following day, about 500 more. The following Monday, another 1,000. By Tuesday— "That was the day the jails were full with no place to put any more," says King—the situation was beyond police control. Five thousand students demonstrated; since there was no place to jail them, the police attempted to disperse them—with fire hoses and police dogs.

The police dogs and the fire hoses of Birmingham have become the symbols of the American Negro revolution—as the knout and the cossack were symbols of the Russian Revolution. When television showed dogs snapping at human beings, when the fire hoses thrashed and flailed at the women and children, whipping up skirts and pounding at bodies with high-pressure streams powerful enough to peel bark off a tree—the entire nation winced as the demonstrators winced. And, as if following the instruction of the old blues song, hundreds of letters were now coming to Martin Luther King, Jr., in his cell, sent "in care of the Birmingham jail." President Kennedy had already intervened to make sure that no harm came to King there. Department of Justice negotiators were frantically trying to work out a truce. By May 8th white and black leaders were involved in talks. By May 10th, King (released from jail) and Shuttlesworth were able to announce a truce. But the bombing of the

---

[4] I remember, most vividly, Bull Connor on the floor of the Democratic National Convention at Atlantic City a year later. The former chief of the inhuman police of Birmingham had come as a delegate with the Alabama delegation and was being expelled from the floor of the convention for refusing to take the oath of allegiance to the Party. Pursued by officers, the quarry of a merciless pack of reporters and television men, he fled like a hare from the chase, harassed and frightened, trying to protect his little grandson from the jostling. Finally he burst through the doors of the Columbia Broadcasting System floor studio, still pursued by reporters and cameramen, to find refuge for himself and grandson under the shelter of the same broadcasting system that he was suing for half a million dollars in libel. I wondered then whether his memory in flight carried him back to the scenes in which hundreds of Negroes had fled from his dogs and his harassment.

home of King's brother and of a Negro motel brought Negro mobs, rioting (with no heed to Martin Luther King's nonviolence) into the street.
The State of Alabama sent in state troopers, their helmets emblazoned
with the Confederate flag, to restore order; and then Kennedy moved
into position battle units of the United States Army, ready to act. The
Federal troops were never used; under their threat, Birmingham calmed
down to live uneasily, by mid-May, with the compromise arrived at: Negroes might eat at lunch counters in a few of Birmingham's downtown
stores; a few of them would be openly employed on the floors of emporia
that depended on their trade.

But the spirit of revolt could not be quenched. ("You got to understand," said a thoughtful Mississippi cotton planter, "that every one of
those Negroes on my land has a television set in his shack, and he sits
there in the evening and watches.") The massive Birmingham protest
had triggered demonstrations all across the nation, and, like firecrackers,
one popping off the next, all through May and June of 1963, Negroes
took to the streets. The National Guard patrolled Cambridge, Maryland;
in Jacksonville, Florida, the police cleared demonstrations with tear gas;
in Memphis, Tennessee, the city fathers closed the municipal pool. And
everywhere from Baton Rouge, Louisiana, to Charlottesville, Virginia,
students manned the lunch-counter front.

The turbulence spread north: in Sacramento, Negroes sat-in at the
State Capitol; in Detroit they invaded City Hall and demanded the city
fire its chief of police and subject him to criminal trial; in New York,
Negro activists dumped garbage on City Hall Plaza; in Philadelphia they
clashed with police at a construction site; in Chicago, at a cemetery that
refused to bury Negroes. In the ten weeks following the Birmingham
uprising, the Department of Justice counted 758 demonstrations across
the nation; during the course of the summer, there were 13,786 arrests of
demonstrators in seventy-five cities of the eleven Southern states alone.

By now, in May of 1963, the Negro revolution had become the
standing excitement of the American news media: magazine articles on
the Black Muslims and Malcolm X; articles of rediscovery of the
condition of the black man; and television programs without end as
learned men and rabble rousers alike competed for public attention. And
in this atmosphere there began to flourish Negro leaders entirely and
ominously different from Martin Luther King. In New York, Adam
Clayton Powell, Harlem's leading black racist, began to invite riots. It
was as if Southern white crime required Northern Negroes to take venge-
ance on Northern whites. And as outrageous statement succeeded outrageous statement, television and press began to require even more dramatic and violent statements in a competition of threat and violence by

which Negro leaders sought the attention of the nation. "Miscalculation of the moment of truth which is upon us," threatened the Reverend Gardner Taylor, leader of a Negro church in Brooklyn, "could plunge New York, Brooklyn, Philadelphia, Chicago, Detroit and Los Angeles into a crimson carnage and a blood bath unparalleled in the history of the nation."

It was out of this turmoil that the Civil Rights Bill of 1964 was born.

Ever since the early days of the Freedom Riders in 1961, the brilliantly staffed Department of Justice had been aware of a growing disturbance in the South. Its services of intervention and conciliation there had long been pressed to the limit of strain. One could matter-of-factly describe as heroes all the tired, ever-traveling men of the Civil Rights Division of the Department, led by Assistant Attorney General Burke Marshall, a man at once so gentle and so tough, so tenacious and sensitive, as to deserve a full essay on the role and responsibility of a creative government servant as he practiced it at his job. Yet with all the conferences, journeys, trips, meetings and private leadership gatherings in Southern cities that the Department could arrange, before Birmingham it remained manacled. The statute books bore no law which the Department could enforce; and without a base in law, its lawyers, marshals and even Federal troops were powerless. "We were like a bunch of firemen," one Justice official was quoted as saying, "trying to put out a big fire at the same time we were trying to set up a permanent code of safety regulations to abolish fire."

The Birmingham uprisings lifted the strain to an emergency level. The Department of Justice's manpower resources in the South were stretched to vanishing: of the no more than 500 Federal marshals (usually middle-aged political appointees), only 150 could be considered young enough to stand the test of violence; the Department of Justice might call on the Border Patrol, but that would strip the borders; or the President might bring to its aid the Department of Defense and call on units of the three divisions of regular Army troops in striking range (82nd and 11th Airborne Divisions and the Second Infantry Division), as was finally done: but that was to invoke force when reason was necessary. Civil unrest in the South had entered a new dimension. "We felt this was an experience like . . . like a galaxy which accelerates when it explodes," is the way Burke Marshall described it. It could not be allowed to explode further.

Yet the explosion itself released the energy of conscience and the attention of public opinion hitherto lacking. On such a crest of feeling, a new law, however it stretched the Constitution, might be possible. Marshall and two Justice aides, Louis F. Oberdorfer and the Director of Public Information, Edwin Guthman, flew back to Washington on May 15th after the final settling of the Birmingham dispute to confer with Attorney General Robert F. Kennedy. He invited them to accompany him on a plane trip to Asheville, North Carolina, so they might talk undisturbed on a small Air Force jet. When they returned to Washington, the talks continued, almost unbroken, for the two days of the following weekend as the Attorney General invited in other concerned officials. The talk centered on the questions: What could a law do? What should it do? What should it stress? Schools? Public accommodations? Voting rights?

The Attorney General brought the result of their discussion to his brother, the President, the day of their return. The President wanted legislation too—he needed the authority of law to pacify his country. His staff—Sorensen, O'Brien, O'Donnell—joined the discussions. Such a bill was inescapable, the President felt; it must come, and he should move it—and better that he do it now, in 1963, than in 1964.

Thus on the evening of June 11th, in one of his most eloquent broadcasts to the nation, John F. Kennedy said:

> This nation was founded by men of many nations and backgrounds. It was founded on the principle that all men are created equal, and that the rights of every man are diminished when the rights of one man are threatened. . . .
>
> It ought to be possible . . . for every American to enjoy the privileges of being American without regard to his race or color. . . . Every American ought to have the right to be treated as he would wish to be treated, as one would wish his children to be treated. But this is not the case. . . . It is better to settle these matters in the courts than on the streets, and new laws are needed at every level. . . .
>
> The old code of equity law under which we live commands for every wrong a remedy, but in too many parts of the country wrongs are inflicted on Negro citizens and there are no remedies at law.
>
> Unless the Congress acts, their only remedy is in the street.

Eight days later, on June 19th, 1963, he placed before the Congress of the United States the text of his Civil Rights Bill to give Negro Americans a remedy for their grievance in law—and asked the Congress to act.

For some men, lawmaking is the most fascinating process of government. New laws call forth the subtlest trickeries, the most brutal pressures, the most corrupt practices of American democracy—from the most bumbling city council wrestling with zoning ordinances, through state legislatures wrestling with highway and labor legislation, to the Senate of the United States, which deals both with follies and with fate. Once a bill is engrossed in law and signed by the Executive, it is an act of government, enforceable by the courts, a leverage for lawyers and petitioners, to be supported by police, marshals and, if necessary, by armed might of state. To invoke this armed might of the law, or to forestall it, men and pressures from all over the country gather in Washington to participate in the talk before the words become concrete in statute. This right of lobbying, so frequently denounced by amateurs, is one of the essential components of American freedom, so deep and fundamental that it is enshrined in the Constitution itself: "Congress shall make no law . . . abridging . . . . the right of the people peaceably to assemble, and to petition the Government for a redress of grievances."

Every proposal of major legislation sets up a tumult of its own, and when it begins, a special climate wraps itself about each Congressman and Senator: pounded by mail and telegrams, assailed by visitors and telephone calls, buttonholed in corridors, at cocktail parties and conclaves, wined and banqueted by the fashionable, seduced occasionally by pleasures of the flesh and sometimes by hard money, pressed by conscience, teased by the desire to avenge personal affronts, hammered at by Executive leadership which wants its will done, worried about the home folks as well as about the greater common good, committed by obligations to other legislators who in the past have helped him with his bills, in debt to campaign contributors whose donations he accepted in good faith, puzzled as well as enlightened by the cascade of testimony before committees, uncertain and usually as ignorant as the rest of us about the complicated stipulations of texts he must vote on, in the hands of expert staffs for analysis of the texts yet pursued by zealots who challenge the analysis with their own experts—in this vast uncertainty, while editors denounce him for delay, the lawmaker wisely shrinks from swift action and, with native instinct, takes refuge in evasion while he listens and mulls words and entertains petitioners, waiting for time and pressure to clarify his thinking and the collective thinking of his colleagues.

In what, to most citizens, seems an agonizingly slow process, major bills are slowly shaped, slowly gutted or slowly strengthened.

In the Civil Rights Bill of 1964, almost eight months of discussion in the House of Representatives had shaped and passed by February 10th, 1964, a bill in many respects wiser and stronger than the original bill sent to Congress by the White House. This bill took the Federal Government further inside the private lives and customs of individual citizens than any Federal legislation in American history. It concerned itself first with voting—guaranteeing that Negroes might vote in every state on the same qualifications as whites; it prohibited discrimination against Negroes in places of public accommodation—hotels, motels, restaurants, with their restrooms and barbershops, gasoline stations and places of amusement. It opened all public tax-supported facilities everywhere—state, county or municipal—to Negro citizens as well as white. It permitted the Attorney General to open suit, on his initiative, to desegregate schools anywhere in the country. It established an Equal Employment Opportunity Commission on a Federal level to see that neither unions nor employers discriminated against Negroes in jobs because of their color. It permitted Federal aid to be cut off from any community discriminating, because of race, against citizens who might share that aid. It established a Community Relations Service to negotiate between racial communities where necessary.

Every one of these proposals presented to the Senate on February 17th, 1964, was anathema maranatha to the Old Guard of Southern Senators who, generation after generation, had here held unbroken the front their grandfathers had lost on the battlefields of the Civil War. Were it not for the titanic importance of the substance of the bill itself, one could say that their defeat in this last stand was the major historic event of the spring of 1964—for their hour had come.

One could sense the tolling of history almost anywhere if one prowled beyond the floor of the debate. For example: the gathering each morning at eleven in the office of the Senate Majority Whip, where there had been organized the greatest Congressional revolt since George W. Norris, in 1910, chained "Uncle Joe" Cannon to majority will. Once a citadel of the Senate's old inner Southern Club, the Majority Whip's office was now the office of Hubert Horatio Humphrey, the first Northern liberal to master the machinery and politics of the inner Senate. It held, of course, the furnishings of that man's puckish sense of humor—a giant-size red panic button, the size of a dinner plate, on the door leading to his lavatory; a name plate on his desk styling him "Chief Leading Feather" (a gift of Minnesota's Chippewa Indians); a red billiard ball (not an "8-ball" but a 7½-ball); and, on a wall, a deerhead trophy of a hunting visit to Lyndon B. Johnson's ranch. But it was engaged in the most serious business: the forging of a controlling bi-partisan alliance in domestic affairs between Republicans and Democrats against the South.

Here each morning gathered the steering committee of the Civil Rights Bill; this was the command staff, where for the Republicans would sit Kuchel, Javits, Keating or Case and, for the Democrats, Humphrey, Clark, Hart or Magnuson. If the Senators could not be present, they were represented by their executive aides, the field-officer corps of the Senate. With them would sit the architects of the bill, the lawyers of the Department of Justice, Marshall or Katzenbach of Robert Kennedy's staff, and usually Charles Ferris of the Democratic Policy Committee. Present, normally, would be two members of Humphrey's personal staff, John Stewart and Robert Jensen, liaison officers with the press, the churchmen, the civil-rights groups. Here, each morning, strategy would be reviewed—psychology of country and floor, response to the growing flood of mail, tactic of filibuster, review of floor assignments and management.

The steering committee was in no hurry. The sole Southern strategy was the rigid and exhausting one of filibuster. Sixty-seven votes would be needed to kill the filibuster, and of these the Humphrey-Kuchel team knew it had 59 or 60 committed by conscience or politics. Lyndon Johnson could always add three or four more by applying the extra last bit of pressure. Yet the key was still Senator Everett Dirksen of Illinois and the four or five Republicans of good conscience he might sway once his own mind was made up. The strategy of the civil-rights Senators was to play the game slowly—to exhaust the formal Southern opposition on the floor, day by day, until gradually the nation might become aware of what was happening and understand the issue; until the unions might educate their members; until the churches might arouse conscience; until the public might grow exasperated by the delay in the nation's business and demand action.

Moreover, their discussions were almost without rancor, even as they discussed the cruelties to be inflicted on old and respected colleagues by the necessity of battle, even as they descended to minutiae like not letting old So-and-So off the floor even to relieve himself. It was almost as if they were in sympathy with the men they must destroy—as if they had heard the comment, "Don't cheer, boys, the poor devils are dying." Hubert Humphrey, normally a zestful gladiator, expressed it best in the first week of debate: "This is no longer a battle of the heart for them—they simply have to die in the trenches; that's what they were sent here for. They're old and they haven't any recruits. They know it—one of them said to me, 'You simply have to overwhelm us.' And so we have to beat them to a pulp. No one can make peace. They have to be destroyed."

The faces of the irreducibles also told what was happening. They numbered as certain, now, no more than 18 or 19—they could no longer count on friendly states like Oklahoma (Monroney and Edmondson) or West Virginia (Randolph) or even Texas (Yarborough). And they were old. Allen Ellender of Louisiana (aged seventy-three) was so tired and faltering that after his first three-hour speech he must take to bed the next day; Willis Robertson of Virginia (seventy-six) could talk for an hour and a half and then was worn out. Neither Byrd of Virginia (seventy-six) nor Stennis of Mississippi (sixty-two) nor Ervin of North Carolina (sixty-seven) nor even Eastland of Mississippi (fifty-nine) possessed any longer the vital energy, lung power and stamina to hold the floor and delay matters as the great, bellowing Southern champions had in times past when a speech might run for six, eight or ten hours. And in the younger, more vigorous Southerners who would vote on their side, the spirit of combat had flagged—not out of lack of vitality but out of lack of conviction. Fulbright of Arkansas, compelled by his electorate to vote against the bill, saw with far vision what this bill meant to America's position in the outer world. Talmadge of Georgia (fifty) and Long of Louisiana (forty-five), bearers of two of the most famous racist names in the South, were young enough to see that even their world was changing and that, sooner or later, they too would have to seek Negro votes as well as white—or lose their seats.

And their chieftain, their floor captain, had lost spirit—this was the armature of their despair. For Richard Russell of Georgia, the greatest of the domestic leaders of the South, was trapped. Tactically, he led divided troops. If he worked out a compromise with Dirksen to restrain the bill by limiting amendments, he might gather four, five, perhaps six wavering Republican Senators to him. But, on the other hand, if he agreed to any compromise, he would be attacked as a sell-out leader by such Southern firebrands as Strom Thurmond of South Carolina. Which left Dirksen open to the cultivation of Hubert Humphrey's almost daily visits as Humphrey sought, by moderation, to bring the Republican leader over. ("We created a hero niche for Dirksen," said one civil-rights tactician, "and we let him walk into it.") Tragically, too, Russell was caught emotionally. He was devoted to the new President, Lyndon B. Johnson, whom he had tutored and shaped in the President's early Senate career. Yet Russell, leader of the South, could see the problem of this man, the first Southern President since Reconstruction. Lyndon Johnson was now President of all the people, forced to act for the will of the majority. Russell could not attack Johnson personally, so he must fight against the President's bill with an impersonal hatred—which crippled him.

Like aging prize-fighters, short of wind and stiffening of muscle, the Southerners were left with no resource but cunning; and cunning told them to delay, day by day, hour by hour, until somewhere, somehow, there might be a turning of national sentiment. It was the strategy, said someone, of "punt and pray."

The Southern hope that Dirksen might become their ally on the floor slowly faded. The debate, which started on Easter Monday, a year after Martin Luther King's Easter arrest in 1963, brought no help from the Illinois Republican. By the end of April the Republican had lined up with the civil-rights advocates. The national sentiment hardened—and then, for the first time in decades, the churchmen began to enter politics.

Slowly at first, then with deeper and deeper commitment, Protestant ministers, Catholic priests, Jewish rabbis began to be aroused; and then, when aroused, began to mobilize the forces which every politician knows to be irresistible. (This involvement of the Religious Establishment was to grow and swell until in 1965, with Martin Luther King's Selma march, the church had entered American politics more decisively than at any time since the Civil War. But even in 1964, in their first tentative entry, the churchmen caused politicians to tremble.) On April 28th the National Interreligious Convocation on Civil Rights assembled in Washington and filled the gymnasium of Georgetown University; then its members diffused through the capital in the next few days, to offices of Senators, pleading personally and directly for the bill with a prestige and authority that few men on the Hill dared ignore.

The Southerners continued to hope: George Wallace's campaign in the spring primaries might uncover a hidden vein of Northern sympathy; or the Negro leaders might overreach themselves; or violence might spurt from some Negro ghetto that would convince other Americans that this bill was more than unworkable, that it was dangerous. But they waited in vain.[5]

[5] Final passage of the bill eventually came on June 19th, at 7:49 P.M., by a vote of 73 to 27, all Senators present and accounted for, a rarity in modern politics. The bill passed on a turbulent Friday. All through that week Hubert Humphrey had striven in Washington, concealing private sorrow and agony. His twenty-year-old son Robert was to be operated on that week in Minnesota for cancer of the lymph glands. Urgently required on the floor of the Senate to captain the bill, he had longed to be at the bedside of the boy. The doctors operated in Minnesota without his presence, removed a malignant lymph gland and then, just before the final vote in Washington, reported to him that the prognosis was excellent—that Robert would recover and thrive. After the vote the leaders gathered in Humphrey's chambers for a moment of relaxation and a celebratory nip; Magnuson observed, "I wonder how many Ph.D. theses are going to be written on what we did tonight." After that a number of the champions in the fight went out to dinner and, while dining, learned that Senators Edward Kennedy and Birch Bayh, who had lingered late in Washington to cast yea votes and had then flown to the Massachusetts Democratic convention in West Springfield, had cracked up in a small plane. They were told that Kennedy was fatally injured (which turned out to be untrue) and they quickly broke up. For almost all concerned, the Civil Rights Bill ended in a week of mingled triumph, hope and sadness.

* * *

A visitor attempting to test this Southern strategy or the purposes of the bill by measuring them against event and sentiment in the nation could not help but be confused.

The morality and need of the bill were so clearly inescapable that no man of good conscience or good sense could oppose it. But what would it do? What would it solve? How much could an Act of Congress do to make the two races of America accept each other as friendly citizens? Whose demands did it satisfy? What other demands lay behind Martin Luther King's simple insistence on dignity and equality of opportunity? And what were all the communities of America, both North and South, ready to concede in compliance with the new law?

Away from Congress, nothing was clear.

One could leave Washington in mid-debate, as this reporter did, and be in the State of Mississippi within five hours. And events wore an entirely different countenance there.

Mississippi seems almost too small a state to torment the conscience of the nation so deeply. Two little communities live there, entirely separated, hating and fearing each other in a condition of total lawlessness and immorality: whites (1,257,000) outnumber Negroes (916,000) by 58 percent to 42 percent. The Negroes are fewer than the Negroes of New York City alone, the whites fewer than the population of New York's Bronx. But for three centuries they have had only animal relations with each other, and all politics, all decision, is magnetized by the primordial fact of race hatred. It is an entirely closed society with an orthodoxy of racism shared by most of its newspapers, its politicians, most of its churches, most of its organs of public opinion; and its clinical immorality has best been described by its most famous son, William Faulkner.[6]

To say that Mississippi is backward is to understate the case. Statistically, of course, Mississippi stands at the bottom of the list in almost every measure of progress: in poverty, in income per family, in teachers' salaries, in expenditure per child in school. It is more realistic to describe Mississippi as crippled than as backward—crippled by a fear which grows worse as the state struggles against fear. The white community is brutal to its Negroes, brutal beyond its own understanding of what this brutality brings. And the more this brutality hardens, the more the Negro flees. Forty-five years ago Mississippi was one of only two states in the Union (the other being South Carolina) with a Negro majority. (Of

[6] For the best political study of Mississippi, one should read Professor James W. Silver's excellent *Mississippi: The Closed Society* (Harcourt, Brace & World, 1964).

Mississippi's 82 counties, 29 still had a Negro majority in 1960.) But the condition of Mississippi life expelled from the state the best, the most energetic, the most able, the most responsible of the Negroes—the seekers, the yearners, the men of hope. They have left behind, by their own out-selection, the most ignorant, most uneducated, most close-to-animal Negro community of the South. White Mississippi has fashioned Black Mississippi by its own hate into what it fears most. Moreover, the general condition of Mississippi life, for white and black alike, is pre-modern: Mississippi is still one of the two most rural states in the nation, where 62 percent of the population lives on farms or in villages. Here, two kinds of human animals live, black and white; the whites have the guns and the machinery of government, and in their semi-beast relationship the white man is the hunter animal, the Negro the prey. One must grasp firmly in mind this root fact: that there is no law to govern the relations between the races in Mississippi. The state is an illegal society, flouting and mocking Federal laws whenever it so chooses; and for about 40 percent of its citizens—the blacks—there is no protection of court, police, law or mercy. From January 1st of 1964 to August of 1964, thirty Negroes were murdered—assailants unknown, but believed to be white.[7]

Jackson, the capital, is one of the more attractive new cities of the South—very new (its population had grown from 62,000 in 1940 to 144,000 in 1960), overwhelmingly middle-class (its white-collar suburbs, with their green lawns and scarlet flowers, are as pleasant as any in southern California) and spankingly clean. And as one enters the city, one passes the small but excellent veterans' memorial, on whose wall is a huge inscription: THOSE WHO SEEK PEACE SHALL HAVE IT.[8]

The fundamental problem of Mississippi is that the Mississippians themselves do not know how to seek peace.

Two men remain in my memory of my spring visit to Mississippi. For it was in them that I first saw and became aware of the new and terrifying rebirth, in our time and in America, of the ancient debate over

---

[7] The judgment above, like those on several pages following, were made in 1964. They may be wrong for 1965—for there begins to be a faint glimmer of light at the end of the tunnel. Mississippi, resisting still in its heart, has moved to comply with Federal law. It has obeyed court orders to admit Negro children to public schools in Biloxi, Jackson and Carthage. It has also heard its Governor say (of voting rights), "It doesn't make sense to turn down a person [Negro] with an M.A. degree when he attempts to register and then register someone who has not been to school. We don't have a leg to stand on."

[8] On Burke Marshall's office in Washington hung another sign, more realistic: BLESSED ARE THE PEACEMAKERS FOR THEY SHALL CATCH HELL FROM BOTH SIDES.

man's relationship to the law, of the individual's responsibility at once for his conscience and for the order that makes progress possible. It was there, in Jackson, that I heard the first sound of what Johnson and Goldwater would contend about in the fall of 1964 and what America will be contending with in the turbulent years ahead—and from two most unlikely men.

The first was Mississippi's Governor Paul B. Johnson: a thin man, slight of build, with a lean face, his countenance crowned by a high Lenin-type bald dome fringed with neat gray hair, and by heavy, bushy, black eyebrows over ice-blue eyes. One realized, with a start, that this was no Northern cartoon of a Mississippi Governor; this was a man of civilization and dignity whose deep, serious voice spoke not cornpone but a cultured English—and spoke at once in fear, perplexity and wistfulness. In his plight one could see half the tragedy of his state.

Johnson had, for a Southerner, a liberal early record: support of Harry Truman against the states'-righters in 1948; support of Adlai Stevenson in 1952; an affectionate reverence for Franklin D. Roosevelt, dating to the days of childhood when his father, a young Congressman (who later became Governor of Mississippi too), had been friends with the then-Assistant Secretary of the Navy and the Johnson children had known the Roosevelt children. Yet, as a liberal, he had recognized that his career in state politics was hopeless; thus, after several unsuccessful tries for the governorship, he had campaigned in 1963 on a platform of segregation and race hatred so inflexibly extreme as to satisfy the most violent white segregationists.

Now, as Governor, Paul Johnson was trapped; and as he attempted to explain himself out of the trap, it grew tighter. Wistfully he hoped the North would understand—would try to understand Mississippi and his position there. He was committed by campaign promise to total inflexibility. He could yield nowhere: he could not admit little Negro children to white schools; he could not admit even one or two Negro students to the university. He wished the President of the United States would understand his position—but he could offer no yield, no give, no tiniest token of compromise in return.

And he was afraid: this was most important. He had only 250 highway patrolmen to maintain peace and order in the state, and was trying to expand their number to 475, with full police powers, to make ready for the summer invasion of civil-rights workers from the North. And he was afraid not so much of the civil-rights workers as of his own white extremists: there was the Ku Klux Klan, already burning crosses from end to end of the state; along the Louisiana border, where matters were worst, there was the operation of the Society for the

Preservation of the White Race; local law officials and sheriffs were unreliable; in context, the White Citizens' Council (which the state officially supported) were moderates. The Klan he recognized as dangerous: "People poor in words who can't express themselves, who don't say 'Let's boycott So-and-So,' but rather, 'So-and-So needs a beating,' or 'So-and-So needs to be killed.' " He himself was dedicated to law and order; his problem for the summer was to preserve this law and order against the retaliation that the Northern civil-rights invasion might set off. (I would think that Governor Johnson probably regards the killing of *only* three civil-rights workers last summer as a triumph of law and order—rather than a disaster and a flame point of the future.) Yet the North would *not* understand—would not understand that the Negroes were making progress (his father, as Governor, had seen to it that they got their first free textbooks thirty years before), that their "carnal wishes and coarse desires" could not be changed overnight.

He had tried to communicate with the Northerners. (He had actually, as soon as he became Governor, journeyed to New York to talk with the heads of the great broadcasting networks to see if somehow the national image of Mississippi could be changed.) He mourned the breakdown of communications between North and South. Yet when one asked him whether he, as Governor of the whites, could communicate with the other community in Mississippi—the blacks—he insisted that he could not. He could not receive Roy Wilkins, the revered head of the NAACP. He could not, indeed, as Governor, talk to any Negro officially on any matter. He could not even communicate with Robert P. Moses, from whose headquarters two miles from where we were talking was being organized that adventure in Christian heroism which was to wake the nation to the squalor of Mississippi as effectively as Martin Luther King's demonstrations had roused the nation to the brutality of Birmingham a year before.

Robert Moses was the second personality to press himself on lasting memory in Mississippi. I found him by driving through the underpass of the Illinois Central Railroad, out of white Jackson, and in a few minutes I was in black Jackson—the shabby houses, the tin roofs, the littered gutters, the peeling paint that tell in the Deep South where Negroes live. Just this side of the forlorn Negro cemetery, overgrown with weeds and uncut grass, the headstones tipped, the pillars leaning and mourning as they bent toward each other over the gate, was the Streamline store block, with the Streamline Billiards and the Streamline Bar and a store without title whose number was 1017 Lynch Street, over whose windows were stuck black-and-green stickers saying, ONE MAN—ONE VOTE.

This was COFO Headquarters—it did not need a sign, for with a shock one realized that the tall white boy lazily tossing a ball so naturally was tossing it to a tall Negro boy—and that they were friends! One needed to be in white Mississippi only a week to be electrified by the sight of a white and a Negro who were companions. Inside, where the mimeographs ran and the pamphlets were stacked, blonde white girls were talking with social ease to Negro youngsters of their own age. Here was headquarters of one of the most perilous and inspiring mobilizations of the year—the summer "invasion" of Mississippi by the civil-rights movement of the North.

Bob Moses was field marshal of the invasion, the commander of troops. Now thirty years old, Moses was as quiet, as shy, as gentle, yet as firm, as Paul B. Johnson was precise, forthright, yet perplexed. A dark, diffident Negro with round eyes and horn-rimmed glasses, slow of speech and soft of tone, he had already in his years in Mississippi endured suffering that had become legend. Born in New York, in a Harlem flat, he had progressed by mind and will and mathematical talent through New York schools to a B.A. at Hamilton College and an M.A. at Stanford. A sweet and luminous person, he had been hired as mathematics teacher in one of New York's better white private schools and then given it up because in conscience he could not live easily in New York while Negroes were in torment in the South. He had, by 1964, been in Mississippi for three years, and the summer invasion of that year was to be the crownpiece of his mission.

Except that Moses did not like the word "invasion." It was, he said, an "educational summer project." The Mississippi State Sovereignty Commission (officially supported by the state legislature) sends missions north to various towns from Wyoming and Wisconsin to Maine to preach segregation, and had sent some 180 such missions to speak in the North in the previous two years. Moses was inviting young Northern white and Negro students to come south to Mississippi to preach another mission, to educate. The laws of the United States said Negroes could vote. Mississippi said no. The thousand students he was gathering from Northern campuses would teach Negroes to read and write, teach them where to go to register, teach them how to fill out the forms. And then, if Mississippi still said no, the Negroes would vote in their own Freedom Democratic Party and send their delegates to the Democratic National Convention—and let that convention decide who would be seated. (See Chapter Nine).

For Moses, the situation in Mississippi started with the fact of illegality. If there was no law—what law bound him? Except that all this was said not in defiance or in flame. It was spoken so softly, so gently, in such a Christian manner, that one quivered in listening. This was a revolutionary of the confessional type, the most difficult for government to handle. *This I will be,* says the confessional revolutionary, *because my conscience tells me to be so; and thus I will die because I must make witness with my conscience.* I have met many revolutionaries in years of reporting—and the most troubling are the confessional revolutionaries, as the Romans discovered when the Christian martyrs bowed their heads before the lions in the arena. Moses had been beaten bloody in his months in Mississippi; he had been fired on at night; he had already spent seventeen days in solitary confinement in a Mississippi jail, working at mathematical theorems in his head; he had braved the police dogs in demonstrations, though he has an almost psychopathic fear of dogs; five of his vote-claiming Negroes had been killed by the Klan in the first two months of the year; more would die that summer; he himself was an almost sure target for a sniper's bullet if he remained. Yet he would stay. One thousand Northern college youngsters were already preparing themselves to come for conscience to teach Mississippi Negroes how to vote —and he would lead them, his head bowed, his life ready to be given.

The Governor's mansion on Capitol Street and Moses' headquarters on Lynch Street were only two miles—but several civilizations— apart. The Governor claimed he defended law and order; Moses insisted there was no law in Misssisippi. They could not speak to each other; and there seemed nothing the civil-rights law could do to bridge the gap between them, until either Federal occupation or a revolution of conscience brought the state into the modern world.

This was Mississippi, which, with Alabama, had somehow drifted out of the Union of American civilization. But it was not the South—for as one moved either west or east from the delta to the newer South that was being reborn, one entered an America in ferment and a South where civilized men could hope. Here the frozen hates were breaking. In the rural counties great ice floes of hate still remained, but in the cities, swelling with new ways of life, the process of accommodation had begun, and now, not only furtively in parlors of good people but in public speech and public utterance, elected officials proclaimed the doctrine of opportunity and defended the civil-rights law of the government's desire.

Atlanta was the center of this new South. It was the center not only of SNCC (the Student Nonviolent Coordinating Committee), feeding from the cluster of six Negro colleges which make that city the only rival of New York in leadership of Negro thinking; the center not only of Martin Luther King's Southern Christian Leadership Conference; but the center, too, of white civilized thought. Atlanta's press—under the influence of its seminal mind, Ralph McGill, publisher of the Atlanta *Constitution*—had in two decades begun to alter the thinking of its leadership.

Atlanta's mayors were elected as much by Negro as by white voters, and dealt with their Negro voters responsibly; and Georgia's Governor, Carl Sanders, was thoroughly a man of the modern world.[9]

Ivan Allen had only two years before succeeded the legendary William B. Hartsfield as Mayor of Atlanta when I visited, and from Allen's conversation one could begin to pluck the threads of reason that had been so tangled in conversations in Mississippi. No Northern liberal in any sense, Allen accepted Negro participation in Atlanta's life (Negroes are 40 percent of its population) as a fact of life. Of 1,800 Negroes in city employ, only 17 had held jobs above the grade of laborer before Allen took office. Some 344 officials above the grade of laborer had been added since Allen had taken office on January 1st, 1962—of whom some 93 were Negro. As he explained, one could see that Atlanta was harvesting what Mississippi had not yet dreamed of sowing. Atlanta's Negroes, said Allen, were stable. The six great Negro colleges had provided them with an educated and self-respecting leadership; Atlanta Negroes had developed a business community as well as college communities, and Negroes had their own daily newspaper, their own radio, their own insurance companies, their own chain of drugstores—their own state senator, freely elected.

Where all this was leading, Allen was sure, was clear: he was for the Civil Rights Bill, unashamedly. Not only that. He had been one of the first to go to Washington and testify that he, as a Southern mayor, *needed* the law. His city, he had proudly said, had made progress: "We accepted the basic truth that the solution which we sought to achieve in every instance granted to our Negro citizens rights which white American citizens and businessmen had previously reserved to themselves as special privileges." For four years his city had bent every effort to open lunch counters, city facilities, theaters, hotels, restaurants, to all kinds of citizens. But Atlanta has "achieved only a measure of success. . . . Significant as is the voluntary elimination of discrimination in our leading restaurants, it affects so far only a small percentage of the hundreds of eating places in our city. . . ." Atlanta leadership had operated not through law but through voluntary civic groups. Now, if

[9] This Atlanta leadership was not able, however, to hold Georgia in the Democratic columns in the November elections. For the first time in its history Georgia went Republican, by a margin of 54.1 percent to 45.9 percent, as Southern rural red-neck counties swung heavily to Barry Goldwater to reward him for voting against the Civil Rights Bill. It is still too early to tell whether this was a final rear-guard protest against the progress Georgia has made in the past ten years—or whether it represented a new counterwave of white resistance to Negro opportunity, which had come faster in Georgia's large cities than anywhere else in the South except North Carolina. The elections of 1966 will give the first reading on the depth of the trend.

Congress failed to pass this law, this progress might be wiped out—"If the Congress should fail to clarify the issue at the present time, then by inference it would be saying that you could begin discrimination under the guise of private business. . . . Cities like Atlanta might slip backwards. Hotels and restaurants that have already taken this issue upon themselves and opened their doors might find it convenient to go back to the old status. . . . Gentlemen, if I had your problem, with the local experience I have had, I would pass a public-accommodation bill."

Thus Allen—who needed the law.

At the level of the governors, the progression became even clearer. Carl Sanders of Georgia and Terry Sanford of North Carolina could easily have taken rank among the dozen best governors of the Union—and, measured by their tasks and the pressures on them, their names must go, permanently, on any honor roll of Southern progress. For they governed not only over large and modern cities like Atlanta and Charlotte, but over backwoods Southern peasants, both black and white, among whom prejudice reached the intensity of Mississippi. They must urge both city and countryside forward, yet if they misjudged the balances, their states, too, might slip back into the moral chaos of the racist twins, Alabama and Mississippi.

They, too, needed the civil-rights law—but what perplexed them was its implementation. How would it be tested? Would it be pinned on Atlanta and Fayette County at the same time with the same severity? Would it be forced on Chapel Hill, North Carolina, with the same intensity as on Hertford County? They wanted it and needed it—but how fast and how quickly would the pace be forced?

It was Sanford who lifted the problem to the theoretical level. I reproduce his thinking only roughly, but it ran like this: His state, like the State of Georgia, he felt, had turned the corner into the new world. It could be held on course only by the strictest adherence to the law. In the name of the law and the Supreme Court, he could apply the laws of the United States because North Carolinians (and Georgians and Texans and Floridians and Virginians), deep in their hearts, *were* Americans and loved the country for which they had fought in so many wars. Its history was their history. If violence came, however—then North Carolina might go either way. If in Mississippi or Alabama the Klan butchered Negroes, then North Carolina and Georgia would stand by the law and would, in disgust, separate from the racists to uphold the law. But if a federal occupation came—or if, anywhere, the Negroes rose and flouted law and butchered whites—then all that had been done in North Carolina and Georgia would be ended and those states would stand by fellow whites everywhere, law or no law. And what seemed to bother

Sanford most was an abstruse theoretical question: How could he, as Governor of North Carolina, insist that his whites obey the laws of the United States if Negro leaders elsewhere insisted that *they* did not have to obey the law?

It was a conundrum. It did not seem to me that this was a conundrum I should impose on Bob Moses, who already had been forced by Mississippi to live beyond the law. His testing time would come later at the Democratic National Convention.

But it was a question forced sharply to attention a few days later when, coming home to New York on the eve of the opening of the World's Fair, I was forced to contemplate the morality and reasoning of an episode that never took place. For in New York, events insisted that I measure the new law against the condition of the Negro in the great cities of the North.

New York is the most tolerant of all the cities of the world—and still the greatest, even in its agony. Of New York's 8 million population, slightly more than one million are Negro. We shall examine the Negro's life in the big city later. But in New York approximately half of the city's budget is spent on this one-seventh Negro population; and, per-capita-per-community, each Negro child receives two and a half times as much budgetary support as a white child. White people leave; more Negroes come. The city must take care of them. But with its limited resources, the city can no longer enlarge what it gives to its Negroes; yet what Negroes have is so far below the threshold of their needs or desires that they writhe in discontent. Among Negro leaders a constant battle of tactics goes on as to how and where satisfaction for their discontent should be sought. All that the Civil Rights Bill promised Negroes in the South had been encoded in law in New York years and years before. And thus, on the eve of the World's Fair, the Negro revolution in New York began to divide in streams of contrary reasoning. Two weeks before the opening on April 22nd, 1964, younger Negroes had begun to break away from the established leadership of the New York Negro community. They would seize the occasion of the World's Fair to paralyze the city; they would "stall-in."

The stall-in was the first touch of the agony that later in the summer was to rowel New York's civic leadership. The stall-in reached for the vitals of the city. The younger Negro leaders urged that Negroes drive their cars to the approaches of the Fair and then stall-in—run out of gas, simulate a breakdown, simply stop. Anyone who has driven a car on the Long Island approaches of New York City knows what this means. In summer, traffic on the Long Island Expressway chokes on a single car with a flat tire; two cars stalled by overheating will clog traffic three miles behind. To plan a deliberate stall-in, with fifty cars clogging the express-ways, meant to reach for the nerve centers of the enormous, delicate megalopolis which is the most technically sensitive in the world. To reach for the Douglaston Interchange, the Van Wyck Interchange or the Triborough Bridge is to grab at the groin of a community of 10 million people. Moreover, even though the police might be able to handle a stall-in without blood, what would happen if, as the Negro radicals urged, protesters jerked the emergency-brake cords in the subways? Subway trains in New York run, at high speed, over thirty-five miles an hour—the emergency cord locks the wheels, and it is as if the hurtling train had hit a stone wall. Innocent people can be killed.

Two days before the stall-in, the Mayor of New York held a secret conference of four full hours with his top administrative leaders in the privacy of his mansion, the beautiful Gracie Mansion, which overlooks from a green lawn the confluence of the waters at Hell's Gate. A slow-moving and benign man who has done as much for civil rights as any elected official of the United States, Robert Wagner could examine the problem before him in these following elements:

First were the demands. The leaflets of the radicals demanded (a) that all construction in the city be halted *immediately* until the construc-tion industry was fully integrated; (b) that an immediate "rent strike" be called in all areas where Negroes lived together; (c) that the school sys-tem produce an integration timetable *immediately;* (d) creation of a public Police Review Board (which would remove control of police from elected authority) that would subject all police action to the veto and investigation of unnamed civil-rights groups. All this, *immediately*—or else chaos.

This was the second element: chaos. New York is at the mercy of anarchy. Its 237 miles of subway and elevated track; its 4,600,000 daily subway riders; its commuters; its ordinary citizens, who need light, elec-tric power, gas, water and roads, depend on the most complicated me-chanical cobweb in the world. If there were to be an enemy attack on the United States, the Triborough Bridge and Hell's Gate would rank among the first five targets; but one hundred anarchists can close it off by simple hell-raising. To govern New York City, it is essential—nay, primordial —that this nerve system be protected.

And, thirdly, the problem of Negro leadership. No government office in the country is more open to Negro leadership than New York's City Hall; indeed, the new chieftain of Tammany Hall, the chief dis-penser of patronage in that city, is a Negro (J. Raymond Jones). But in all the colloquy over the years between elected Negroes, self-appointed Negroes and TV-appointed Negro spokesmen, the city has had to select and decide for itself who speaks for Negroes and who presents their demands. And now, under the stall-in threat, the men gathered at Gracie Mansion had to decide who spoke for the Negroes—or how they might negotiate with wild men who threatened the city with catastrophe.

For the self-appointed leader of the stall-in was Isiah Brunson. Isiah Brunson operated from exactly the same kind of store-front headquarters (on Van Nostrand Avenue, Brooklyn) as Bob Moses (on Lynch Street, Jackson, Mississippi). But they were different. Bob Moses was a New York Negro who, in New York, had earned the right to a soft and easy existence as a respected mathematician and had chosen instead to go to Mississippi and face death. But Isiah Brunson was a twenty-two-year-old from South Carolina who had fled the terror of his state years before and chosen to come to New York, where he was safe, and then decided to paralyze New York.

I drove out to the Negro ghetto in Brooklyn to visit Brunson's headquarters the night before the scheduled stall-in with one of the ablest Negro writers in American journalism. Perhaps because the writer resented my joining him, he was almost explosive in his remarks. I, who had just come from Mississippi, was as fearful of him as I had been of the Mississippi Klansmen. He quivered in anticipatory ecstasy: tomorrow, when the stall-in took place, Martin Luther King would be through; so would Jim Farmer of CORE; so would Roy Wilkins; so would all the other Negro leaders who made the bridge of contact between white and black. Brunson, he said, would open the new world. The whites had broken the social contract between men; the United States owed the Negroes for three hundred years of unpaid labor as slaves; the reparations bill was going to be presented tomorrow; New York would be paralyzed. It was too bad that white people might be killed—but if that was the way it had to be, it had to be. "If a thousand innocent people have to be killed . . . if that's the way it has to be," he said, "well, that's the way it has to be." He respected me, he said ("I have respect for you, White")—but it was obvious that if I had to be shot, he would merely add regret to his respect.

Mr. Brunson treated the Negro writer with scant courtesy when we entered. A tall, shambling youth, rather handsome, dressed in a black leather jacket, Brunson dominated his command post. (NO TALKING TO REPORTERS, read one sign on the walls of his office, BY ORDER OF ISIAH. NO STATEMENT WILL BE GIVEN WITHOUT PERMISSION OF ISIAH, read another. NO OUTGOING TELEPHONE CALLS AFTER TEN A.M. MONDAY, BY ORDER OF ISIAH, read a third.) Tomorrow, and two years out of a South Carolina village, he could be king—if he could indeed paralyze New York.

All this had been as obvious to the men who gathered in conclave at Gracie Mansion with the Mayor as it was to this observer. The Mayor, all of New York City's nineteen Congressmen, both of New York's Senators were supporters of the Civil Rights Bill. But there was no contact between them and this new leader of the Negro revolution. Paul B. Johnson of Mississippi could not find it in his conscience to talk to the leaders of the Negro revolution in his state on any subject; but Robert Wagner of New York could find no established leader of the Negro community willing to sit with him and mediate between Isiah Brunson and the chaos he threatened. And if he negotiated with Isiah Brunson alone, on demands which were beyond his power to satisfy, who among the younger Negroes would repudiate Brunson the next day and declare that whatever agreement or pledge Brunson and the Mayor made was not binding on *him?* Somberly they decided they could not negotiate with Brunson—and prepared for the challenge.

Never has a finer police mobilization taken place than that staged for the opening of the World's Fair.

For this, finally, was the Mayor's decision: he must protect the city. Under the command of Michael J. Murphy, as fair and firm a police commissioner as any city boasts, tow trucks by the dozen lined up on the Triborough Bridge; three command posts were set up to protect the approaches and communications of the city; helicopters hovered overhead; and the city waited for the confrontation.

And then—nothing happened. Most of the august officials, state, Federal and municipal, decided to avoid the expected stall-in by flying into the Fair by helicopter. Jacob Javits, a leader of the civil-rights fight in Congress, alone bravely decided to travel by subway and confront the expected ring of Negroes as a champion of their battle, which they were undermining. He did indeed travel by subway, but no wall of flesh confronted him. For the Negro community of New York simply walked away from Isiah Brunson. Those who owned cars among the Negroes and civil-rights advocates refused, for the moment, to risk them by flouting the law; they were content to wait on the working out of their problem in Congress. It was a victory for the law, among the Negroes of New York as among the whites. How fragile this victory was would become apparent only three months later (see Chapter Eight, on the Harlem riots).

But for the moment, in the spring of 1964, the Negroes of the cities, like the Negroes of the South, were content to let the legislative process in Washington determine how far the law should go in guaranteeing open opportunity to all people. It was this that Lyndon Johnson favored; and this that Barry Goldwater, who felt that no laws could solve matters of conscience, opposed.

Under these circumstances, as Barry Goldwater voted against the unsettling law, the Republican Party faced both its past and its future and prepared to gather to choose a possible President in San Francisco.

CHAPTER SEVEN

BARRY GOLDWATER'S CONVENTION: COUP AT THE COW PALACE

S AN FRANCISCO is a magic city.

Whereas the new cities of the West face out on a cubist future from a rootless past, San Francisco magically marries both new and old as only Washington and Boston do elsewhere in the country. New bridges and newer freeways plow the approaches of the city; the sterile curtain walls of the new architecture rise on the slopes of the hills; but the city remains the same. The cable cars still clank up the old hills; in Chinatown there are Chinese who still speak Chinese; the restaurants still serve food cooked by people who care about cooking; and through the Golden Gate steam the deep-prowed vessels, trailing the scent of the Orient, to tie up at the Embarcadero.

At evening, as one sits in the glass eyries on the towers of Nob Hill watching the sun go down over the Pacific, watching the strands of light gradually weave golden shawls over the shoulders of the mountains, reverie takes one back to wagon trains and gold rush, and coolie labor and nabobs. There are the coastal ranges on the one side, and beyond them the Sierras, and then the Rockies and the journey of three centuries of Americans; on the other side is the ocean, the angry Pacific with its far rumbling of clash and menace. From the hills one looks down on an amphitheater of history past and history yet to come. What is real and what is unreal blend; past becomes present, and so, too, does future.

The Convention of the Republicans in San Francisco was a pivot in history; like San Francisco itself, it meshed the oldest and the newest— the most complete electronic and technological political controls with the oldest of American ideas. The sound of its dominant voice bore the wisdom of the grandfathers, rising from the history of an older America; the method of its masters was entirely new and of tomorrow. No one can yet define accurately what happened to the Republican Party at San Francisco—whether the forces that seized it were ephemeral or were to become permanently a majority that would alter and perhaps end the Republican Party as known through a century of American history. This will become clear only as the years throw perspective. But it is certain that few, if any, of the 1,308 delegates and 5,400 reporters from all over the world who gathered in San Francisco from July 13th to July 16th ever correctly sensed what we were living through as we lived through it.

In retrospect, it seems, we kept climbing, both physically and reportorially, in search of a jugular, and the stages of our journeying were quite neatly set out geographically:

At the foot of the great hills rose the new San Francisco Hilton, opened for occupancy only three weeks before the Convention—in which modern hotel architecture had reached a new efficiency in squeezing the smallest possible rooms in the most economical possible fashion about the newest plumbing and the most convenient floor-level garage space. Here was the press and communications headquarters where thousands of newspapermen and electronic journalists churned constantly for two weeks, waiting for the news conferences at which the champions of various candidates would volley back and forth at one another; where the central press room stacked its mounds of releases, statements, resolutions, calendars; where the mill of gossip, rumor, whisper and occasionally fact ground twenty-four hours a day.

The press and electronic media had come prepared to report Armageddon. Trained in the analysis of political struggle, cocked for a battle of giants, prepared to report the great forces here locking once again as they had locked so many times before in mortal combat, we could not do otherwise than report what was happening as combat, and anticipate the supreme moment. "There is something about a national convention," wrote H. L. Mencken, "that makes it as fascinating as a revival or a hanging. . . . One sits through long sessions wishing heartily that all the delegates and alternates were dead and in hell—and then suddenly there comes a show so gaudy and hilarious, so melodramatic and obscene, so unimaginably exhilarating and preposterous that one lives a gorgeous year in an hour."

In stalking the gorgeous hour, one first climbed three blocks from the Hilton Hotel to the gracious old St. Francis facing out on Union Square, set aside for the genro of the Republican Party—Eisenhower, Nixon, Lodge and others of yesterday's heroes. Here were the proceedings of the Platform Committee of the Convention—the first skirmishing ground where the opposing armies might clash. In 1952 the Taft-Eisenhower war had broken out at such proceedings; in 1960 the Rockefeller-Nixon forces had chosen these proceedings as their field of engagement.

One attended the proceedings in the Colonial Ballroom of the St. Francis, sitting on little red-cushioned gilt-frame chairs, and observed the hundred members of the Platform Committee sitting in similar chairs in a block of ten by ten, peering dutifully up at their Executive Committee and a great blue curtain on which a legend read FOR THE PEOPLE. On successive days Governor Nelson A. Rockefeller appeared with a legalistic summary of his mainstream philosophy; then Governor George Romney; then Governor William Warren Scranton of Pennsylvania with a precision-tooled fifteen-minute analysis of the complexity of American life and the role of the Republican Party. Scranton's was the best of the statements—a gemlike little speech, a model of clarity, lucidity and analysis, it fell, nonetheless, on the ears of the composed ladies and gentlemen as had the words of Mr. Rockefeller and Mr. Romney before him: with the near-soundless thud of a cracked egg dropping in sawdust.

On Friday, Senator Barry Goldwater appeared before the Platform Committee and spoke with that mixture of subdued wrath and moral fervor at which he reaches his best—and the Platform Committee roared with excitement and delight. Preceded into the room by the chanting of "We want Barry, we want Barry," followed out of the room by the same surge of sound, Goldwater might have read a page out of the Sears-Roebuck catalogue and received the same cheering of approval. The Platform Committee was his; there were to be no subcommittees, contrary to previous Republican tradition, to divide in majority and minority reports that might focus argument. It had been well arranged beforehand—the Platform Committee would divide in "panels" to report to the Executive Committee, already dominated by the Goldwater men, and the Executive Committee would, alone, write the platform of the Party.

Outside the St. Francis all of the mammoth apparatus of television had been assembled to show, visually, the opening of war. Giant white cranes had been wheeled in from Los Angeles; technicians by the score manned the roof-high mobile cameras on the crane platforms and guarded the ropes, entanglements and cables that fed the eyes of the networks which waited to show the nation the struggle over its future. But the harvest of drama was slim: inside, the stolid gentlemen and ladies, selected in advance from each state by the dominant Goldwater leaders and their allies to avoid drama, offered only a disciplined decorum. Outside, on one of the days when I counted, the score read: 130-

odd TV technicians waiting for dramatis personae to appear and perform; 24 pickets for the Congress of Racial Equality; 8 chubby teenagers celebrating the shaggy Beatles, urging: RINGO FOR PRESIDENT; 3 dour pickets parading with white-lettered red placards bearing messages that praised God and damned Earl Warren, Chief Justice of the Supreme Court. For Scranton: zero. For Goldwater: zero.

Since there was no real combat here, however much all of us tried to make it appear like combat, we proceeded by clanking cable car a few more blocks up Powell Street to Nob Hill, where two of the finest hotels in America face each other: the Fairmont and the Mark Hopkins. The Fairmont, owned by Benjamin Swig, is *the* Democratic hotel in San Francisco. As always, graciously elegant, the Fairmont was an island of tranquillity from which to observe proceedings across the street at the Mark Hopkins, the Republican hotel, owned by film cowboy Gene Autry. The coat of arms of the Mark Hopkins displays two golden horses, upreared and facing each other, and bears the motto: WHERE GOOD FRIENDS GET TOGETHER.

No motto could have been further from the truth during the Republican Convention of 1964—for by some bizarre misjudgment months and months before the convention, both Governor Scranton of Pennsylvania and Senator Barry Goldwater of Arizona had unwittingly been assigned to the same hotel. Now, layer upon layer, like a club sandwich of ice cream and pickles, the Goldwater and the Scranton forces were interwoven: on the 12th floor were the command elements of the Scranton combat team; directly above on the 14th and 15th floors (there is no 13th at the Mark) was the command layout of the Goldwater team; above that, on the 16th floor in the Royal Suite, the personal quarters of the Governor of Pennsylvania; and above that, again, on the 17th floor in the Presidential Suite, the personal quarters of the Senator from Arizona.

Goldwater and Scranton had once been good friends; Goldwater had been Scranton's senior officer in the Air Force Reserve unit on Capitol Hill; they corresponded normally, and with some warmth, by handwritten letters; they had gone overseas together on their Reserve duty in the NATO exercises of 1959. Even today Goldwater recollects in bitterness that Scranton had been his first choice for running mate as Vice-President all through the spring of 1964. But now they were enemies. Pinkerton agents guarded the Goldwater floors from both press and Scranton spies; electronic experts checked the interweaving floors for listening and bugging devices; in between the floors, on the staircases, hardy newspapermen staked out their posts, insisting to the police and

guards that staircases were neutral territory. Below, in the lobby of the
Mark, zealots, partisans, lobbyists and recruits churned in permanent congestion.

Here, if anywhere, there should be combat and clash—and the political setting was magnificent. Precisely on the green summit of Nob Hill
lies the Pacific Union Club, housed in the old brownstone mansion of
mining magnate Jim Flood (an underground passage is said to run
from the club to the more modest house nearby where Flood, an original
nabob of Nob Hill, kept his mistress), whose members cling nostalgically to the kind of California that Hiram Johnson destroyed. (Even the
great earthquake of 1906 failed to shake down their mansion, one of the
few buildings on the hill undestroyed.) Once when Democrat Harry Truman arrived in San Francisco and stayed at the Fairmont across the
street, the club drew its shades down so as not to see him on his
morning stroll. But most of those who live in the modern apartments that
flank the central enclosure of the Pacific Union Club are upper-class
executive families, college people and sophisticated Republicans of the
Scranton persuasion. Thus, from a pictorial point of view, Nob Hill, the
Pacific Union Club, the Fairmont and, above all, the Mark Hopkins,
with its constant comings and goings, should have been the finest setting
for drama in the Convention city.

The fact that even here no real drama could be observed should
have impressed on all of us the realization that the drama of this Convention lay not in episode but in perspective—in what had taken place
long before the delegates had assembled in this Convention city, and
what that portended. The long-range perspectives were amusing. Here
the forces of Eastern liberalism were captained by the grandsons of three
men who, seventy-five years before, had set their granite faces against
the future: grandsons, respectively, (Rockefeller) of Senate Majority
Leader Nelson Aldrich, the most truculent defender of turn-of-the-century big industry against any government regulation or control whatso-
ever; (Lodge) of Senator Henry Cabot Lodge, the arch-isolationist,
architect and progenitor of the exclusionist immigration acts; and
(Scranton) of William Walker Scranton, a burly industrial primitive who,
in his home town, had helped organize armed posses to shoot down, in
hot blood, workers trying to unionize. While their opponent, true spiritual
descendant of their own grandfathers, was the grandson of a Democrat,
an immigrant from Eastern Europe.

The shorter-range perspective, however, was more exciting than
amusing. For the Mark Hopkins, if one could get inside its guarded 16th
and 17th floors, was the stage for two separate exercises—one of futile

gallantry on the part of William Warren Scranton, and the other of futile conquest on the part of Barry Morris Goldwater.

The Scranton exercise was either folly—or absolute gallantry. One could understand it only by understanding the man. For Scranton, in a mute and profound way, loves the Republican Party; it is as much a part of his breeding and spirit as the Christmas gatherings at his home or the flag of the United States. He had entered the race belatedly, moved by his personal humiliation at Cleveland but propelled more profoundly by the Goldwater vote on the Civil Rights Bill. (See Chapter Five.)

Where or when hope fades in a man is as difficult to make precise in date as when hope is born. Better schooled in practical politics than any of the young men on his staff, Scranton had known from the outset of his mad gallop that his chances of heading off the raiders at the pass were less than one in ten. That slim chance had evaporated in the June adventure (see Chapter Five). To win, he would have had to bring over to his banners the old Taft conservatives of the Midwest, with whom, underneath his polish, he has a deep philosophical kinship. Thus, Illinois, Ohio, Indiana, Minnesota had been the key states in his forlorn strategy of June. In Indiana he had won surreptitious second-ballot strength, available, however, only if Goldwater were stopped on the first ballot. But he had failed, despite his concentration on the state of Lincoln, to shake any Illinois delegates in June. And by the time the convention opened, Ohio leadership had betrayed him. Governor James Rhodes of the Buckeye State, a courthouse politician whose promises and commitments are now held at general discount among high-level Republican politicians, had been among the leaders of the Stop Goldwater movement in Cleveland. Rhodes had not only pledged himself to Scranton to hold his delegation firm, but had made a similar covenant of faith with Ray Bliss, his partner in Ohio politics. But, breaking faith with both Scranton and Bliss, Rhodes announced from Ohio on July 9th that he was releasing his delegates— and then, arriving in San Francisco, proceeded to Scranton's apartment to explain why: Rather than seeing Goldwater as a disaster, as he had at Cleveland four weeks before, Rhodes was now convinced that Goldwater would be the Truman of 1964—an underdog victor. Goldwater's vote on civil rights which had horrified Scranton had not upset Rhodes; the "backlash" of white votes would carry Goldwater to victory in Ohio, predicted Governor Rhodes. (Goldwater was to lose Ohio, that sturdy Republican state, by over one million votes in November.) Rhodes urged Scranton to run for Vice-President as Goldwater's running mate.

When Rhodes had left late that night, Scranton knew the fight was over; and shortly after the Ohioan's departure he gathered his young men and several of his allies of the Lodge-Rockefeller wing. It was probably hopeless now, said Scranton. If any of them wanted to get off the Scranton bandwagon and go over, now was the time to do it; he would accept their departures with no reservations of affection or feeling. But as for him and for Mary—they were going right on with it, down to defeat on the convention floor if necessary, hopeless as it was. At which all in the room rose and applauded; no one would leave.

There was, thus, by the weekend no strategy for Scranton except the gallantry of hopelessness: to appeal to the delegates by going over their heads to the rank and file of the Republican Party; by rousing the nation to respond by mail and telephone calls and telegrams and thus exert pressure on the delegates already committed either by law or conviction to vote for the Arizona Senator.

Now gathered all the grave and eminent chieftains of the older Republican Party of the East to discuss this last strategy of desperation. Their unending councils of war all through Friday, Saturday and Sunday, futile as they may have been and at times comic, included some of the greatest names of the Party. Rockefeller, Javits and Keating of New York; Case of New Jersey; Lodge of Massachusetts; Scott of Pennsylvania—all gathered with Scranton, his wife and his young men to plot strategy. A confrontation area must be found, some vivid area of deep division on which they might force the Party on the Convention floor to reverse the Goldwater-dominated Platform Committee and thus make a first breach in the discipline that seemed so inflexible. All were willing to serve as lieutenants of the young Pennsylvania Governor or anyone else who would carry the banner ("I'm willing to knock myself out as John the Baptist," said Hugh Scott, "if only someone else will be Jesus Christ"). But what tactical area could they choose as confrontation ground?

On this the eminent Easterners disagreed. Some, like Scott, wanted to put the Republican Party on record against a national right-to-work law, thus forcing the Convention to choose between repudiating Goldwater and repudiating the vote of the workingman. Others felt that civil rights was the key issue. Still others, like Rockefeller, felt that a denunciation of "extremist groups" would finally force Goldwater to define his stand on the John Birch Society. And yet others, notably Javits, felt that the critical concern of the nation was war and peace, and that the Republican Party should reaffirm the national conviction that only the President, not subordinate military commanders, should be allowed to detonate nuclear weapons.

Scranton himself hovered between the nuclear issue and the civil-rights issue. Briefly, before the Convention opened and over the weekend, as all expected Eisenhower's arrival on Sunday, it was felt they might have Eisenhower's support on the nuclear resolution. Eighty-odd members of his administration had signed a denunciation of Goldwater —including seven former Cabinet members. If Eisenhower could lend his overwhelming military authority and Presidential aura to a denunciation of Major General Goldwater's strategies, it might be the miracle that could sway the floor. On Monday at 10:15 A.M., before the Convention opened, Tom Gates—Eisenhower's last Secretary of Defense, who had flown overnight from the East Coast to make a last plea to his old chief in San Francisco—joined Malcolm Moos in Eisenhower's suite and came away with an apparent approval of the text of the Scranton resolution on nuclear control. All were gay. But at lunch they heard that Eisenhower had repudiated the statement on which they felt they had his agreement; he had decided an ex-President should not discuss the secret technology of American national defense in public politics; and by late Monday there was little time for the frantic straightening out of the lines of communication in which the Eastern Republicans had snarled themselves in their desperation.

As the Eastern Republicans debated their strategy, each and every proposed resolution had been celebrated by the press as a new challenge. One of the drafters, with a foretaste of what they were doing to Goldwater, summed up their efforts thus: "What we were looking for was something that would put the nation and the rank and file of the Party on the alert to the fact that our leading candidate was impetuous, irresponsible and slightly stupid." But the capping effort to reach the rank and file over the weekend was an extrapolation of a simple suggestion by former National Chairman Meade Alcorn into Scranton's famous denunciatory letter. Alcorn had suggested that Scranton challenge Goldwater to open debate before the Convention itself. The strategy board had approved; Scranton, busy appearing at as many delegation caucuses as would give him a hearing, could not find time to write the challenge himself. He left instructions that the letter be composed on guidelines he had generalized, while he continued in the exhausting routine of beseeching delegates to turn back the inevitable. In Scranton's suite on the 16th floor, Scranton's young men took the Alcorn suggestion and encased it in rhetoric; a secretary signed the letter, and a messenger carried it to Barry Goldwater. Goldwater, stunned by the vehemence of the letter, examined the signature briefly and, knowing Scranton's handwriting from previous exchanges, declared immediately that the signature was spurious—and that the Pennsylvanian had probably never read it.

Goldwater was right. The young men of the Scranton staff on the floor below, fighting desperately to provoke the explosion that could tear the Convention apart, had composed—without their chief's perusal—a letter that was less a challenge to debate than an indictment, a summons to Goldwater to stand trial before the Convention delegates. It read:

> Your organization . . . feel they have bought, beaten and compromised enough delegate support to make the result a foregone conclusion. With open contempt for the dignity, integrity and common sense of the convention, your managers say in effect that *the delegates are little more than a flock of chickens whose necks will be wrung at will.* . . .
> You have too often casually prescribed nuclear war as a solution to a troubled world.
> You have too often allowed the radical extremists to use you.
> You have too often stood for irresponsibility in the serious question of racial holocaust.
> You have too often read Taft and Eisenhower and Lincoln out of the Republican Party. . . .
> *In short, Goldwaterism has come to stand for a whole crazy-quilt collection of absurd and dangerous positions that would be soundly repudiated by the American people in November.*

Goldwater, of course, refused debate and grandly returned the letter. Scranton later admitted he had never read the letter and disapproved of much of its language. The dedicated Goldwater delegates and the Midwestern conservatives reacted almost predictably to the description of themselves as chickens whose necks had been wrung. The Goldwater command had circulated the text of the letter through San Francisco on Monday before the Convention opened at 10:20 A.M., by which time it was obvious the letter had hardened the Goldwater Convention against Scranton, rather than divided it.

But the letter and the weekend seeking of confrontation ground had done something else, something more important. It had made the Republican Convention the stage for the destruction of the leading Republican candidate. What Rockefeller had begun in spring, Scranton finished in June and at the Convention: the painting for the American people of a half-crazed leader indifferent to the needs of American society at home and eager to plunge the nation into war abroad.

There was, from the Thursday before the Convention opened, no reality whatever to the Scranton hope of victory. There was only a historical perspective that required him to go down bloody on the floor, in defeat. History would have to record that the Republican Party had not

submitted docilely to this new leadership, but had resisted to the end—
so that from this resistance and defeat, others, later, might take heart and
resume the battle.

If one unreality wrapped itself about the Scranton leaders, a similar
yet more important unreality enveloped the Goldwater camp. Where the
Scranton leaders confused the applause and the press and the reading of
the polls with the will of the Republican Party, the Goldwater leadership
confused the applause and the praise of the San Francisco Convention
with the will of the nation.

It was only on Monday afternoon, after the Convention opened, as
one proceeded from the Mark Hopkins to the fourth and last of the stations on the symbolic peregrination of observation, that one could begin
to understand how the Goldwater leaders could have been so swayed to
misjudgment.

The Cow Palace of San Francisco, built by the early New Deal in
one of those social give-away programs so odious to Barry Goldwater,
sits on sixty-seven acres, a spacious but not overpowering arena, good in
acoustics, an auditorium with complete visibility, generous of access and
entry, perfect for conventions in all respects except for the clogged roads
and highways that lead from downtown San Francisco to its site six miles
away.

And here, in the plaza before the great Cow Palace, all the neat,
well-barbered men and prosperous women of the Goldwater delegate
army and gallery support could see the face of the enemy—as if stage-
managed to rouse their anger.

First the demonstrating civil-rightsers—girls with dank blond hair,
parading in dirty blue jeans; college boys in sweat shirts and Beatle hair-
cuts; shaggy and unkempt intellectuals; bearded Negro men and chanting
Negro women. They paraded in an endless circle, shouting, singing,
sometimes clapping their hands to the rhythm "Aunt Jemima—*clap clap*
—must go!" "Barry Goldwater—*clap clap*—must go!" "Jim Crow—
clap clap—must go!" "Uncle Tom—*clap clap*—must go!" DEFOLIATE
GOLDWATER, read one sign, and GOLDWATER FOR PRESIDENT—JEF-
FERSON DAVIS FOR VICE-PRESIDENT; VOTE FOR GOLDWATER—COURAGE,
INTEGRITY, BIGOTRY; WILL 100 MORE YEARS CHANGE WHITE HEARTS?;
GOLDWATER—'64, HOT WATER—'65, BREAD AND WATER—'66; and still
the chant: "Freedom—*clap clap*—Freedom—*clap clap*—Freedom—
clap clap."

Simultaneously, on another patch of the fore plaza paraded the
peace-niks. WOMEN FOR PEACE read their placards, and NO MORE WAR;
PEOPLE WHO LIVE IN GLASS HOUSES SHOULD OPPOSE MLF; NATO MEANS

NUCLEAR REARMAMENT; GERMANY HAS FORTY PERCENT CONTROL OF NATO; KEEP EIGHT NATO FINGERS OFF THE NUCLEAR BUTTON.

Between these two groups stood a large white bus, a silent demonstration in itself with a half-dozen yellow posters: JUSTICE FOR HOFFA; INVESTIGATE BOBBY KENNEDY.

Through all these demonstrations the well-dressed and well-mannered Goldwater delegates made their way. All these—the civil-rightsers, the peace-niks, the Hoffas, as well as the Kennedys—were what they had mobilized to oppose. But they had been sternly admonished to good behavior. Discipline urged them to restraint. Thus, they sat out the keynote address on Monday, as well as the first writhing effort of the Easterners to find (and fail to find) a loophole in the Convention rules, exactly as instructed.

But on Tuesday they began to roar. It was Eisenhower who stirred them first. The General's speechwriters had written a conciliatory, states-manlike address reviewing the entire history of the Republican Party, its mighty achievements, its struggle to defend liberty against centralization in Washington, the ravages on liberty in the previous three years by the Democrats in the White House. So far, so good. But with a personal desire to express more human yearnings, the General had inserted a few touches of his own: "Let us," he said as he wore to the end of his remarks, "particularly scorn the divisive efforts of those outside our family, including sensation-seeking columnists and commentators, because, my friends, I assure you that these are people who couldn't care less about the good of our party"—at which point the Convention exploded in applause, shouts, boos, catcalls, horns, klaxons and glory. Here at last was someone cutting a bit of raw flesh from the hitherto unnamed enemies, and the delegates could vent emotion. (On the floor, one delegate from North Dakota, jumping up and down, was heard yelling: "Down with Walter Lippmann! Down with Walter Lippmann!") But Eisenhower was not through. He had another penciled addition to his discourse: ". . . let us not be guilty of maudlin sympathy for the criminal who, roaming the streets with switchblade knife and illegal firearms seeking a helpless prey, suddenly becomes upon apprehension a poor, underprivileged person who counts upon the compassion of our society and the laxness or weaknesses of too many courts to forgive his offense." A second time a nerve was twinged: an ex-President of the United States was lifting to national discourse a matter of intimate concern to the delegates, creating there before them an issue which touched all fears, North and South. The Convention howled.

It subsided for the ninety-minute reading of the platform and then, after prime television time had thus been blotted out in the East, Nelson

Rockefeller was permitted to rise and defend the first of the three draft minority resolutions which the Easterners had prepared as a challenge to the text of the majority in the Platform Committee. The slight ripple of normal applause as he rose was almost immediately drowned by a wave of boos, and then a billow of howling, and then by explosion. "We want Barry! We want Barry!" the galleries shouted. Squeeze-horns and cowbells, toy trumpets and hooters began to clang. Somewhere in the auditorium someone began to pound a bass drum in rhythm with "We want Barry!" Chairman Morton with difficulty gaveled the auditorium to silence, and Rockefeller began. His five minutes were allotted to the topic of extremism—and as, with absolute zest in the first minute, he swung into his call, the audience exploded again. It was as if Rockefeller were poking with a long lance and prodding a den of hungry lions— they roared back at him. This was the face of the enemy—more important than columnists or switchblade hoodlums; this was the man who had savaged Barry from New Hampshire to California all through the spring. This was the man who called them kooks, and now, like kooks, they responded to prove his point. This reporter was sitting in the Goldwater galleries to savor the moment, but suddenly found two men peering over his shoulder, noting every word written in the notebook—and commenting angrily as they read. As Rockefeller progressed and the roars grew, his tone alternated between defiance and mockery; he smiled; the audience yelled and roared, and the bass drum thumped; and Rockefeller taunted them all. In a passion that he had rarely achieved in his entire spring primary campaign, he was reaching emotion—and delighting in it. And as he taunted them, they raged. Nor did they, apparently, know what they were raging at: the East; or New York; or Communists; or liberals. As this reporter uncomfortably left the Goldwater galleries for the safety of the floor, he noticed a tall, thin, blonde woman, her fists upraised and shaking, screaming at the top of her lungs: "You lousy lover, you lousy lover, you lousy lover!"

The floor itself was comparatively quiet, Goldwater discipline holding firm; and here one had the full contrast that plagued all reporters throughout the year—the contrast between the Goldwater movement and the Goldwater organization. There is not, and was not, anywhere in the entire high command, in the brains trust or in the organizational structure of the Goldwater campaign, anyone who remotely qualified for the title "kook." Nor was there evident any "kook" on the floor. But the "kooks" dominated the galleries, hating and screaming and reveling in their own frenzy.

As Rockefeller, enjoying the spectacle and combat, a lock of his full hair tumbling over his forehead, taunted them ("This is still a free country, ladies and gentlemen"), they yelled even louder. "These things," called Rockefeller above the tumult, "have no place in America. But I can personally testify to their existence," he went on, recalling his California primary campaign. "And so can countless others who have also experienced anonymous midnight and early morning telephone calls, unsigned threatening letters, smear and hate literature, strong-arm and goon tactics, bomb threats and bombings"—the audience halted him for long seconds here with their roaring—"infiltration and take-over of established political organizations by Communist and Nazi methods." They yelled back at him and strained in their seats as if trying to grab him and get him, and he shot back jauntily, "Some of you don't like to hear it, ladies and gentlemen, but it's the truth." And more booing and yelling and roaring. And as the TV cameras translated their wrath and fury to the national audience, they pressed on the viewers that indelible impression of savagery which no Goldwater leader or wordsmith could later erase.

It was impossible, for example, for TV to show the contrast between organization and movement. The gallery seethed—which was visible. What was invisible was the instant attempt of the Goldwater command to squelch the demonstration. Floor control of the convention, in anticipation of battle, had been given to Clifton White; and White, with Goldwater's enthusiastic participation, had meticulously studied the Kennedy communications planning of the 1960 Democratic Convention —and bettered it. In his green-and-white trailer behind the convention hall, White sat at a web of some sixteen walkie-talkies and thirty telephone lines. From the twin consoles of his telephone board he could reach seventeen phones strategically placed at the aisle seats of his floor managers; two more were "hot" lines to Goldwater's suite at the Mark Hopkins; two more to headquarters of the Southern delegates at the Jack Tar Hotel; two more to the main Goldwater switchboard. One white button (the "all-call" button) could reach every phone with its attendant monitor all through the convention hall and the city. An antenna, located jam-proof at the very pinnacle of the auditorium, reached the walkie-talkies, deploying men to any necessary point instantly. Higher-powered walkie-talkies of newest manufacture, delivered just before the convention, were on standby at the control wagon, able to communicate for twenty miles from base. (Goldwater, a devoted ham radio operator and fascinated by electronics, regretted later that if they had had just a few more weeks to prepare, they could have put a pocket telephone in the shirt pocket of every single delegate on the floor.)

The gallery and the emotions it showed were what the nation could see; what it could not see was Goldwater delegates and lieutenants leaving the floor and deploying through the galleries to hush, to control, to silence the anger.

Rockefeller had his five minutes; so did each of the other sponsors of minority resolutions; so did George Romney with a compromise set of resolutions. And each resolution as it came up was "debated," then gaveled or voted down. This Convention had been locked up before it started; now, almost effortlessly, the machinery built up through the spring months was showing the true nature of the Convention.

For this was a new thing in American conventions—not a meeting, not a clash, but a *coup d'état*. What was evident was the sight of demonstrators and the sound of galleries; but the troops and their captains had been moved into place months before.

Politics needs demonstrations. Demonstrations vent emotions. They satisfy the citizens' desire to participate, to have a share in their own fate, to feel themselves part of something larger and greater than themselves. The next day, Wednesday, as nominations were called, the Goldwater delegates and demonstrators were more classically managed. At 3:25 P.M., after one of the most tasteless nominating speeches ever made, Senator Everett Dirksen of Illinois nominated Barry Morris Goldwater, "the peddler's grandson," for President of the United States. Down from the rafters cascaded flakes of inch-square gold foil; up from the pit boomed the bass drum that beat in rhythm to "We want Barry"; from behind the rostrum a band blared "When the Saints Come Marching In." Onto the floor poured the demonstrators: the Californians in golden bibs, the Nevadans in red silk shirts, Texans carrying longhorn insignia. Gold balloons rose from the floor. Delegates in white hats with black ribbons, blue ribbons, gold ribbons—to distinguish themselves by states—squeezed into parade. The cowbells and police whistles and squeeze-horns joined the bass drum. Signs were brandished: BETTER BRINKSMANSHIP THAN CHICKENSHIP; PEGGY, INDIANA'S FAVORITE DAUGHTER; OSCEOLA, ARKANSAS, TIRED OF PUMP WATER WANTS GOLDWATER; LAND OF LINCOLN GOES GOLDWATER. And the old favorites: WE WANT BARRY; A CHOICE NOT AN ECHO; $AU + H_2O = 1964$. And the favorite blue-and-gold portrait of Goldwater himself, jut-jawed in profile, bravery and integrity facing the world; and the state banners, state flags and spoils of victory wrenched from the Easterners. (One burly Californian, in his golden bib, had broken off the staff which bore the legend MASSACHUSETTS, having wrested it from the devoutly anti-Goldwater Bay-Staters. He had lashed it to a new stave with his own belt, and as he jiggled in the serpentine on the floor his pants kept slipping down, so that he half danced and half wriggled, holding his combat trophy up with one hand and his falling pants with the other.)

Then, as suddenly as it had been called forth, the demonstration ended. With none of the pandemonium and pleading exhortation from the platform that normally accompany uncontrolled demonstrations, the Goldwater delegates returned to their seats, and the demonstrators to their galleries. Clif White had passed the word: Order. Order in the seats again, quiet orderliness and courtesy, to hear the other candidates nominated. And once again the contrast: decorum on the floor where Goldwater control could be exercised, but the exhalation of hate on the fringes where no man controlled. At the gateway to the Convention floor, as the Scranton demonstration was getting under way, a towering lug wearing a Goldwater button was yelling at two pretty teen-age girls trying to join the Scranton demonstration: "Get the hell out of here, you little sons of bitches."

One could savor the contrast best from Clif White's command post in his trailer truck just outside the convention hall, where this reporter proceeded shortly before the roll call on the nominees.

Antiseptically clean, its cream tile floors glistening under the overhead fluorescent lights, its beige curtains drawn against intruders, it had the same quiet hush, broken by flat, laconic interjections, as a chamber in surgery. At one end, behind a glass-frosted panel, sat White, alone before the console panels of the communications system, his blue bow tie sagging and his pale face even paler than usual. Before him along the wall were three booths separated by glass panels, and in each booth two men monitored the communications that ran to the floor captains of the regions they commanded. Some, like John Grenier (South) and Wayne Hood (Midwest), had plotted this hour through years of secret meetings and planning.

A handsome young lady named Pamela Reymer, in green dress and glowing white pearls, stood at a delegate chart hung on the wall and waited, grease pencil poised. White spoke sharply and softly, calling his team to attention for their last vote projection of the states in their charge before roll call. As White read the states, the young lady wrote figures in the projection column, adjacent to the column left blank for the official roll call. The adding machine clicked mechanically as figures were added, and the final total—seconds before the gavel fell on the TV screen—read 890; they had made their calls technically, quickly, in flat voices. Ed Failor said once, "Tell Pam to make it twenty for New Jersey," and they jeered at him. But then their eyes were on the television sets. What had been planned three years ago was now to become apparent to the nation from the calling of names one hundred and fifty feet away on the floor of the Cow Palace.

As the TV carried the voice "Alabama?" somebody in the trailer yelled "Yea Barry!" and was instantly hushed.

Tense and soundless, they watched: Alabama—20 for Goldwater in the column next to the prediction which read, similarly, 20. Alaska—a stutter; they had predicted 9 for Scranton, and Scranton had 8. Down the line: Arizona—16 predicted, 16 for Goldwater. Arkansas, on the nose. Through the Cs and Ds and on down the alphabet, each prediction true. Georgia paused to poll its own delegation. Silence in the trailer. Illinois: 56 for Goldwater, 2 for Rockefeller, none for Scranton. As predicted, but someone in the room muttered, "What a slap in the face!" No change from prediction all the way. At Kansas (Goldwater, 18; Scranton, 1; Romney, 1) someone asked in the silence, "How about that?" and White said, "Exactly two above the April projection." Maryland was up one above the last posting—a plus.

At Mississippi the count read 430 to 87.

At New Jersey—Goldwater, 20; Scranton, 20, and someone said: "You did it, you did it, my God, that's great!" To which came Ed Failor's response, "You didn't believe me," and a handclapping, then a hush again.

It was 561 to 122 before Ohio—and Ohio called 57 for Goldwater, 1 for Margaret Chase Smith. In the command trailer, a girl's squeal and a voice: "How beautiful!"

The adding machine was clacking, and White pressed a hot-line button to the Goldwater suite: "Just for your information, on this projection you'll make it on South Carolina." One could not hear Goldwater's reply.

The roll call was approaching South Carolina now—Oregon committed to Rockefeller, Pennsylvania sure for Scranton. It was 647 to 193 as South Carolina's vote was called. On the floor Wirt Yerger of Mississippi was floor agent of John Grenier's Southern command, including South Carolina, and Grenier said softly, "Go get 'em—Come on, Wirt!" A rebel yell yodeled in the trailer as the camera shifted to South Carolina's chairman, Drake Edens, and Eden's voice came off the floor: "Mr. Chairman, we are humbly grateful that we can do this for America. South Carolina casts sixteen votes for Senator Barry Goldwater." The TV screen flashed 663 votes for Goldwater, 8 more than needed for the nomination—and the trailer command suddenly went soft. It was as if tension had been released by the turning of a valve. "If that ain't a twist of fate," said one, "South Carolina nominated him in 1960."

They rose now, stretched, yawned, began to chatter. White hushed them. They watched the screen, then White interrupted and the voice of Denison Kitchel came over the squawk box from on high. Then came another voice, metallic, martial, dry, familiar to all of them: "Clif, you did a wonderful job, all of you fellows. I can't thank you enough. See you down here a little later." "Thank you, boss," said White. "O.K.," said Barry Goldwater, and the squawk silenced itself.

The TV screen continued to show the roll. It was, finally, 883 for Goldwater—7 less than the last posting.[1] A walkie-talkie reported that Scranton had just left his trailer and was entering the hall. They lounged against the wall, the room now full of smoke and stale air, to comment on Scranton's march down the aisle, his manner on the rostrum.

White watched with them this, the end of his three-year plan, then pressed another button. "All call," he said, and repeated, "All call. Close your switches. I want to remind you we're having a party for all the delegates and alternates at the Mark Hopkins, and the Southern region is having one at the Jack Tar for those who can't get to the Mark. Please check in your headsets and all radios. I will talk to you tomorrow. Congratulations. Out."

With that, White closed his switch and was, not knowing it, *Out* himself. He had misjudged his boss as much as his boss misjudged him.

Like every professional politician's, White's dreams had been enlivened by the hope that someday he might be National Chairman of his party. Though a National Chairman is fool's gold in terms of power, it glitters like real gold for many an artist of the political technology. Goldwater believes they had discussed the post several months before, and that he had then told White: no. White's account differs—he has no recollection of such a conversation. He felt that the candidate should be free to choose his own man, but that he, White, should be consulted for his pains. White says the matter was left open. However it was, the National Chairmanship was the reward White had sought for his labors in organizing the nomination.

But it was not to be. After the nomination Goldwater greeted White briefly in his suite at the Mark Hopkins for mutual congratulation. Then festivity swept them apart. They were not to see each other again for several weeks. Friday, at the conclusion of the Convention, riding down in an elevator in the Mark Hopkins, White overheard in a chance conversation between two men that Goldwater had chosen Dean Burch, not himself. No formal notice or personal message came to him. Friday evening, at the Fairmont, White's band of original plotters held a party of their own as farewell to him; several cried; some suggested there was still

[1] Erroneously, in their last posting, they had included 6 Goldwater votes from Oregon, promised to them as breakaways on the second ballot as soon as the delegates were free by Oregon law to vote individually. But for this technical error, the final prediction would have been only one off the mark.

time to dump Goldwater and go with a candidate of their own. White killed the insane idea.

In August, White was named national director of the Citizens for Goldwater-Miller, a job of amateur organization. But his heart was not in it. Like so many professional politicians, and blinded by his own technical virtuosity, he had misread reality: power in America goes where the voters put it, not to the men who organize the votes. Jim Farley and Carmine De Sapio might have told him so from their experience. But White's chieftain, Goldwater, had misread reality too. He had misread the nature of America in the modern world.

I have seen a privately made film which shows Goldwater in his paneled suite atop the Mark Hopkins at the precise moment of victory, as Clif White and the command team were watching from their trailer command post.

Behind Goldwater is a TV screen showing proceedings on the Convention floor, and Drake Edens of South Carolina is finishing: ". . . humbly grateful that we can do this for America. South Carolina casts sixteen votes for Senator Barry Goldwater." Goldwater, jacketless, a knee slung over the floral-covered armchair, sits immobile, his white shirt unbuttoned at the neck. He sits thus for seconds as the words bray from the screen, saying nothing, peering out directly before him, his fingers plucking at his chin, and behind in the window is the black fall of San Francisco, studded with lights far below.

Half a dozen men are in the room with him—Denison Kitchel, Bill Baroody, Tony Smith, Ed McCabe, his brother Bob—and Kitchel is the first to reach him with the outstretched hand of congratulation. Goldwater mutters almost to himself as he takes the hand, "I never thought it would be that close." Others shake his hand. He is lost to them, but says perfunctorily, "I couldn't have done it without you fellows." The TV screen behind him scrambles the picture; he raps it with a knuckle, says, "Come on, stop that"; and the screen obediently comes to attention and shows a picture of Peggy Goldwater.

As the others draw away to leave him alone, he sits pensive, the body entirely relaxed but the fingers continually stroking, plucking, toying with the chin and cheeks. Scranton appears on the television screen, and as Scranton talks of his home in the Republican Party, Goldwater growls, "He should've been there all along." [2] His eyes glazed, he stares

---

[2] The public telegram of congratulation, however, read: I APPRECIATE MORE THAN I CAN TELL YOU YOUR FINE AND GENEROUS SPEECH TONIGHT. IT IS MY DESIRE AND I KNOW IT IS YOURS TO NOW UNITE FOR A DRIVE TO DEFEAT JOHNSON IN NOVEMBER.

in silence at another television screen. The telephone rings. He ignores it. He rouses himself and says, "I want to call Hoover first thing in the morning . . . then Ike . . . and Nixon," relapses into reverie. From across the room he is reminded by someone that the press is waiting for him; does he want to take questions? He ponders briefly, assents, wonders what to say, and muses, "Something will come to me . . . ," hesitates, wryly smiles and says, "Ain't the white folks got rhythm"; he laughs, then lapses again into silence. A young man enters the room, bends over, kisses his hand; and then Goldwater sits alone again, pensive, peering through the black horn-rimmed glasses, and one is conscious of how handsome a man this is, with the white hair and the firm lines—and how perplexed, how utterly joyless is his mien at this hour of victory.

For he was as he sat there, if not the captain, at least the emblem of a major *coup d'état* in American politics.

But in the next twenty-four hours he was to convert that coup into a rupture—the first such rupture in fifty years in the great and sophisticated spectrum of opinion that blends both American political parties together in shades that differ gradually, from extreme to extreme, only by the faintest hues. Whether this rupture of the following twenty-four hours was inevitable will be debated for years. Whether the man Barry Goldwater had been chosen as the vessel of a wordless indignation across the country, or whether Barry Goldwater had roused this indignation, is of interest chiefly to philosophers who debate whether man is prisoner of his fate or its master. But the man, prisoner or master, had perhaps wanted his triumph more, yet wanted the Presidency less, than any man before him in Presidential politics. His life and past had cast him as a minor prophet of the desert rather than as a leader of American government; in the next few months, as the American people tried to accustom themselves to his idiom, he was to carve a forever unerasable mark on the thinking of the country.

But first were the questions: Who was this man? Where had he come from? What ideas did he bring with him? Could he win? And if he won what would his victory mean?

First, the man, the nominee: tall, six foot even; a muscular 182-185 pounds; not lithe, with dancing step, as was John F. Kennedy, but slow and dignified of walk; the face a deep tan, as is common among people who live in the sunlands; the frame of the face all sharp planes—the nose clean and sharp, the jaw pinch-pointed with one vertical dimple, the lips thin, the cheeks flat, no hint of the sagging jowl that normally overtakes men at his age, fifty-five. All this topped off by the silvery white hair set off by black-horn-rimmed eyeglasses that made him instantly recognizable.

Then the sound: the dry voice of the Southwest, far lower pitched than the Eastern voices that have dominated American politics for so long. It drawls in private questioning, but becomes a snap when offended. Man-to-man conversation with Goldwater is a pleasure—his is a crisp, creative mixture of cussing, eloquence and vivid phrasing that comes from the secular rhetoric of the field command of the American military forces, a distinctive substream of American vernacular. But in press conferences, when he is provoked, the voice starts with a defiance, passes through a range of grievance to reach a bitterness that can become a snarl. He is at his best in the formal public address where, with the audience booming applause, his voice rises to a rhythmic roar (Goldwater has a good ear for the cadence of English prose) and develops the pounding wrath of an Isaiah. And, altogether, all qualities combine to give a sense of hard virility and barely controlled tension.

However often one listened to him at any time in 1964, there was always this tension—an exhalation of sincerity which could rise almost instantaneously to fury. One puzzled over the peculiar quality of outrage one could find in almost any Goldwater utterance—and then, after a long while, it became curiously resonant of another world: the world of the frustrated intellectual.

For Goldwater, on examination, was indeed a frustrated intellectual come late in life to the wonder of books and ideas. Thus, ideas, for him, seemed to have a vigor and validity and virulence strange to those inoculated by learning earlier in life. His outrage was that of a man who could perceive all things with the brittle certainty of the frustrated intellectual —with mechanical precisions and fixes entirely unreal, as if he were a Trotsky of the far right. Where his conqueror, Lyndon B. Johnson, knows there are only pressures and directions, Goldwater is a man who believes there are certainties.

From this outraged certainty one could pick up the two threads that led directly on to the campaign he was about to wage against this Lyndon Johnson. The threads were difficult to perceive at the beginning of the year, but they led inescapably to twin Goldwaters—Goldwater the Patriot, and Goldwater the Prophet, both of them utterly certain of the way the country must go.

His peculiar fascination with mechanics, much written about during the campaign, was the point from which to follow one thread—to Goldwater the Patriot.

His home, for example. Goldwater lives in a flat, angular sandstone house atop the slope of one of Arizona's bare hills. Behind the house rises a twenty-five-foot flagpole controlled by a photoelectric mechanism. As dawn rises over the desert, light activates the motor and the Stars and Stripes mount the staff; as dusk descends, the flag automatically comes down. A little waterfall ripples out of the pool behind the house—and as dark settles, a microphone picks up the sound and pipes it into the house, where it soothes sleep in the bedroom.

"Gadgetry" is an awkward and demeaning word for all that fascinating panel of wares which American technology spreads out to enchant the questing amateur trying to grope his way through a bewildering world of electronics, supersonic speeds and third-generation computers. Gadgetry gives a hobbyist a sense of fixed responses—and Goldwater is a gifted gadgeteer. A radio fan since the age of twelve, he is now an expert tinkerer and manipulator of all the dials, tubes and transistors that the most advanced radio "hams" of this country know. One complete set sits in his apartment in Washington, another in his home in Phoenix; another set was taken to the Convention at San Francisco. But Goldwater is also an accomplished photographer, a man whose superlative photos of his native Southwest approach art. (Indeed, so handsome are they that a collection of Goldwater photographs was sold at $1,500 per volume of fifty to rich contributors in his campaign, and no one was cheated in the purchase.) He is also an *aficionado* of airplanes and aerial technology; a new plane delights him as a new continental racing car delights a young hot-rodder.

Goldwater's love of planes leads us on. During the war he was one of the brave ones who flew the early C47s, with service ceilings of 6,000 feet, over the passes of the Hump, where the clouds hold rocks in them at altitudes of 15,000. (During the Convention at San Francisco, Goldwater took to the air to weave and cross the skies over the Cow Palace, where he was about to be nominated.) Planes still fascinate him, but not only as a hobbyist—even more as Major General Barry Goldwater of the United States Air Force Reserve. The phrases of the Air Force High Command pepper his thinking: distances, tonnages, mileages, maps, settings, communications, ranges, payloads, missiles and numbers. No recent civilian candidate for President has had wider and more intimate connection with the high military brass than Barry Goldwater—and his talk reflects it. His sound in conversation echoes the sound of alert-shacks where pilots chatter about their trade, or, as he ages, the select dining rooms of the Pentagon where generals generalize on combat.

It is here that one is led to the first of the political Goldwaters— Goldwater the Patriot. Unaware that his technical military musings lead to an enormous sense of public alarm, devoutly convinced that the fear of nuclear weapons must be lifted from the American mind ("You mention the word nuclear," he once said, "and all they can think of is the big mushroom cloud, the red blast and twenty million dead"), Major General Goldwater is a Patriot with an incandescent sense of mission. His inner convictions tell him that America is in danger. Although he is never compared to Winston Churchill, there is in conversation with Goldwater a genuine echo of Churchill's alarm for England in the late thirties when Churchill counted the handful of Hurricanes and the timetable of the Spitfires and measured them against the growing numbers of Goering's *Luftwaffe*. Goldwater deeply believes the United States to be in mortal peril, approaching a decade of the seventies when Russian missile strength will outweigh American strike strength; and that the nation must be roused to face the peril.

What makes Goldwater so different from Churchill is that Churchill alone in England counted Spitfires and Hurricanes and measured them against the *Luftwaffe*, while in the United States the Defense Department is directed by the greatest counter of all time—Robert Strange McNamara, whose devotion to the defense of the United States is as great as was Churchill's to England, and who possesses the authority to act. (Goldwater's bitterness at McNamara, whom he considers a liar, is far more intense than his bitterness at Johnson. But, in trying to arouse other patriots against McNamara by super-zeal and super-militancy, Goldwater was to wrap tightly about himself the cloak of firebrand that Rockefeller and Scranton had so colorfully tailored for him in the spring.)

This military Patriot might have had his roots anywhere in America. He might have been almost any man in whom the ecstasy of war experience had been deepened by continuing devotion and permanent association with the armed-forces reserves—and who, from the eminence of his political position, could be alarmed and upset by the inner debate on American military strategy.

But the other Goldwater, the Prophet, can only be understood against the background of his time and place, for his roots twine back to the pioneer epic of American life.

If there is an authentic pioneering family in the American Southwest today, it is the Goldwater family; its history is the story of the opening of the American desert—and, inescapably also, of the story of the Jews in American life.

Those who think of American Jews in stereotypes—as the scholars and scientists of the universities, the self-questioning intellectuals of New York and Hollywood, the muscular businessmen of Chicago, the prosperous middle class of the suburbs, the diminishing toilers of the garment industry—think of the children and grandchildren of the great turn-of-the-century migration which brought two million Jews from the

ghettos of Eastern Europe to the great cities of America in one generation. But these immigrants had been preceded fifty years earlier by more
lonesome, solitary Jews who, sifting through the South, Midwest and Far
West, struck root and fathered such great families as the publishing Ochs-
Sulzbergers, the merchant princes of the Straus, Gimbel and Rosenwald
clans, the mining dynasty of the Guggenheims—whose descendants only
later moved back to New York to form the aristocracy of the Jewish
community on the East Coast.

These people, who arrived more than a century ago, came generally
as roving merchants and peddlers, wanderers on an open frontier—like
Roman traders who clustered where the legions staked out civilization.
They came by foot and by covered wagon; as they trickled West they
fought Indians and marauders and joined the loose frontier communities
of strangers, not as aliens but simply as other strangers, accepted or
rejected as men and individuals. Of these, Michel Goldwasser of Konin,
Poland, the Republican candidate's grandfather, who arrived in San
Franciso during the California gold rush in 1852, is an archetype. Drifting from San Francisco to Los Angeles, thence along the gold strikes to
Arizona, "Big Mike" Goldwater (he had anglicized his name to Goldwater in England en route) and his brother arrived in that Indian territory when Apaches and Navajos were still on the warpath. They settled
in a time of violence: the Civil War, drawing Federal troops away from
their Indian outposts, left the first settlers to defend themselves and establish their own law and order. Hip-shooting, a charge made so frequently against Goldwater, terrifies those who think of the trigger as a
nuclear weapon. But in the Old West, hip-shooting was a masculine virtue—a man had to be quick on the draw, else he were dead.

The Goldwaters helped build the State of Arizona—not only as
merchants (they developed the two best department stores in the state)
but as fighters, school-board members, territorial legislators, citizens involved in every public affair of the territory still not a state when Barry
Morris Goldwater was born in 1909.

This is, of course, remote history. Goldwater's own father, Baron,
married a Protestant, as did Goldwater himself. He is a member of the
Episcopal Church. Yet the saga of his pioneering Jewish ancestry adds a
tingle of pride to every speech he makes recalling Big Mike and the past.
And though he knows neither Hebrew nor Yiddish nor the ritual of the
synagogue, his dry Southwestern tonalities still cloak a biblical rhythm of
speech and the endless Jewish capacity for righteous indignation.

To this family background must be added the experience of a man
growing up in boom country. The Arizona boom started with an act of
distant paternalism that fully qualifies in the standard Goldwater lexicon

of politics as "socialism"—the building of the first great Federal dam in Arizona in 1906-1911, which gave water to the parched valley of Phoenix and let the town grow. In Goldwater's boyhood, however, Phoenix was still an overgrown egalitarian town of about 30,000 people, an open society where the boy could learn boxing from a Negro master, roam the Indian reservations, tinker with radios, be a student leader (though a poor scholar) at high school. He spent one year at the University of Arizona, which he left after his freshman year, following his father's death, to go into the family department store. War took Goldwater away to be a gunnery instructor in the Air Force, later a pilot over the spurs of the Himalayas on the run over the crumpled Hump of the China-Burma-India theater; and he returned to find himself posed with a problem of conscience.

Pre-war Phoenix had had a population of about 65,000. But it was growing rapidly, and during the war years had been governed by as corrupt a band of politicians as the Western states knew, with gambling and prostitution wide open. (Phoenix still has one of the highest crime rates among American cities.) In a campaign of vigilante virtue, Barry Goldwater, the leading merchant, joined other civic leaders to reform and purify the little city's government in one of those traditional victories of the Old West whereby the "good guys" appeal to the citizens to clean out the "bad guys." "There's always been one and sometimes two Goldwaters damned fools enough to get into politics," he wrote, ". . . it ain't for life and it may be fun."

Thereafter, as the postwar westering impulse seized millions of Americans, Phoenix exploded in a growth phenomenal even for the incredible Southwest. By 1950 the population of its metropolitan area had swollen to 332,000; by 1960—664,000; 1963—816,000. It is essential to feel the psychology that grasps boom areas of the United States— southern California, the Arizona sunlands, southern Florida. Here, in a world of exuberant growth, dreams come true. Those who buy land early see the land quadruple and quintuple its value in a decade. Those who open small businesses thrive in an endlessly expanding market. Those who flee the crowded, regulated cities of the East, with their noise, congestion and friction, find here that all the virtues of private enterprise extolled by John Stuart Mill are still valid. Pluck, hard work and a bit of luck bring their rewards quickly and tangibly. And no man over the past eighteen years, from City Council of Phoenix to the highest forum of his party's offering, has spoken more eloquently, more loudly, more consistently for these simple virtues than Barry Goldwater of the Old Frontier.

In the life of Goldwater, in the development of his thinking, there have been no sudden conversions such as Senator Vandenberg's to internationalism. It is still the individual that is central to his philosophy; it is still good guys against bad guys. His world is the world of desert illumination, like his Arizona—white and black, blazing sun cut sharp by slanting dark shadows. It is totally different from the thought world of Lyndon Johnson; in Lyndon Johnson's world of striving and denials and triumphs and compromises, there is no such thing as an all-good man or an all-bad man. In every saint Johnson can find the hidden sinner; in every villain Johnson can detect a native good. The immense and phenomenal range of Lyndon Johnson, the political artist, extends octaves above and below the Goldwater range—from Bobby Baker at the one extreme to Hubert Humphrey at the other. This makes Goldwater a helpless prey for a man like Johnson.

For Goldwater's credo—that the individual must be free to do and build as he will, free to organize his life without the interference of government or suffer without the mercy of government, free either to make a fortune or to provide his own Social Security in withering age—is sincere, heartfelt and unblemished by any compromise of instinct and thought. The Prophet Goldwater insists that men must be good, that it is their moral duty to be good—and that thereafter they and their country will thrive. Lyndon Johnson regards all men as potential sinners—who will, however, be good if government makes it possible, or even profitable, for them to be good rather than bad.

Both the Prophet Goldwater and the Patriot Goldwater were lucky in their choice of moment in American politics. For in the summer of 1964 Goldwater had arrived center stage in American history at a time when the intellectual vitality had apparently run out of the generations-old liberal orthodoxy. This liberal orthodoxy had dominated American politics, unbroken for the thirty-two years since Franklin D. Roosevelt's first election, through Republican and Democratic administrations alike. All that liberals sought had been won—and each victory had generated new problems. Great trade unions had been created by heroes of the workingmen—and had, with few exceptions, ossified into ponderous labor bureaucracies like the Steelworkers, or festered with the hoodlumism that controlled the Teamsters, largest trade union in the country. Federal financing had given more men ownership of their homes than ever in history in any society—and created the ungovernable sprawl of megalopolis suburbs. Rigid legislation and scientific progress had raised farm income to new peaks—but the number of farms had shrunk from 6,812,-000 (in 1935) to 3,703,000 (in 1959) under this protection. All through the North and West, the American Negro of the big city had been freed for every civil right; yet he stood at the edge of violence. Americans had given some $100 billion in aid to foreign lands after the war, taxed from their own incomes; but that once-brilliant policy had degenerated into a mechanical orthodoxy which created as many enemies as it did friends, and in some cases made enemies out of old friends.

The creative American liberal thinking of a generation earlier had succeeded—and in success had become dogma. The essential tragedy of the death of John Fitzgerald Kennedy is that he left his work incomplete. No man for orthodoxies, Kennedy had, in his brief power span, begun to examine the very dogmas that had brought him to power. The great political watershed that his administration marks in American history was his invitation to all men and good thinkers to examine the dogmas and see America afresh, as it was. And then he was dead.

And at that point entered Barry Goldwater to challenge the dogmas too. But not with newer ideas, only with the older dogmas of the Old Frontier.

From the entire Kennedy experience the leaders of the Goldwater movement took only one new chapter—the mechanics of seizure of power in an open democracy. Those who had mastered the mechanics sat in the trailer at the Cow Palace watching the same scenery as did Goldwater on television—dismissed from further influence, as mechanics, without their knowledge.

Those who sat with Goldwater in his suite at the Mark Hopkins on nomination night were men who believed, like the Prophet, that to guide the power they had seized and the further power they hoped to have, it was necessary only that Baal be cast down and Jehovah restored.

For the men who sat with him were the chief ideologues—William Baroody, Edward McCabe, Denison Kitchel. Together these men, along with speechwriter Karl Hess and the scholars of his doctrine, had already composed the speech of acceptance to be given at the Cow Palace the next day. It was to be no ordinary speech; it had been worked on for over two weeks; it was thoroughly thought out. The Convention had been a leisurely time for the Goldwater staff; the Prophet himself had, as he put it once, "sat on his duff" most of the time and talked to friends or toyed with the ham radio in his suite. Five times he himself had gone over the speech with all its key phrases; no word was unmeasured; the speech was to proclaim a new morality, an uncompromising challenge to the course American politics had followed for the past thirty years—within his own party as well as in the opposition party. The candidate himself had considered tossing, as he once put it, "a rose or two" to the defeated—for, in the politics of democracy, compromise is the fee the victor pays to the vanquished to earn the vanquished's consent on the road the victor must go. But the vanquished at San Francisco had embittered Goldwater beyond compromise—they had painted him as crazy, stupid and bloodthirsty. His speech would offer no compromise; it was Goldwater's decision to go it alone.

Technically, of course, he was not quite "alone." That Wednesday night Goldwater finally confirmed the choice of Vice-President he had decided several days before. He would have as his running mate Congressman William E. Miller of upstate New York. An arch-conservative, Miller is a handsome dark-haired man of fifty whose qualifications, in Goldwater's eyes, were two: he was a Northeastern Catholic, who thus gave the ticket proper geographic and ethnic balance; and he was a quick and partisan debater of no mean skill. It was Goldwater's thought that Miller could savage Johnson with abuse better than any other man in the Party and provoke some outburst that would have the President on the defensive during the campaign. Miller tried quite hard; but, though he is a gut-puncher, he does not foul his blows. He never got within striking range of Johnson or Humphrey throughout the campaign; and his combativeness later oozed away week by week on the road.[8]

It was Thursday, the next day, that Barry Goldwater accepted his nomination at the Cow Palace. Richard M. Nixon introduced him in an excellent speech, himself looking now, at long last, nostalgically attractive. Proclaiming himself "a simple soldier in the ranks" of the Party he had led four years before, Nixon pointed to the uplands where the Republicans must go, urged them to follow their new and great American leader, and concluded as he pointed, turning to the flag-draped catwalk that led to the speaker's rostrum, "Down this corridor will walk a man into the pages of history."

For a moment the thousands gathered in the Cow Palace held themselves in check—like a wave curling to surf. And then, as Barry Goldwater appeared, the surf burst.

From the rear pit the band blared the crusaders' song of American politics: "Glory, Glory Hallelujah." Down from the rafters tumbled red, white and blue balloons, cascading about the new candidate and his wife. From the far left of the hall, Texans waved a huge banner of the Lone Star State. Tier by tier down from the roof the hands fluttered and waved, making the hall ripple in pale white-of-palm shades. "We want Barry, we want Barry," chanted the deeper voices, while as an overtone

[8] Miller, an accomplished bridge player, played bridge frequently with the correspondents who accompanied him on his campaign tour, and won consistently. One of the correspondents with the Miller party relates that toward the end of the campaign he asked Miller whether the candidate, as a sporting man, would give him a chance to get even by betting on the outcome of the election. Miller, according to the story, replied that he might seem stupid—but he wasn't crazy enough to bet on the outcome of *this* election.

came the female voices moaning, "Ba-a-a-a-arry, Ba-a-a-a-a-a-arry."
The noise grew, the sound roared, the balloons popped now as they
bubbled and bounced off the rostrum to delegates who cigarette-touched
them to clear them away. The gavel banged and banged, and slowly the
crowds subsided to listen to the hoarse-voiced man.

He began with a ceremonial salute to the dignitaries of the Party. He
cleared his throat to accept his nomination with a "deep sense of humility," then swung into his text, and it was he and the Good Lord alone, co-
pilots in a rescue mission to save America at the edge of Gehenna: "The
Good Lord raised this mighty Republic . . . not to stagnate in the
swamplands of collectivism, not to cringe before the bully of Communism."

Gradually the dry voice picked up the force of an incantation:
"Our people have followed false prophets. . . ."

And the speaker was leading his audience way out there into a new
world, a crusader's world unexpressed in American politics for generations—the visionary Prophet and the martial Patriot alternating, first the
Prophet, then the Patriot, over and over again:

"This Party, with its every action, every word, every breath and
every heartbeat, has but a single resolve and that is:

"Freedom!

"Freedom—made orderly for this nation by our Constitutional government.

"Freedom—under a government limited by the laws of nature and
of nature's God.

"Freedom—*balanced* so that order, lacking liberty, will not become
the slavery of the prison cell; *balanced* so that liberty, lacking order, will
not become the license of the mob and of the jungle."

Then the Patriot reviewed the failures of the incumbent administration:

". . . failure cements the wall of shame in Berlin; failures blot the
sands of shame at the Bay of Pigs; failures marked the slow death of
freedom in Laos; failures infest the jungles of Vietnam; and failures
haunt the houses of our once great alliances and undermine the greatest
bulwark ever erected by free nations—the NATO community."

Then the Prophet again, turning homeward:

"Rather than useful jobs . . . people have been offered bureaucratic make-work; rather than moral leadership, they have been given
bread and circuses; they have been given spectacles and, yes, they've
even been given scandals.

"Tonight, there is violence in our streets, corruption in our highest
offices, aimlessness among our youth, anxiety among our elderly, and

there's a virtual despair among the many who look beyond material successes toward the inner meaning of their lives. . . .

"We Republicans see all this as more—*much* more—than the results of mere political differences or mere political mistakes. We see this as the result of a fundamentally and absolutely wrong view of man, his nature and his destiny. . . .

"It is the cause of Republicanism to insure that power remains in the hands of the people—and, so help us God, that is exactly what a Republican President will do with the help of a Republican Congress."

Then the Patriot again, talking as Cromwell:

"It is . . . the cause of Republicanism to remind ourselves, and the world, that only the strong *can* remain free—that only the strong *can* keep the peace!"

Then the call:

"The Republican cause demands that we brand Communism as the principal disturber of peace in the world today. Indeed, we should brand it as the only significant disturber of the peace. And we must make clear that until its goals of conquest are absolutely renounced, and its relations with all nations tempered, Communism and the governments it now controls are enemies of every man on earth who is or wants to be free."

Then the message to the Party itself:

"Anyone who joins us in all sincerity we welcome. Those who do not care for our cause, we don't expect to enter our ranks in any case.

"And let our Republicanism, so focused and so dedicated, not be made fuzzy and futile by unthinking and stupid labels. . . ."

And then the final, unforgettable thrust at the Party moderates:

"*Extremism in the defense of liberty is no vice!* . . . *Moderation in the pursuit of justice is no virtue!*" (The italics represent underlinings in the candidate's own text.)

With this, the Republican Convention of 1964 was over.

The delegates and gallery audience wandered back to town through the gates where Negroes softly sang "We Shall Overcome," and then proceeded to celebration on the hills of the old city. All night, on Nob Hill, a projector illuminated the front of the Mark Hopkins with a huge picture of Barry Goldwater and the sound of singing rose to the sky, while overnight analysts took themselves to their typewriters to begin the examination of the Republican Party that has continued ever since.

The best quick reaction I remember is that of another reporter who, halfway through Goldwater's speech, slowly became aware of the politics of no compromise, then turned and remarked, "My God, he's going to run as Barry Goldwater."

Indeed he was. But what Barry Goldwater was, or represented, was best summed up several weeks later by the house-thinker of the Scranton headquarters, James Reichley, who produced a mordant internal house memorandum which I wish were my own and from which I now borrow heavily.[4]

The Republican Convention of 1964, thought Reichley, writing from the bruised and melancholy Scranton headquarters in Harrisburg a few weeks later, ended not only the Eisenhower era of American politics, but something more important: the reign of pragmatism in the Republican Party.

Pragmatism had begun to make its way in American political thinking at the turn of the century when a group of notable American philosophers (chiefly at the Harvard Yard) laid the intellectual basis for a re-examination of the twentieth-century world. The somber confusion of a Henry Adams confronted by the full turbulence of the American industrial revolution, and his tragic view of the destruction of the old values of American life and the Republican Party, which had been born out of moral indignation—these were no part of the pragmatists' thinking. The test these men set for policy and purpose was: *Does it work?* "The old idealisms, along with the old religions, were found to be mere words, verbal concoctions to fuddle the senses," says Reichley.

It was only, however, after the First World War that pragmatism became the dominant political code of both parties: first in the Republicanism of Harding, of Coolidge and of Hoover; and then, when obviously Hoover could not make it work, in the inspired pragmatism of Franklin D. Roosevelt.

Overwhelmed by the magnificent success of Roosevelt's achievements, both in hard fact and in the hearts of voters, the Republican Party in the wilderness was briefly swayed by the encapsulated idealism of Robert Taft. "In the late forties," says Reichley, "Robert A. Taft—the most representative American Puritan since Cotton Mather—sought to lead the Grand Old Party out of the wilderness of electoral defeat behind the standard of conservative individualism. Taft was not, as a matter of fact, very conservative—he advocated Federal programs in the fields of health, education and welfare which, in 1964, have yet to be realized—but his career was based on application of a philosophy of Christian individualism. His followers tended to extract the emotion from his Christianity, and the economics from his individualism."

---

[4] Where, as in the first few paragraphs below, this chapter encloses certain thinking in quotes, the thoughts are from Reichley. Elsewhere, where the thinking of this writer diverges from that of Reichley, it would be unfair to hold him responsible for my opinions.

With the triumph of Eisenhower in 1952, however, what triumphed was neither a moral purpose of Taft extraction nor the commercial pragmatism of Harding, Coolidge and Hoover. What triumphed was an extended pragmatism of the Roosevelt-Truman descent, directed by a man who had been made great under Democratic administrations and whose thinking, with a few budgetary changes, continued their policies at a period when those once vital policies were becoming obsolete.

Thus the violent, almost explosive appeal of the Goldwater slogan: "A Choice, Not an Echo."

The ground was ready. Others had plowed it before. Joseph McCarthy, one of the most immoral characters ever to terrify good men in government, could nonetheless reach and communicate with millions of people of moral fervor. In the sober Midwest, scores of little politicians had plowed the ground and found a harvest of votes ready for anyone who claimed spiritual descent from Robert A. Taft. In the South the white-collar segregationists of the newer cities, trapped by the ugly racism of the wool-hat Democrats and the, to them, more dangerous race-mixing notions of the Northern Democrats, were likewise ready to respond to a moral and emotional appeal.

To this Reichley analysis one must add other ingredients of emotion at hand: simple geography, for one, and the Far West's inherited suspicion of the men of a mythical Wall Street and Eastern Seaboard who had financed, but enslaved, their development.

One of the finest governors in the Union is Robert E. Smylie of Idaho, a tall, lanky, easy-going Westerner. As, finally, he was resigning himself to vote for Goldwater, Smylie—a liberal and a Rockefeller man at the beginning—described best for me the emotional content of the Western revolt: "This continent tilts," he said, "but Easterners don't recognize it. You have an 'overflight' complex—you think the first stop after Idlewild is Los Angeles. You don't recognize the condition of our explosive growth and development—and Kennedy did more to emphasize the fact that you don't care than anybody. What did he care about the little mountain states of the West? Our state of Idaho has only one trade mission abroad—in Tokyo. It's Idaho wheat that makes the Japanese kids grow so fast—and we sell it, we don't give it away. We can get a pretty good-sized loan in San Francisco these days; we don't have to go to New York. If I buy stock, I'm more apt to do it through Schwabacher and Company in San Francisco than with Merrill, Lynch. Harvard may still be the biggest shareholder in Idaho Power and Light—but the board is local. Boise Cascade is local. Morrison-Knudsen is local and it operates all around the world. Our kids go to school in the Mountain West, or perhaps to Stanford or California—they don't have to go East any

longer." Then, explaining what made so many Westerners simple conservatives while he was of another view, he concluded, "We're growing,
we're growing. And as we grow, these new governments will realize that
we have to have conditions of growth; that they, too, need a government
that governs. But now it's too early."

None of these emotional ingredients was necessarily locked up for
Goldwater at the beginning of his Presidential campaign. But no one else
could tap the hidden streams of morality and fervor as did he. Rockefeller, whose fervent political moralities are deep but not easily recognized, expressed privately his own drives in terms of "problem solving";
but in public, through his campaign, he tried to define himself as "moderate"—a dead word. Crippled further by his divorce and remarriage, he
shrank from making the issue between himself and Goldwater one of
morality. So, too, did Scranton. Trapped by a moral ambiguity of his
own—he believed Goldwater to be dangerous for the nation, yet felt
compelled to repeat again and again that he would nonetheless support
any candidate of the Party for the Presidency—Scranton campaigned
with a bellicosity unnatural to him; his campaign was based on the ancient pragmatism of: *Will it work?* His call to the delegates, founded on
the declaration that Goldwater could not win (which was true), had appealed deeply to no one's conscience and fervor.

Thus Goldwater was left, from beginning to end of the Republican
civil war of 1964, in command of the moral high ground. The fervor, the
frenzy and the excesses of the dedicated Goldwater troops appalled a
huge and unknowable percentage of Republicans of other moralities; but
there was no banner to which they could commit themselves. And so
Goldwater and his Puritans were left alone to face the greatest pragma-
tist of all time—Lyndon B. Johnson.

This reporter went to sleep in the darkness of the early morning
after Goldwater's nomination, bothered by the victorious singing beneath
his windows. The strains were familiar but cacophonous until he separated them. Some of the Goldwater jubilants were singing the "Battle
Hymn of the Republic" and others were singing "Dixie."

But they were united at last.

CHAPTER EIGHT

RIOT IN THE STREET:
THE POLITICS OF CHAOS

EW of those who, that Thursday evening in San Francisco, cheered or ululated for the Republican Party in their hotel rooms had paid much attention to an early-afternoon wire report from New York. The paragraphs that reached the West Coast seemed to repeat, with a distant staccato, the vivid theme of Law and Order which the Prophet from Arizona had just laid before the Republican Party and the nation. But they gave little fore-echo of what was to come.

The wires stated simply that a Negro lad of fifteen, James Powell, had been shot and killed by a New York police lieutenant, and that a protest by Negro students had immediately thereafter been dispersed.

The death of James Powell on July 16th, 1964, was not, however, so simply to be disposed of.

It was to torment all of New York for a week; kindle rioting worse than the great metropolis had known in modern times; inflame Negro communities to riot up and down the Eastern Seaboard of the nation; place directly before the city of New York—for the first time in any American city—a fundamental consideration of the strategy of domestic tranquillity; and it continues even now, in retrospect, to summon the nation to ask itself, in agony of conscience, what kind of civilization is being bred in its great and changing cities. For in James Powell's brief life one could find gathered all the perplexities, all the dogmas, all the dangers, all the sorrows, all the unanswered questions of the darkest area of American life.

James Powell was, by the police record, already a potentially dangerous person: the child of a widowed mother, he had been charged earlier that year with attempted robbery; three cards in his previous police record were for minor instances of juvenile delinquency; in school he had been unruly and violent. Yet in his short life he had shown other qualities too: evidently he was one who strived. A year before, he had gone down to Washington, at the age of fourteen, to join in the great Freedom March. He was, by all report, reaching for something. And he was shot while he was reaching both for an education to make him better—and a knife to slash with.

The shooting of James Powell is a classic example of how, in an unstable situation, dogma and accident can escalate into explosion.

James Powell had volunteered to go to summer school for remedial reading in the anti-poverty program which the distant Federal Government and the city government of New York were jointly launching. The confused School Board of New York City, whether by the dogma of integration or the dogma of "quality education," had decided that the program required youngsters of Harlem tenements to journey down to a strange neighborhood to attend classes at the Robert Wagner Junior High School. Children of varying social climates are as different in manner as their parents. The normal swagger, noise and natural boisterousness of youngsters, black or white, take different forms. But on East 76th Street—where the frontier of the Perfumed Stockade meets the Bohemian-German neighborhood of Yorkville in a block visibly and rapidly becoming fashionable—the apartment superintendents and the white families are not used to the manners of Harlem, the stoop-sitting and the shouting laughter. Nor can they distinguish what may be the innocent high spirits of black youngsters from the violence they read about in the papers.

An apartment-house superintendent on 76th Street was, thus, on this Thursday morning, July 16th, hosing the sidewalk, the flowers and the trees in front of one of the buildings. And whether annoyed by the lounging Negro youngsters across the street, or whether by accident, he turned his hose on them.

To Negro youths, effervescent with the growing élan of the Negro revolt, and particularly to those striving to better themselves, the fire hose is the symbol of Birmingham—as inflammatory as a swastika to a Jewish youth.

The youngsters thus, almost by instinct, escalated from the hose to the Harlem response: off with garbage-can covers. From the garbage cans they extracted natural weapons, bottles; and with garbage-can lids as shields and bottles as weapons, they charged the superintendent, who fled. Dropping his hose, he ran—a bottle hitting him as he went, a garbage-can cover breaking the glass-paned door through which he dashed. Powell, apparently a natural leader, still incensed and unappeased, de-

cided, according to later testimony, that he was "going to cut that ——."
He walked across the street after the superintendent, escalating the response from garbage can to cold steel as he opened and shut his knife.

By this time an off-duty policeman had come on the scene—escalating to a gun as he challenged Powell and ordered him to drop the knife. The policeman fired a warning shot; Powell lunged, slashing and cutting the policeman's right forearm; the policeman fired twice more—fatally. And Powell was dead a few minutes later in a grotesque tragedy of misunderstanding, of reflex hatreds and unplanned response. A child trying to better himself; an apartment-house superintendent trying to patrol the manners and fashion of his neighborhood against strangers imposed on him by distant goodwill; a policeman doing his duty—these were the ingredients.

But these ingredients were only the fuse: seventy-two hours later the entire police force of New York City was on emergency stand-by, Harlem rattled to gunfire, and raging Negroes by the thousands were in the streets, smashing, looting, burning in a series of riots that were to convulse New York for five successive nights and then sweep like crown fire across the Eastern ghettos.

What had happened? What hidden dynamite had been touched off by this sudden spark? Why had the Negroes chosen to disrupt New York first, that city of the nation which had gone furthest—to the edge of bankruptcy—in trying to open opportunity and equality for its colored citizens? Were these truly race riots? Or something more sinister—absolute chaos? In how many other cities would events take a similar course? Would the bloodshed and shock affect the campaign of 1964 just about to start? What lay ahead in the future—beyond 1964—as all of America's cities faced the same problems of Negro growth and adjustment?

History can signal its changes to wise men without violence, but for those who prefer to ignore its message, it then raps with gunfire to call attention.

The riots of 1964 were episodes far larger than campaign episodes; they were the first violent call upon American politics and civilization to pay attention, before too late, to a condition which may result in a transformation not only of political forms but of American life itself.

Starkly put, the gross fact is that the great cities of America are becoming Negro cities. Today only one major American city—Washington—has a Negro majority. But by 1980—*if the arithmetical projections of present population trends come to pass*—Negroes will be the majority in Detroit, Cleveland, Baltimore, Chicago and St. Louis; and in the decade following, in Philadelphia. By 1990, then—which is almost tomorrow in the eyes of history—these trends, *if unchanged*, will give America a civilization in which seven of her ten largest cities (all except New York, Los Angeles and Houston) will have Negro majorities; and the civilization of this country will be one of metropolitan clusters with Negroes congested in turmoil in the central cities and whites defending their ramparts in the suburbs.

One must look back to grasp the full dimension of this change in the nation's life in our time—and the direction in which it points.

Until 1940, 75 percent of United States Negroes lived in the South, the great majority of them functionally illiterate, primitive, excluded from society, as they are still excluded in Mississippi and Alabama, and so cruelly policed as to have no understanding of the law but fear. But World War II made the North into a great industrial furnace—and the draft of the furnace sucked Negro labor by the scores of thousands up the Mississippi Valley, up the East Coast to the forges, foundries and factories of the war effort. The postwar boom and the Korean War kept the furnaces blazing, and the migration northward continued. In some years of the 1950s as many as 200,000 Southern Negroes moved north or west to the big cities in search of jobs and freedom.

Where they pooled, Negro death rates fell and birth rates rose. In the twenty years from 1940 to 1960 the Negro populations of New York and Philadelphia doubled (to 1,100,000 and 529,000 respectively); the Negro populations of Chicago and Detroit tripled (to 813,000 and 482,-000 respectively), and that of Los Angeles multiplied by five (from 63,-000 to 335,000).

Though migration from the South has now slowed, the Negroes' urban birth rate—approximately 40 percent higher than that of whites—continues to swell their numbers. In the cities, with social services available, with welfare and hospital care provided, with family responsibility assumed by municipality, Negroes grow in numbers at a rate faster than ever before. In Chicago, for example, where Negroes have risen from 278,000 in 1940 to approximately 900,000 (1964 estimate), their natural growth rate is nine times faster than it was before the war. The fertility rate of Chicago Negroes increased in the 1950-1960 decade three times as fast as that of Negroes generally across the country. Americans are, in short, witnessing in the black ghettos of the American cities one of the great population explosions of all time.

Something has to give. And what gives is the neighborhood pattern in big-city living as Negroes, fleeing the smell and the rats and the cackle of inhumanly crowded slums, burst out like flood waters under pressure, spurting and spilling into adjacent neighborhoods. Street by street, block by block, neighborhood by neighborhood, solid black precincts crunch their way year by year through the hearts of our cities.

Some ten years ago this reporter made an early survey of the Negro communities of the big cities. In Chicago, white people then hoped to hold "them" (the blacks) to a few small beachheads north of the Chicago River, which would then, as an internal Mason-Dixon line, divide the city into a white north and a black south. By 1964 the beachheads had enlarged into total breakthrough. In New York, 96th Street was to be the line protecting the East Side of Manhattan. But Negroes are now crossing this line into the Eighties. In Boston "they" were supposed to stop at Franklin Park; "they" have by now completely outflanked the park in Dorchester and have closed in on Blue Hill Avenue. In Los Angeles, in Philadelphia, in Cleveland—everywhere the FOR RENT and FOR SALE signs on lawns and apartment houses mark the moving edge of a Negro frontier. And everywhere the problem is discussed by citizens of goodwill, by school committees, by real-estate boards, by philanthropic foundations and academies, in terms of localities, in terms of villainous black blockbusters or villainous white mortgage banks. What is pursued is a word—"integration"—and a goal—"stability." What is ignored is the fact that the sheer thrust of population growth, as elemental as the tides, can be contained by no dikes erected by social workers.

"Where are the whites going to stop running when they know they can't hide?" ask some Negro leaders.

And the answer is, of course, as clear as the maps that chart Negro expansion: "they," the whites, are going to the suburbs. There a more modest white population explosion closes a ring about the changing central city. In Chicago, in Philadelphia, in Detroit—everywhere except New York City—the same pattern repeats itself: the suburban "white noose" tightens about an inner city becoming more and more rapidly Negro, while within the strangulating city a turmoil begins that feeds on fear and hate.

It was within the strangulating city that the riots of 1964 were bred. For, unlike the previous race riots of American history, these were not riots of black against white, or white against black, in an area of expansion and contest on a changing frontier. These were riots of black men and adolescents against the conditions of life they have found—and, in large part, made—within the ghettos of the big city.

One must explore deep within the city ghetto to understand the full nature of the awful Negro tragedy—and the dilemma of Negro leadership over the next decade. For the city is beyond the reach of the new Civil Rights Act. That bill was designed, as law, to remedy illegal and inhuman injustices subject to definition. But all that bill did was to impose on the South measures of decency which had already been adopted voluntarily by every large Northern city years and sometimes decades before. In the North the great Civil Rights Act is not so much inapplicable as irrelevant. It cannot meet the conditions of life as it is lived in the ghetto, where "Freedom Now" has long been granted—and where, in this "Freedom," more complicated problems fester.

In the city, American Negroes have found both success and disaster at the same time; in the big city, for many Negroes, life is better than ever before in all their ancestry—but also, for many others, worse.

The success of the Negro in so many of the great American cities is one of the breath-taking stories of human progress. To have been freed from slavery only one hundred years ago; to have learned to create, cherish and hold dear a family after having been abused as animals for two previous centuries; to have absorbed education and developed leadership in a civilization as complex and technical as America's today—all this is an achievement which nations of independent tradition in Asia, Africa and Latin America have not yet realized.

Negro success has many faces. Success is the political power to claim in one recent year over thirty judgeships in New York City. Success is as sleek as Baldwin Hills in Los Angeles, with its sparkling modern homes and their ground ivy, their fiber palms, their dichondra lawns. Success is Pasadena with its small colony of Negro scholars, engineers, scientists. Success is education—in 1947, United States colleges and professional schools counted only 124,000 non-whites; in 1963 they enrolled 286,000. In the decade of the 1950s alone, the number of Negro engineers jumped fivefold. Success is Negroes directing embassies, leading orchestras, commanding Army units. Since 1940, Negroes have claimed government jobs more and more insistently—and won them, particularly in the big cities and in Washington. In the past two and a half decades, Negroes employed in government have jumped from some 4.5 percent to almost 13 percent (in Philadelphia to about 50 percent). Success is a sparkling crown of Negro writers, artists, musicians, athletes, political leaders. But success is even more than this: quiet, pleasant, stable, hardworking Negroes bringing up their children, or trying to bring them up, in conditions of modern decency.

But disaster, too, has found the Negro in the city. The Negro who migrated to the North came chiefly to work at the heavy jobs—the steel mills, the packing houses, the assembly lines. And for the past ten years these jobs have been evaporating as automation has condemned scores of thousands of middle-aged Negroes to discard and scores of thousands of their youngsters to a future that holds nothing. For the Negro had arrived in the American big city at a turning of American industrial life—when the electronic wafer and the engineer's skill were, increasingly, designing away functions and skills for which nature originally shaped human beings.

In Detroit proper, for example, the largest employer of labor is Chrysler. And in the late-summer layoff of the boom year 1963, before the new-model changeover of that year, layoffs in the automated metal division of Chrysler had pushed out onto the street all men hired after February, 1948. Thus, automatically and without prejudice, a huge number of Negroes who had journeyed to Detroit in the 1950s was eliminated. These were temporary layoffs. But automation changes some industries permanently. In Chicago, erstwhile "Hog Butcher for the World," the packing houses once employed 20,000 workers—predominantly Negro. Today Chicago no longer butchers hogs for the world, for the packers have moved to new and more efficient plants elsewhere, and their South Side workers will never be employed again. In Chicago in 1963 the Urban League estimated 18 percent of employable Negroes as unemployed. In New York the Mayor quite rightly boasts that New York thrives: the largest employment center in the world swings from new prosperity to newer prosperity. But it is a prosperity not equally shared. Manufacturing jobs, to which undereducated Negroes are usually directed, shrink in New York (in the past fifteen years factory jobs in the city have dropped from 1,013,700 in 1949 to 868,700 in 1964). The jobs that open are construction jobs, skilled jobs, white-collar jobs; and for these the great majority of Negroes are still unprepared; or else they are excluded from them because of union discrimination. Generally, in 1964, American unemployment averaged 4.6 percent—but among Negroes it was 9.2 percent. And what is worst is the impact of this unemployment among Negro teen-agers; for although 15 percent of all American teen-agers seeking work were unemployed in 1964, 23 percent of Negro teen-age boys and 31 percent of Negro teen-age girls could find no work—beginning adult life, thus, with a perspective of uselessness or hopelessness.

One cannot speak, therefore, of a single Negro community in a big city. Each has two Negro communities: one that is beginning to achieve, and another that is threatened with collapse of all human values, all dignity, all function; they are almost as different as two separate ethnic groups.

And it is in the second of these communities that one must seek the sources of the uprisings of 1964; for the second, or collapsing, Negro community threatens both its black kin and its white neighbors with the greatest of all disasters: biological anarchy—a decomposition of family life and family discipline which simply cannot be contained in the traditional forms of American democracy or orderly politics.

Here, alas, a writer must contend with an issue that, when raised, so embitters Negro leaders as almost to hush friendly dialogue: the rate of illegitimate births among Negroes. Yet it is inescapable. For neither the stable elements of the decent Negro community, nor the cities as corporate bodies, nor America as a whole can absorb a pattern of mating and breeding which imposes so large and growing a population of illegitimate, yet innocent, children who mature from indifference to violence, and whose care must be charged against other families, black and white, which adhere to the common standards of American culture and Western morality.

One must begin this morbid subject with a word that is common currency among Negro intellectuals—"castration."

When Negro intellectuals use the word "castration," they mean the lack of pride of so many Negro men, which has come down from the days of slavery. Marauding white bandits and African black chieftains conspired, centuries ago, to tear apart Negro family life by buying and selling human bodies. For two hundred years thereafter in the New World white men defiled the bodies of Negro women and violated the spirit of Negro men powerless to protect the women and children they loved. Responsibility was stripped from Negro men; they could only agonize as they watched what white men did to people of their skin. Too many American Negroes bear on their faces the signature of the white man's lust and ancestry for anyone to ignore what happened. And so, the theory goes, the responsibilities of parenthood and family have become less important to Negro males than to white males.

To this historic "castration," say Negro intellectuals, has been added the newer "castration" of industrial city life. Most of the brawn-and-muscle jobs for which Negroes originally came north are disappearing; in an automated world of steel and transistors, the robust, masculine heave of shoulder and back earns less. Negro women can become schoolteachers, collect tolls on highways, become clerks or secretaries. In the United States, three Negro women attend college for every two Negro men; among whites the proportion is just the reverse. But for the uneducated healthy male the doors are closing—and the uneducated healthy male is disproportionate among Negroes. In the big city, too many Negro fathers are less and less able to support a family; and too many no longer associate copulation with family. Fatherless babies grow up in a world with no family standards of decency—a burden and menace to all about them, white and black alike.

With this, one arrives at the terrifying figures of family decomposition. In general, across the country, Negroes bear ten times as many babies out of wedlock as do white families. Nationally, one fifth of all Negro children are illegitimate. But in the big city this figure soars—so that in central Harlem, according to the last available figures, more than two fifths, or 43.7 percent, of all children were born out of wedlock in 1964. The pace of decomposition in that community can be indicated by noting that as recently as 1956 illegitimacy was already a frightening phenomenon at a rate of 33.2 percent. *More than half of all Negro children* in central Harlem today come from homes where one of the two parents is permanently absent.

What transfixes the student of the figures is the dynamics of the change—for it was not always this way, and now it gets worse. During the days of depression, white women and Negro women bore children at equal rates in the big city. Today Negro women bear children in the big city at a rate 40 percent faster than white women; and the differential between them is composed almost exactly of children born out of wedlock. And thus decent Negro families and decent white families alike find themselves pursued by the clutch of a different kind of culture in which it is accepted that children grow up wild. And both Negro and white families are haunted by the biological potential of the despairing Negro for upsetting the entire course of American urban civilization.

The development and pace of this biological anarchy in the past few years in large American cities reads like this: between 1950 and 1962, a twelve-year period, the rate of illegitimacy per thousand Negroes rose by 30 percent in Chicago, by 43 percent in Cleveland, by 70 percent in Houston—and by 78 percent in Minneapolis! One can see the charts steepen in an ominous climb as eighteen-year-old uneducated boys and girls are added to the work force. In the last five years 800,000 Negro teen-agers joined the available work force; in the next five years this number will jump to 1,500,000; and beyond that the projection rises more steeply, with an ever greater percentage of undisciplined and fatherless children in the number. Another extrapolation can be made. When Kennedy was elected in 1960, one of ten Americans was non-white. Today one of nine is non-white. When Johnson's Presidency ends in 1972, one of eight will be non-white. Today, one of seven American children under fourteen is non-white; of infants under a year, one in six is non-white.

These figures could be considered without alarm if they meant simply a numerical increase in the proportion of Negroes making up our urban population. But it is the kind of Negroes as well as their number that counts. Those who come from the zoological tenements, deprived by birth of mercy and kindness, offer no mercy or kindness to others either; and a civilization that has lost its capacity for mercy is no civilization at all, a dictum as true for central Harlem as for the Klan-controlled villages of Alabama or Mississippi. It is the city that has finished the work of the slaver who actively broke up marriages and love among Negroes; the city by its impersonality deprives Negroes of their sense of manhood—and simultaneously nourishes breeding without responsibility. "You have to realize," said George Edwards in 1963, when he was the police commissioner of the city of Detroit and the most humane and sensitive police commissioner in any great city, "that in many cases we are dealing with children who never, in all their lives, have met a decent man—a kind man, an honest man, a responsible man."

One can roam the streets of the Negro core ghettos and the eye can, of course, pick up the classic symbols of despair: the mid-morning wino on the corner staggering about with his bottle in a brown-paper sack; the cluster of half a dozen unemployed sitting on a curb watching the sun move across the sky, with nothing to do today or tomorrow or the next day; and the sluts. But nothing creates more despair than to see the teen-agers loping and loafing and exploring what little there is of life and kicks in their concrete jungle. Out of these broken homes and loveless breeding warrens come, first, the delinquents—and then the criminals.[1] Thus as they mature, like the postwar *besprizorni* of Russia, they make a larger and larger social problem. Their terrorizing of the New York subways is a condition generally referred to statistically—as, for example, in the report of the New York Transit Authority that in the year 1964 alone major crime on its underground system had jumped 52 percent from the previous year. No figures on adult crime are broken down in New York by race; but unofficially the mayor's office makes available the estimate that 80 percent of all crime in New York is committed by what are called "non-whites"—Negroes and Puerto Ricans, generally of the late teens and early twenties.[2]

[1] For an exceptionally fine study of the correlation between broken homes, illegitimate births and the problems posed to New York City by its biological crisis, the reader should go to "The Arithmetic of Delinquency," by Julius Horwitz, in *The New York Times Magazine* of January 31st, 1965. For a classic and tragic longer study of the plight and menace of these children, the reader is referred to "Youth in the Ghetto," distributed by HARYOU. HARYOU, a Negro youth center in Harlem, was perhaps the most promising exploration of the anguish of life in the ghetto before its selfless and brilliant leaders were driven out and the project gutted at the end of 1964 by a political deal which gave control of the enterprise to racist Adam Clayton Powell.

[2] Few Negro leaders care to discuss the situation publicly, but among the most courageous of them is Roy Wilkins, executive secretary of the NAACP. In an eloquent outburst of dismay, he wrote, "The teen-age Negro hoodlums in New York City are undercutting and wrecking the gains made by the hundreds of Negroes and white youngsters who went to jail for human rights. These punks . . .

These teen-agers are an element that no one knows how to handle. In New York City, 5,481 illegitimate Negro children were born out of wedlock in the year 1954. By the year 1963 this number had jumped to 11,809. There are, thus, today in New York well over 100,000 Negro youngsters, ranging in age from infants to the late teens and early twenties, denied love or dignity or patrimony or tradition or any culture but television—and they rock around on the deck of an unstable society, their bread given to them by underpaid welfare workers, their hopes zero, their mothers despised, their hearth the gutter, a subculture in a general American urban culture which itself does not know where it is going.

Let there be no mistake about it: these junior savages are a menace most of all to decent Negro families, penned by white prejudice into the same ghettos with the savages; it is the good Negro child who is first beaten up by the savage, the decent black family which suffers from the depredations of the wild ones. But they are a menace to everyone else too. And it is they who made the riots of 1964—planlessly, aimlessly, without purpose.

For this is what the politics of 1964 faced in midsummer: a revolt against the nature of life in the American city. All summer, and then slowly through the fall, the two candidates were to give their campaigns a hitherto new dimension in American politics: a discussion of the quality of American life. Two men as remote from the libraries and studies of philosophers as any who ever vied for American leadership were to begin a political discussion never before held in any country: What is man's relationship to, and responsibility for, his fellow man? But one must mark for future historians that, long before the major candidates began this discussion, the Negroes of the black ghettos had raised the matter in blood and made it central on the American agenda of the next generation.

The riots of the summer of 1964 were not race riots.

They were worse: they were anarchy, a revolt led by wild youth against authority, against discipline, against the orderly government of a society that had taken too long to pay them heed.

In almost every case the riots began in the same manner: a police episode (the killing of Powell was the only fatal initiatory step) or a

---

these hot-shots, tearing up subway cars and attacking innocent people, are selling the Freedom Riders down the river. They are helping Mississippi. These foul-mouthed smart alecks are really beating up the six-year-old Negro girls who dared to go to school in New Orleans through a screaming crowd of white mothers. The Harlem and Brooklyn morons . . . are cutting and slashing at the race's self-respect, something they can never rebuild with their knives, their baseball bats, their brass knuckles and their filthy language."

police arrest, and then a boiling over of a mob against the police, then against its own leaders, then against any symbol of authority.

Nowhere was there any organization to the riots; even the FBI, normally as quick as a hound on a hare in snuffing conspiracy, could find none. Instead, said the FBI report:

> A common characteristic of the riots was a senseless attack on all constituted authority without purpose or objective. While in the cities racial tensions were a contributing factor, none of the nine occurrences was a "race riot" in the accepted meaning of the phrase. They were not riots of Negroes against whites or whites against Negroes. And they were not a direct outgrowth of conventional civil rights protest. Victims of the rioting were often Negro storeowners as well as whites. The assaults were aimed at Negro as well as white police officers struggling to restore order. While adult troublemakers often incited the riots, the mob violence was dominated by the acts of youths ranging in age to the middle twenties. They were variously characterized by responsible people as "school dropouts," "young punks," "common hoodlums," and "drunken kids."

The rioting in New York began shortly before ten P.M. on Saturday, July 18th, after an obscure rabble rouser on a street corner at 125th Street and Seventh Avenue suddenly called for a protest march on the nearest police station. Met by barricades, the Negroes surrounded the station. When the police attempted to break up the ring, the first bottles were flung; within minutes, as the police attempted to clear the streets, the bottle throwers were joined by allies from the rooftops; and within an hour mobs were surging up and down a seven-block area of Harlem's main commercial thoroughfare, breaking windows, looting stores, smashing cars, hauling away television sets, cameras, fur coats, radios. The rioters took to the streets of Harlem again on a second night; and the next night the violence spread to the larger Negro ghetto in the Bedford-Stuyvesant sector of Brooklyn and continued there for three days.[3]

---

[3] It is important at this point to stress for the record how splendid was the police work in New York City during this week of violence, for it was almost completely ignored—and the studied indifference of the great television media to the heroism of the city's police was as significant as any other factor in the spread of the senseless rioting to other cities. For though one brick-thrower was shot to death and 92 civilians and 48 police were injured, never at any time did the police of New York lose control of the streets, even in the heart of the ghetto.

But television, with its insatiable appetite for live drama, found in the riots gorgeous spectacle. Protected by the police from the mobs, cameramen could thus catch the police at work in the ugly business of grappling with rioters, subduing

The New York riots were the trigger for the summer's discontent. In previous days, news was separated from impact and decision simply by the filter of time. A swift horse would bring Caesar's news from Gaul to Rome in ten days—and the Roman Senate thus had time to reflect on strategy and reinforcement. In Napoleonic times the Rothschilds were able to peg their financial decision on Waterloo by news brought by carrier pigeon. But television has erased the filter of time; one sees the event in the *now*, in all its stunning immediacy, without the overnight filter that the morning newspapers once provided. One sees the cop beating the rioter over the head, two helmeted policemen manhandling a Negro to a wagon; there is a jazz to the immediate news which is unsettling, and this jazz dominated the last weeks of July and early August, 1964.

On Friday evening Rochester, New York, had its riot. Two police-

---

them by clubs when necessary. For months previously, all through spring, television had displayed out of New York a procession of brilliant and eloquent Negro intellectuals who, each surpassing the other, had denounced unsparingly every facet and institution of the city's and country's power structure. New York, which has drawn to its freedoms the most brilliant Negro minds from across the country, thus provided a forum for denunciation of the life of the city by Negro thinkers which gradually, through the months, provided a moral absolution to juveniles for any form of Negro violence against any form of law and order. This moral absolution from restraint had already eroded any internal discipline in Harlem by the time the riots broke. And during the week of violence the police, pictured with their clubs and helmets over and over again, were denied the support of public opinion so essential to quiet civil disturbance. The TV media capped their summations of the riots with statements and charges of Negro leaders and interviews as inflammatory and provocative as any ever heard in so delicate a situation: charges that the police had moved up bazookas to blast Harlem (completely untrue); charges that police had shot a Negro woman in cold blood through the groin (untrue); and other charges of similar nature.

At no time, to this writer's knowledge, has television ever shown to the nation the spectacle of a Negro community of decency. All across this country, from Baldwin Hills and South Los Angeles, through Conant Gardens in Detroit, through the suburbs of Atlanta, through the finer parts of Negro Jamaica in New York, and elsewhere are Negro communities—working-class as well as middle-class—of dignity, tranquillity, quiet and decency far above the average of their white neighbors. The national audience has never been shown these neat homes, these children playing in pleasant happy schoolyards. Television, reaching for a distorted dramatic effect, has ignored the triumphant achievements of such Negro communities as it ignores, generally, Negro political leadership of goodwill. Negroes are generally shown living in trash and slums, in garbage and in filth—a condition which embitters those who try so hard to live decently and discourages those in the slums who are led to believe that this is their only end. The New York riots were capped by a national network's half-hour show of one of the foulest and worst blocks in Harlem, as if indicting the city and holding the city responsible for the riots. Completely ignoring the fact that the City of New York has built more public housing in the ghetto of Harlem than most other cities in the nation have built for all their people, or that Harlem, to an overwhelming degree, is composed of decent people living in decency, TV gave its implied absolution to the rioters, further embittering the best of Harlem's Negro leadership and terrifying the whites already simmering and ready for counterexplosion of their own.

men interfered in a dispute between two Negroes at a street dance, and the town came apart; abandoning the Negro area, the police called on the National Guard, and more than 1,000 troops were mobilized to restore order.[4] On Sunday, August 2nd, the rioting had spread to northern New Jersey, where decaying Jersey City saw the first outbreaks and a week later, Patterson and Elizabeth; then, again with a jump, to Dixmoor, a Chicago suburb, and finally to Philadelphia.

These events thus suddenly force-fed the earliest catchword of the campaign of 1964: "backlash." "Backlash" was a word that had been coined in the summer of 1963 by a New York economic columnist, Eliot Janeway, to describe what he feared might happen if automation and economic downturn combined to squeeze factory employment down in the near future. In any competition between Negro and white working-men for jobs in a shrinking market, Janeway feared that white workers might "lash back" at Negro competitors. Thus the word "backlash"—a fashionable word among political commentators as the year 1964 opened, but one whose political impact was unproved.

By spring, however, the word had become a term critical to political judgment as Governor George Wallace of Alabama invaded the Northern Democratic primaries to test whether racism could magnetize votes in the North as well as in the South. Campaigning in the Democratic primaries of Wisconsin, Indiana and Maryland, Wallace astounded political observers not so much by the percentage of votes he could draw for simple bigotry (34 percent of the Democratic vote in Wisconsin, 30 percent in Indiana, 43 percent in Maryland) as by the groups from whom he drew his votes. For he demonstrated pragmatically and for the first time the fear that white working-class Americans have of Negroes. In Wisconsin he scored heavily in the predominately Italian, Polish and Serb working-class neighborhoods of Milwaukee's south side; in the mill town of Gary, Indiana, he actually carried every white precinct in the city among Democratic voters; and in Maryland he did better, running almost as strong among the steelworkers of Baltimore as among the hereditary racists of the Eastern Shore.

Backlash, as we shall see shortly, was a midsummer political thunderhead—frightfully black and dangerous as it approached, but then over very quickly.

[4] Rochester, a town of 319,000, is a town of industries requiring high skills, its chief employer being Eastman Kodak. In Rochester the Negro population is now about 24,000–a sevenfold jump since the census of 1940. These Negroes, who originally arrived in upstate New York as unskilled migrant farm labor from the South, have been drawn into the Negro settlement of the city without either the skills or the education required to give them hope or career in Rochester, which, without inviting them, has now become responsible for their lives.

Yet what was included in the vague term had the most somber implications for Democratic leaders. For years—ever since the triumph of Franklin D. Roosevelt—the Democratic power base in the big cities had been an alliance of the workingmen in their unions, the minority ethnic communities and the Negroes. Backlash implied that this power base could be dissolved at the polls if the workingman were examined realistically—both as a union member and a community member.

For backlash provoked a long-range re-examination of the term "integration," which is supposedly the modern extrapolation of the old melting-pot theory. But the fact is that the melting pot melts different people at different rates. From Boston to Los Angeles, most urban Americans live in neighborhoods of their own "kind," with people of their own origins, traditions, tastes and holidays. Americans usually leave the "old neighborhood" to move up, and as they move, most ethnic groups in America sort themselves out by income and class. In the new neighborhoods they "mix" with "strangers" only as their education grows and they find themselves closer to people of kindred interest than to people of kindred roots. Businessmen mix with businessmen, professionals with professionals, lawyers with lawyers, politicians with politicians. The lower the educational level, the more people tend to cluster with their "own kind," seeking social life with those who live nearby.

What makes a neighborhood different from a ghetto is simply that a man can always choose to leave his neighborhood—but in a ghetto he is locked in. When Negro leaders call their neighborhoods ghettos, they mean they are locked in. And thus the black ghettos become, for sensitive and aspiring Negroes, prisons. Many Negro leaders are convinced that if Negroes were free to move anywhere they wanted, they would, like most ethnic groups, prefer to live with their own and sort themselves out, as the other groups have done, into the quiet and the noisy, the diligent and the shiftless. But they are given no elbow room.

They expand, therefore, only by crude pressure of numbers—the resistance outside usually determining the rate and direction of growth. Across the country, Negroes usually gobble up Jewish neighborhoods first, then Italian neighborhoods. In the folklore of the big cities, Negroes meet their toughest resistance in Polish or Irish neighborhoods, so that police watch such frontiers everywhere for flashes of violence.

In the big cities, thus, two primordial emotions confront each other: one is that of the Negro who "wants out" from the violence of the ghetto, who wants room and dreaming space. The other is that of the white workingman who has an equally deep emotional need and hunger to live comfortably with neighbors like himself. For many a white workingman, his two-family house is the dream-end symbol of life's striving.

He plans to live on the lower floor, rent the upper floor, and add the rent to Social Security or pension, finishing his life in modest contentment near his friends, watching his grandchildren grow up, feeding them tidbits from the food of the old country at the old delicatessen. For him, the Negro neighbor ravages the dream.

Most American workingmen will work on the job with and, frequently, befriend Negro fellow workers. As working people, the unions and the Negroes have made their progress together. But when he goes home the white workingman joins another community—the home neighborhood community. It is this community which the Negro population explosion threatens; and as it threatens, it scissors the ethnic minorities and the working people into a schizophrenia. And thus—backlash, as George Wallace discovered it in the spring of 1964.

There was never any reality to George Wallace's threat to mobilize the backlash in the spring and summer of 1964; what emotions this narrow-minded, grotesquely provincial man aroused soon swung to Barry Goldwater.

But the riots were important to Democratic strategy. In New York, as Negro pickets picketed police headquarters, several hundred Italian-American youngsters demonstrated against them, bearing GOLDWATER FOR PRESIDENT signs. In Chicago the *Sun-Times*, in July, took a plant-gate poll at the South Chicago steel mills, whose workers are overwhelmingly Democratic—and discovered that they were voting for Johnson by the perilously narrow margin of 53 percent for Johnson and 47 percent for Goldwater. Could Goldwater exploit the riots? Would he? For a brief but anguishing moment, leadership on both sides pondered whether an American Presidential election could indeed be made to turn on the race issue.

In politics the things that do not happen are frequently as significant as those that do. For the riots surged, reached their crest, alarmed the white cities—and then evaporated. The fact that they were contained, then sterilized, then removed from public attention, their rhythm somehow broken, is in all respects a negative fact—but a principal one. For it marked the triumph of all concerned—white *and* black—who strove to bring Negro fury back into normal American political channels.

Exactly how the riots were eliminated from the politics of 1964 is not yet known—and perhaps will not be known for many years, the sequence of events that capped them being secret.

One part of the major decision was admirably shared by the two men about to contest the Presidency—Goldwater and Johnson. Both agreed that the election must not turn on the violence inherent in the confrontation of the races. And it should be recorded that it was Barry Goldwater who, on his own initiative, approached the President at the height of the rioting (on the afternoon of July 24th) and volunteered to eliminate entirely any appeal to passion of race in the fall campaign, to which the President agreed in private compact. In so doing, Goldwater yielded certainly the strongest emotional appeal his campaign might have aroused. Nor did he later, even in certain disaster, break his agreement: flatly, he vetoed his staff when they brought to him, as a desperate solution to the widening gap between him and Johnson, the most electrifying campaign film of street violence ever prepared. (See page 332, footnote.)

It is difficult to trace or define all the activities of the President of the United States (see Chapter Nine) during this period. Yet his prodigious efforts, coupled with those of Attorney General Kennedy, were everywhere. Every form of pressure—political, financial, investigative and persuasive—was quietly applied to the problem in maneuvers directed from the White House. White liberals who normally finance Negro movements in the North were brought to heel and persuaded to threaten a cut-off of all moneys. The large sums of money appropriated for registration of Negro voters by the Democratic National Committee were held up long enough to make sure that every penny so directed would go to registration and not into hell-raising. At one point in the spring, civil-rights leaders had discussed bringing 50,000 Negro demonstrators into Atlantic City to overawe the Democratic National Convention. Now Federal surveillance skeined every big Northern city, exerting itself down to such minor details as pinpointing chartered buses which might be carrying known troublemakers to Atlantic City and the Democratic Convention—and halting them. Civil-rights leaders were persuaded to issue a call to halt all demonstrations; all national leaders except James Farmer of CORE and John Lewis of SNCC concurred. And slowly, as word got through to the streets to "cool it," the pace of the demonstrations slackened, so that by late August Johnson was free to propose and find accepted with enthusiasm the name of Hubert Humphrey, one of the greatest champions of Negro rights in American history, for the Vice-Presidency. By fall, as the children went back to school, the riots came to an end (the last: Philadelphia, August 28th-31st, over $2,500,000 of damages, 2 killed, 339 injured).

With quiet restored, politics could return to the formality of issues, personalities and cross-firing denunciations. Yet the impact of the riots was certain to last for years—if, indeed, they were not to repeat themselves in 1965, 1966 or 1967—and we should examine the perspectives that framed them. For if the characteristics of the riots were senselessness and anarchy, they were only reflections of two larger aspects of the greater problem: the problem of dogma and the problem of leadership that push up as anyone attempts to think through to a solution of the race problem in our cities.

Despair incubated the riots: but dogma created the thought climate which released them: the dogma that all ills within the Negro big-city community are the fault of white men alone; the dogma that all will be solved by the acceptance of a semantic cure embodied in the word "integration."

The first of these dogmas is best summed up in the phrase which dominates Negro thinking about American life: "power structure." To Negro intellectuals, the term "power structure" means not only government but every system or combination of decision-making in American life and every crevice of privilege within it. It means the men who control the banks, industry and the great insurance companies; the men who control newspapers, magazines and television; it means the schools and their administrators; it means the churches and their hierarchies; it means labor leaders and their hiring practices and seniorities.

In these elements of the power structure, Negroes have leverage in only one: government. They have won this leverage by translating the growth of Negro population in the big cities of the North into solid blocs of votes that swing six to eight of our largest states one way or the other. Exploiting this strength, Negro leaders insist that government use *its* leverage on all the other elements of the power structure, public and private. They propose that government press itself into *every* area of decision, that it penetrate, dominate and purge the most private areas of American life until discrimination in every form is abolished.

When used soberly, the term "power structure" has great descriptive validity. For, although Negroes have rarely successfully explored, as other ethnic groups have, the opportunity of creating their own industries and enterprises to make a power structure of their own, it is rather late in American history to do so. Giant corporations grow larger, small businesses perish, and Negroes cannot create the industries and leadership groups that the Irish, the Italians, the Germans and the Jews all created when opportunity seemed closed to those groups, too. Unless government, playing its balancing role against the new and giant corporate bureaucracies, insists that these bureaucracies open opportunity to Negroes, too, they will remain forever closed. And unless government, in the personality of its leadership, accepts Negroes and promotes them, then the example of its indifference will be followed everywhere.

But the term "power structure" when translated down to the level of the street becomes the term "whitey"—and "whitey" in the Negro ghetto is as contemptuous a word as "nigger" is in a Southern village. The dogma of "power structure," as translated down, means that "whitey" made it this way—that "whitey" filled the streets with trash, bred the babies with no fathers, soiled the schools and houses, as well as closed off the jobs. Translated out to where it reaches juvenile hoodlums in the streets, it lifts from them the responsibility for anything—"whitey" made it this way, so "whitey" has got to fix it. ("You're supposed to be so smart, man, you fix it—you made it this way.") And in making "whitey" the enemy, and government responsible for all evil, the dogma paralyzes both the development of order in the Negro communities and the assumption of responsibility for order by Negro leaders. The dogma requires Negro leaders, too, to denounce the entire power structure; it forbids them to recognize its awkward, ponderous but sincere effort to wipe out prejudice. They must stand by and watch rioting and looting—and explain to their children and followers that it is wrongful, yet that the power structure caused it.

Dogma, most of all, paralyzes goals and thinking—thus the words "segregation" and "integration" are equally powerful as motivating forces. But while segregation is a specific and illegal violence to the human spirit of man—insisting that the color of skin must lock him off from opportunity—"integration" is a term never defined but always sanctified. And, at level after level, the blind application of the word "integration" defeats the purpose it proposes to serve. After some ten years of reporting on integration in housing and schooling in American cities, this writer must insist that in no area of American public life does greater hypocrisy exist. In housing, one of the Negroes' great champions, Saul Alinsky of Chicago, has defined "integration" as the period in a neighborhood's life "between the day the first Negro family moves in and the last white family moves out." Almost all Negro leaders agree with him—for the Negro population explosion is so great that almost any crevice of entry into a white neighborhood becomes, within years, a breach through which thousands of other Negroes pour to create another black or "segregated" neighborhood to be denounced.

Overwhelmingly, Negro leaders agree that stable neighborhoods of mixed ethnic groups can be created at this transition stage of life only by what is euphemistically called "controlled integration" or "benign quotas." Leaders of NAACP quietly and realistically negotiate for entry into new housing developments on a controlled basis—but must deny, publicly, that they are doing so, for to do so would be to flout "integration." In New York, leaders in its public-housing administration will quietly tell a reporter how desperately, and by what specific measures, they attempt to control the entry of Negroes into public-housing projects to keep some sort of integrated balance—but must officially deny they are doing so.[5] To eliminate the hard-core Negro ghettos, the most meticulous and detailed planning must take place. But dogma prevents such planning; and thus catastrophe is always imminent. No sadder event took place in the election of 1964 than the repeal of the Rumford Fair Housing Act by the voters of California; yet, faced with the all-or-nothing choice between "integration" by an overflooding Negro population or a system of white communities locked up for their own protection, California chose the lock-up.[6]

The catchword "integration" serves its purpose worst, however, in public schooling. The dogma of "integration" requires that "neighborhoods" be broken up so as to mix and homogenize ethnic groups whether they wish it or not, whether they are good or bad. The prime examples of result mocking purpose lie, of course, in Washington and New York. Unplanned "integration" in Washington, D.C.—the first great city in America to live with a Negro majority—has brought about the most "segregated" school system outside of the South. And a similar erratic and unthinking bowing to dogma is in the process of destroying community support of the public school system in New York. On Manhattan Island the proportion of white children in the public elementary schools of that New York borough has declined from about 60 percent in 1945 to 45 percent in 1954 to 22.8 percent in 1964.[7] Yet perhaps worst of all effects of dogma on school "integration" is to create the thought among

[5] Only one public agency, the New Jersey Advisory Committee to the U.S. Commission on Civil Rights, has ever had the courage to suggest publicly what all housing students know privately—that Negroes can be introduced into new communities and these communities kept stable only on an orderly basis of "benign quotas."

[6] One could not help but be perplexed later on in 1964, after the elections, when the students of the University of California rioted on their campus. The Bay Area students had been among the vanguard of the invasion of Mississippi, and their campus had been a base for recruiting civil-rights missionaries for the South. Yet only one year before, Berkeley, the town in which they dwelled, had called a referendum on whether to repeal its open-housing ordinance or not; and the vibrant students, so readily excited by discrimination in Mississippi, had refrained from any participation in a test of discrimination in their home town. The ordinance was repealed by a vote of 22,720 to 20,325 thus barring Negroes from free housing in the university city. Within a year, ironically, the students destroyed the peace of their campus to protest their right to help Negroes in Mississippi.

[7] The figures for 1945 and 1954 are only estimates. "We weren't supposed to count them by color until 1958," said one sad spokesman of the New York City Board of Education. "Up until then a kid was a kid, no matter what his color, and we tried to educate them all as best we could." The speed of change has been, of course, phenomenal. In 1954, in an early survey, I was told that white children would remain a majority in the city-wide elementary schools as long as could be seen ahead—or at least a half-century. But on January 15th, 1965, the Board of Education reported they were now a minority—49.2 percent—on a New York city-wide basis.

Negroes that they *are* inferior in ability to white schoolchildren—and that only if white children are forcibly transported or directed to schools of Negro majorities can some contagion of learning be ignited among them.

Such dogmas have paralyzed and divided Negro leadership, and in their dilemma and its resolution the fate of American cities will in large measure rest.

For no more delicate or exciting or difficult adventure goes on in American life than the effort of Negroes to find leadership among their own. Their problem is uniquely complicated. They cannot measure their leaders either by concrete accomplishment or concrete blunder in a self-governing, independent community. They must measure their leaders' efforts within the greater, many-purposed efforts of the embracing white community, and blurred by white men's reports and black intellectuals' critique, where achievement and rhetoric can never be separated.

Partly, and to a larger degree, the problem of Negro judgment on its own leadership is complicated by white men's contempt—and the spill-over of that contempt to Negroes themselves. Roy Wilkins explains it by saying that the white man has so leveled all Negroes, making them all so cheap and mean in each other's eyes, that they measure their heroes by the degree of white bitterness they arouse. (Wilkins remembers discussing Ralph Bunche with a porter in a railway station and the porter saying in conclusion, "Well, that Ralph Bunche still has to get his hair cut in the same barbershop I get mine.") Partly, too, the search for leadership is complicated by television's search for the dramatic artist—the fiery Negro who can inflame passion and make good shows, which leaves the responsible Negro at a disadvantage as he tries to define in public the main thrust of the complex demands of his people and adjust it to the greater responsibility he bears to the American people. And, lastly, by the fact that the long-range strategists who consider the plight of their people are divided: Should they tear apart all American life, if that can be done, and exact revenge in chaos? Or should they try to open it, preparing their community to join in the opportunities, if opened, by adherence to the common standards of Western civilization?

It is this last, the strategic cleavage in the thinking of Negro leadership, that reflects best the muted civil war that goes on within every black community of the North. For, tormented as they are in their daily lives in the city, Negroes are divided between those who wish to earn their way to a place in American civilization as it is, and those who, in despair, prefer to make chaos. A terror reigns in most Negro communities of the North—not only a physical terror in those streets where the decent are prey for the savages, but an intellectual terror which condemns as Uncle Toms or traitors all who try to participate in the general community or lead the way to better life. Nor is this internal struggle anywhere near as simple as the phrase "civil war" might indicate, for it is overlain with what in effect is a guerilla war expressed in demonstrations, boycotts, muggings and violence led by self-appointed leaders. Like all guerilla wars, it brings the wrath of uniformed authority down on an entire un-identified group, innocent and guilty alike; so that the more spasmodic the guerilla actions, the more harsh is the reaction of nervous police. And thus the entire black community finds itself all too frequently locked together in support of black psychopaths whom so many Negroes, in-dividually, despise.

No city had made a greater effort to include Negroes in its commu-nity life—or succeeded better—than had New York by 1964. The New York City low-rent housing program in two Negro districts of the city alone—Harlem and Bedford-Stuyvesant—had come to more than 33,-340 units, or more than the low-rent housing programs for all people in the state of California, Michigan, Ohio or Massachusetts. Nor did these figures include the thousands of Negro families housed in "integrated" projects elsewhere in the city. No office of any mayor was more con-stantly in communication with what it considered to be the best Negro leadership than the office of Mayor Robert Wagner in New York when the riots broke out. Yet, as one of the Mayor's aides says, "Not one of us or not one of them had any idea how completely out of touch they were with the undermuck of Harlem, or what could happen on a hot summer night." None of the rabble rousers who took to the streets of the big cities in 1964 had any credentials for leadership except for hatred of the white man; none had risked life or dignity, as had the summer students under Robert Moses in the perils of Mississippi Yet the conditions of city life and family decomposition had provided for them adolescent troops whose moral restraint had been entirely eaten away by dramatic producers and eloquent intellectuals on television, who somehow per-suaded them that revenge for Mississippi and Alabama could be taken by looting and violence in the cities of the North.

The riots in New York—as in Philadelphia, New Jersey, Rochester and elsewhere—were finally stilled not only by force but by intercession of the political leadership that Negroes themselves had elected.

One should linger lastly, in this chapter, to look at these elected Negro leaders. For, by and large, Negroes elected by the civil vote of the Negro communities are among the finest men elected by free politics anywhere in the country. Headlines are claimed by such egregious and frightening elected exceptions to this rule as Adam Clayton Powell of

New York. Yet for every Powell there are five others in every city as well educated and responsible as the finest white legislators of the country. From Earl Brown and J. Raymond Jones of New York (who have done more for their people in New York by democratic process than all the demonstrations of the streets put together), to William Dawson of Chicago, to Edward W. Brooke of Massachusetts, to William Patrick, Jr., of Detroit, to Thomas Bradley of Los Angeles, to that pungent and perceptive thinker Louis Martin of the Democratic National Committee, these were the men most challenged by the expression of the turbulence inside the Negro ghettos which first vented in the summer riots of 1964. None of these men make headlines or draw national attention. Nor can any white man help them, for praise from the white society weakens their base in the civil war within their communities.

It was men like these who, in first line of battle, hushed the riots of 1964, which they styled "Goldwater rallies." The first order of politics for Democrats in the summer of 1964, as they approached their Convention, was somehow to preserve the leadership and authority of these men in their own communities. Thereafter it was up to such Negro leaders to mobilize their people so that the anger might be brought to the polls as votes without provoking or intensifying the backlash that might cancel out their efforts in the vote totals.

Their perspectives for the summer and the fall and the thereafter of the elections were immensely complicated. There were the Freedom Democratic Party delegates of Robert Moses' creation, arriving from Mississippi, determined with the faith of martyrs to tear the Democratic Party apart on the Convention floor. The Convention would have to be negotiated without violence at whatever necessary compromise. Thereafter Lyndon Johnson would have to be elected, with the fattest plurality of Negro votes in history, to give him the biggest margin ever. Thereafter the Federal Government might, they hoped, give them the resources to start helping their communities to help themselves. In their perspectives they had only two allies—Barry Goldwater and Lyndon Johnson; and now it was time for Lyndon Johnson's Convention.

CHAPTER NINE

LYNDON JOHNSON'S
CONVENTION

T HE Republicans had claimed the major political reporting of the
spring and summer months; the Negroes had claimed the attention
of political thinkers. Politics had been a-simmer and a-boiling
from February on.

But at the White House a pious innocence ruled. From the lowliest
White House courtier to the President himself, a question on politics
drew a reply of hurt and embarrassment, as if one were guilty of tasteless
vulgarity: President Johnson was President of all the people; when the
time came, so ran the standard response, he would shoulder his political
burdens and reluctantly descend to the marketplace of partisan politics.
Meanwhile he was too busy healing the wounds and governing to pay
attention to the sad and melancholy spectacle of those Republicans
chasing around the country saying such awful things.

All this was true. Yet it was only partly true.

The President *was* busy. He was learning his job. He was governing
well. The world was pressing at his time. He was traveling the country.
The best politics for any President is to be a good President—and Lyndon Baines Johnson was being a very good President. But not only that:
with an artfulness of performance that ranged from country-boy shyness
to the most sophisticated use of every public instrument of persuasion,
he was making sure the country knew what he was doing and how well.

For the President, alone among men, can set the climate of politics;
and as Johnson proceeded about his work it was impossible for a close-
range observer at the White House to measure what the master was doing to the weather over the land.

What made it so difficult to measure the master's work from the

White House was not only that one was too close to the volcanic personality who now presided there. Although that was part of it, the greater part of the difficulty was that no one, any longer, at the White House could say exactly what was going on in the President's mind or how he saw politics. Only the President could talk for the President— and although he was the most loquacious of Presidents, he used talk as a smoke screen, weaving in and out of his own clouds like a destroyer flotilla behind its own smoke curtains. Not even the White House staff knew where he would emerge next.

The changing of the White House guard had come gradually; but as it proceeded, it was not only revealing of the man—it was fascinating as a study of the political process at a court of power.

Under Kennedy, the undisputed prince gallant of his own team, a romantic glow bound his staff not only to his person but to each other. The White House was a community. The politics of America were discussed not only at the White House and in its offices but were chewed over for hours in the evenings and over the weekends by men who were transfixed by their participation in the thrill of power. They sharpened ideas on each other; they knew what their companions thought; they talked and talked the long nights through, so that a common body of opinion slowly shaped ideas and alternatives to be presented to the appellate court of the President's decision. If in Kennedy days one spoke, say, to Robert F. Kennedy or to Theodore Sorensen or to Kenneth O'Donnell or to Lawrence O'Brien or to Pierre Salinger, one could learn, in the round, the whole outline of the President's thinking and the color of his mood. Lesser communities contributed to the high community of the White House, slowly creating doctrine, each in its own sphere. One could speak to Rusk or Bundy or Schlesinger or Ball or Harriman and have a whole rounded view of foreign policy; one could speak to a Heller, a Dillon or a Kaysen and have a whole rounded view of economic policy; one could speak to Nicholas Katzenbach or Burke Marshall or Edwin Guthman and have an immediate reading on civil rights. One could not, indeed, find out what was happening on the Hill, where Congress met—but the White House itself was a constant seminar in national policy which, if it opened at all, opened on illumination. It was as if, under Kennedy, a fourth branch had been added to the traditional trio of executive, legislative and judicial branches. The new fourth branch was the policy-making branch of government.

To some of those on the outside, there was something almost too precious about the court of the Kennedys—too gay, too glamorous, too elegant. Outsiders sneered at the seminars of the Hickory Hill group, where several earnest Cabinet members gathered to discuss great thoughts and hear great thinkers. Outsiders denounced as Babylonian revels the parties and the swimming-pool gaieties that the press reported.[1] But a collective body of thought, opinion and intertwined loyalties and trusts grew up which, in such a great time of testing as the Cuba missile crisis, fused the body into magnificence.

Under Lyndon B. Johnson, however, it became gradually apparent that the White House staff was a staff—not a community. More of the President's staff, probably, had more immediate access to their chief than had been true in Kennedy's day, but only as staff servants. Lyndon Johnson's men operated on a radial principle—with no rim of companionship to hold them together. When one spoke to any one of them, he replied, if he replied at all, only in his own area of competence; and following the thread from such an individual led nowhere else but to the Oval Office and the President himself. Lyndon Johnson was his own Lyndon Johnson, his own Robert Kennedy, his own Sorensen, his own O'Donnell, his own Salinger.

Good men, strong men, weak men intermingled as the new staff members joined or replaced the old Kennedy staff at the White House. Correspondents tried to fit them into familiar pigeonholes. Walter Jenkins, obviously, was closest to the President, but he was no Robert Kennedy; when one could see him, one found a gentle, quiet, kindly man; he gave no insights or illuminations to outsiders—but within the walls he was the steady centripetal force that held things together; his weakness was to become public later. Jack Valenti was next—he was pegged originally as another Kenneth O'Donnell, master of the body. Whirled up overnight from Dallas on November 22, 1963, by Johnson's personal command, with no clothes and no bags, and carried to the White House

---

[1] The largest, gayest and one of the most useful of these gatherings was the most publicly deplored and the last—the Robert F. Kennedy swimming-pool party of June, 1962, described as an orgy in which drunken members of the American government one by one tossed themselves fully clad into the Kennedy pool. For those who were there it was entirely otherwise: the Kennedy children in night-gowns and bare feet peeped over the hedgerows to see old friends and great men arrive; the dogs barked as the Attorney General tried to present his mother (in whose honor the party was given); in one corner Ambassador Charles Bohlen talked earnestly in Russian about the Berlin crisis with Khrushchev's personal emissary to the Kennedy court; in another corner astronaut John Glenn talked with several Cabinet members; an earnest lady buttonholed Secretary of HEW Abraham Ribicoff for more funds for retarded children; Lyndon Johnson danced gaily among the rest. Mrs. Robert F. Kennedy did indeed slip into the pool as she walked over the moist catwalk that spanned it; Presidential adviser Arthur Schlesinger gallantly jumped in, fully clad, to help her out; and later in the evening one other member of the party slipped or stumbled into the pool from the glistening tile border where people danced. There was no drunkenness, and the party ended at three in the morning. It was good for all of these people to have talked to each other and learned from each other away from their offices.

for the first time on Air Force One, Valenti must have been staggered within those twenty-four hours by an explosive change in his life greater than in anyone else's except Lyndon Johnson's. Valenti was to grow into a man of dimension and genuine strength as the year wore on, but in the spring months he was described as the master's "choreographer" rather than the guide O'Donnell had been to Kennedy. Bill Moyers, a young man of twenty-nine, was as shy yet as important as Sorensen had been; witty, thoughtful, of definitely superior quality, he was, however, difficult to get to and so dedicated to the President's privacy that one could only vaguely guess where, in his all-embracing functions, was the main thrust of his influence. Johnson had established a pedestrian political errand-doer named Clif Carter at the Democratic National Committee to oversee the political management of John Bailey, Chairman of the Committee; but Carter seemed suspicious of anyone who came from east of the Mississippi or north of the Mason-Dixon line, and when he talked, it became apparent that he had little feeling for politics outside of the Southern courthouse states.

Conversations with all these men ended on the same note: the President was not interested in politics. The President would wait until the Republicans held their Convention. The President would see whom they nominated and then decide strategy. The President was busy governing his country. And so, too, the President himself in his sporadic bursts of press conferences and his sporadic stretches of silence.

One could learn best what was happening from the President's new Press Secretary, George Reedy—and perhaps because Reedy was so far removed from inner access to the court and to the President's thinking, he could observe what was going on as a thoughtful and veteran outsider. Reedy had succeeded Pierre Salinger, a Press Secretary of another school. Salinger had had the complete run of the White House, action authority greater than that of several Cabinet members, participation in every matter from civil rights to Cuban missile crisis. Salinger's role changed under Johnson—along with McGeorge Bundy, he was one of the two members of the old team in closest operating contact every day with the President; but he found himself demoted to servant, like others of the operating staff. It was not easy for Salinger, a sensitive man, to smile when Johnson teased him and made him the butt of public jokes. Johnson bore no ill will to Salinger, nor Salinger to Johnson; they simply were not meant for each other. On the afternoon of his quick decision to run for the Senate in California, Salinger had had difficulty getting in to see the President to tell him so; Moyers interceded and arranged for some time so that Salinger could break the news to the President. When Salinger explained his decision, the President entered no demurrer at all, no plea that Salinger stay, such as had been made three months before. Instead, in a characteristic Johnson gesture of generosity, he asked Salinger what the filing fee for the Senate was in California. Salinger said $450. The President reached in his pocket, pulled out a large wad of bills and carefully counted out $450 in cash to Salinger as the first campaign contribution—and wished him well.

With Reedy the President was more at home—Johnson had alternately cared for and abused George Reedy for twelve full years of service and had molded Reedy to his will. Reedy could not prevent Johnson from making mistakes, could not restrain him in dealing with the press; all too frequently he was left out of high decision and minor detail alike. But on those rare occasions when Reedy permitted himself to give a full-dress analysis of his master, he could explain the President better than anyone else, making him seem big and wise, cosmetizing away the warts on the giant he served.

The warts had begun to be apparent by the end of January. A president is like a horse in a horse market—experts in political livestock examine him daily, poke him, pinch him, stare at him, prod him. The gait, the stride, the breathing, the sound are all examined minutely hour by hour, day by day, until one wonders whether any human being can stand up to the strain. Even a normal man entering the White House becomes abnormally sensitive under such attention, and Lyndon Johnson was certainly no exception to the rule.

The press had begun to chip at him in January. He had received and accepted as a gift during his Vice-Presidency a $542 stereo set (retail value perhaps $800) from Bobby Baker. How could a Vice-President accept such a gift from a character like Baker? The *Wall Street Journal* in March published the first tentative exploration of the President's personal finances. How had a poor Texas boy living on government salary acquired all those millions? He had invited the press to his ranch in Texas at the end of March—and the dispatches blossomed with stories of Lyndon Johnson, beer can in one hand, hat over the speedometer, pressing on the throttle of his car and letting her rip up to 75, 80—or was it 90 miles an hour? None of the terrified correspondents who accompanied him or tried to follow him knew—but they had been frightened white by the crazy spin, and as they retold the story it became ever more hair-raising, while the President felt they had betrayed his personal hospitality. A few weeks later on the White House lawn he had pulled his two pet beagle dogs by their ears up onto their hind legs—they had yipped and yelped; dog lovers the country over were indignant; the President of the United States said it didn't hurt at all; a spokesman of the American Kennel Club disagreed. Privately, the President told a staff member he was going to lift those beagles by the ears until he trained them not to yelp.

The public President did his best to adjust to the attention of the press, to court it and please it. The inner President was adjusting, simultaneously, to an enormous range of problems—to the whole pressure of the clangorous, disturbing outer world which took up more than half his time; to the crescent and swelling problems of civil rights; to the Congress, which remained the focus of his attention; to the daily administration and guidance of his executives. But it was the press that occupied the largest part of his time. He was wooing the entire nation through it.

"In the last seven days," James Reston of *The New York Times* wrote at the end of April, "President Johnson has held two formal and two informal press conferences, made seven impromptu statements after White House meetings, delivered a major foreign policy address to the Associated Press, helped settle the national railroad dispute, opened the New York World's Fair and talked at a political rally in Chicago. The White House is now more open than any residence in Washington. 'Welcome to *your* house,' he tells his visitors, and that's the way it is. No majestic aloofness for him. He regards himself as the temporary occupant of a public building and simply does what comes naturally."

Yet the courtship did not work. In the first week of May, still trying, the President had the genuinely hospitable idea of inviting all the newsmen with their wives and children to a press conference on the White House lawn—an idea similar to an enlightened corporation executive's display of the plant where the daddies work to the children. Several hundred people showed up—not only wives and children, but parents, and sometimes grandparents, of Washington correspondents.

The event produced one of the memorable pieces of reporting of the year from Russell Baker of *The New York Times,* a gaiety so good it deserves reproducing in full:

> *Washington,* May 6—In his search for the ideal news conference, President Johnson is like a woman buying a hat. He is trying on everything.
>
> Today he tried the al fresco family frolic, a format that blends pink lemonade and "Pop Goes the Weasel" with heavy data about corporate profits and South Vietnam.
>
> While the reporters sat in the May sunshine on the south lawn of the White House and did their familiar jumping-and-asking routine, their children mashed cookies into the grass and played tag under the candy-striped awnings of the lemonade stands.

For a finale, the President let the toddlers mob him for mass photographs and, just like Captain Kangaroo on the morning kiddies' hour, showed them the proper way to pet a beagle.

When the television cameras had shut down and the happily perspiring President had finished his fourth cup of punch, Jack Valenti, a White House aide, turned to Malcolm Kilduff, an assistant press secretary, and said, "Well, Mac, how do we top this one?"

More than a thousand newsmen and their families were sprawled over the lawn when the President strode out of the White House at 4:30 P.M. and stood on a temporary bandshell platform for his opening statement.

Until that moment the occasion had seemed more like a Buckingham Palace garden party without ascots than a news conference. Two red-and-white canopied stands, manned by a dozen waiters in black, dispensed pink punch and cookies.

Under the elms, the Marine Band in dress scarlet and blue entertained with patriotic and children's airs, opening with "Merrily We Roll Along" and "Jingle Bells" and ending with "America." Children in various sizes, most of them in Sunday best, wandered around or sat still doing the following:

Sucking their fingers, scratching their necks, inspecting trash cans, grinding cookies in the lawn, taking snapshots, looking at each other through the wrong ends of opera glasses, calling "Mommy," looking awed about being dressed up on a Wednesday, sneaking glances at the White House policemen's pistols and squirming on chairs.

A Fast Opening

The President came on with a fast opening. "I thought you children deserved a press conference," he said, after everyone had settled down. "I want to prove to you that your fathers are really on the job sometimes."

That was appreciated with laughter, but then the President raced through a 12-minute monologue on affairs of state that left the youngsters with that glassy-eyed look they get when a good movie comes to "the love part."

The first squeals of protest came from the back row as the President announced that he had "told Senator Byrd we were going to do everything in the world to hold down" spending. They then ran through some figures on corporation profits while a lost two-year-old surrounded by photographers called, "Mommy! Mommy! Where are you, Mommy?"

"I asked labor to hold wage increases within bounds of . . ." the President was saying, and another small boy began systematically tearing out fistfuls of White House grass and showering it over his mother.

The President finally asked for questions, but his first answer was lost in the commotion as Mrs. Johnson walked onto the platform and took a seat. For the President, it was also a family affair. In addition to his wife and his beagles, his older daughter, Lynda Bird, also attended.

While Mr. Johnson assessed politics in Alabama, a girl of about 6 sidled up to Lynda and asked, "How old are you?"

Miss Johnson said, "I'm 20."

The little girl pondered that while the President moved on to Vietnam. Then she asked, "How old is Luci?"

"She's 16," Lynda said.

The little girl spun quickly and dashed away shouting, "Mommy! She's Lynda!"

The President went on, disposing of Indiana politics, dogs, his health, Cuba, economics. The children began drifting away, squealing, playing tag, crawling under the chairs.

A little girl in blue with a white bow in her hair sat in the shadow of McGeorge Bundy, Special Presidential Assistant for Foreign Affairs, contemplatively eating a fistful of grass.

Frustrated Crasher

The first tantrum occurred at 4:58 as the President was winding up. A small boy had made a desperate dash to crash the White House mansion and had been restrained by his mother. He howled and raged.

"We weren't invited to go inside," his mother explained. "Just be outside and see the lawns and fountains. You just don't go into other people's homes uninvited." Unpersuaded, John broke and tried again, and again was restrained, howling.

The news conference was followed by a half hour of chaos as Mr. Johnson posed in a swarm of several hundred children, showed the beagles, downed lemonade, shook hands and traded small-fry talk.

A serious 9-year-old who had come with a notebook, under teacher's orders to report to his third-grade class, was asked as he trudged out what the news conference had been about.

"Oh," he said, "India, war on Vietnam, and all that stuff."

Everyone enjoyed the carnival atmosphere of the conference and left in high spirits—except the President. He had made a real effort of friendship—but they had pushed him too far. Example—*Q:* "Many of the young people here have dogs. Now that you have brought the subject up, perhaps you would tell them the story of your beagles?" *A:* "Well, the story of my beagles are that they are very nice dogs, and I enjoy them, and I think they enjoy me, and I would like for the people to enjoy both of us." The beagle question was followed by a meaner one. Was not, asked one correspondent, the President worried about overexposure, about being seen too often on TV and in the papers? The President went into a slow burn. "Well, I strive to please," he replied, "and if you'll give me any indication of how you feel about the matter, I'll try to work it into my plans in the future. . . . I'm trying my best to accommodate them [the press]. . . . I hope all of you are enjoying it today. I sometimes think that these press conferences can be conducted just as accurately and perhaps as effectively in the President's office . . . [but] I always want to remain accessible. And I hope the press will never criticize me for being overaccessible." Shortly thereafter the Presidential lawn conference closed; he invited the children forward to have their picture taken with him; but this was his last appearance before the press for a full four weeks, during which he sulked at their ingratitude.

All the while—no word of politics, no attention, apparently, to the arena.

It was here that Reedy could explain best what was going on. For Reedy invited all who would listen to pay attention not to the personality but to the public President. And the public President was creating with such broad strokes, with such heavy reiteration, such a picture of the President-in-action as to be incomparably more effective than any particular campaign tactic or strategy.

First, the President was Getting Things Done. He was conferring on Vietnam almost weekly. He was flying over the Ohio River floods to see what the Federal government could do to help. He had settled the Panama problem and resumed diplomatic relations with that country by the beginning of April. The next week he was deep in the four-and-a-half-year-old railroad dispute and had postponed a strike for fifteen days while he sweated and beseeched and cajoled to bring labor and management together. He opened the World's Fair in New York on the morning of April 22nd, and that evening told the nation over television that the railroad strike would not happen—a settlement had been reached. And two weeks later he was off for Appalachia to grapple with poverty in a fourteen-hour non-stop tour.

For the second element of his Presidential strategy was to be visible. He was traveling—north, south, east and west—at a rate and pace that not even John F. Kennedy had matched. And as he traveled, he trailed behind him a set of speeches and headlines that drummed over and over again on the public ear.

To the press these travels were boring, the speeches repetitive.

But, away from the President, as one studied the speeches day after day they gradually assumed a quality of grand simplicity. Lyndon Johnson was not, like John F. Kennedy or Richard M. Nixon, to find his themes in mid-campaign, in the fall, on the road. He had found his themes in the first few weeks of tragedy; and, long before Convention and campaign, day after day he pounded them at Americans in every walk of life. Long before the campaign orators began to adorn the same speeches with campaign rhetoric, Lyndon Johnson's themes were as clear in his own mind as any President's ever were before.

There were the two grand themes—Prosperity and Harmony. (Peace was to supersede these as a theme after Barry Goldwater was nominated.) Week after week, month after month, the President would open a press conference with the latest economic statistics; and in formal addresses he was to package the statistics into a perspective and picture of a nation living at the peak of an all-time boom—which indeed it was. In no city he traveled, in no hotel to which the correspondents followed him, could they sleep late without being wakened by the sound of jackhammers as new buildings rose. Nowhere he paused but the throngs seemed at least half composed of platoons of amateur photographers whose intricate and expensive cameras and lenses were trinkets of the nation's carnival of buying. Month by month, each industrial index passed its predecessor—and the country, oozing fat at every pore, its bays and lakes speckled with new watercraft, its highways clogged with the all-time outpouring of the automobile industry, its vacation resorts already overbooked from Maine to Florida, was made aware of it. There was no need to politic about the boom; it was necessary only to point it out.

From his November accession to the Presidency until late May, Johnson made only two "political" speeches—both at fund-raising dinners to which John F. Kennedy had previously committed the President. But the non-political Johnson, bearer of glad tidings, was never absent from the news for more than a few days. The Dow Jones stock-market average stood at 733 on November 21st, the day before he became President; it reached 827 in April, 837 in May, 851 in July. And within the theme of prosperity lay, like the yin within the yang, the subordinate theme of poverty. For the first of his own measures before the Congress was the anti-poverty crusade; and as he traveled through Appalachia, the Midwest, the South, his presence established that the Field Marshal of Prosperity was also Commander in Chief of the War Against Poverty.

The second theme was graver—for Harmony meant civil rights. Historically, it was high drama to find Johnson, now, the great champion of Negroes and civil rights. He had come a long way from the Texas Congressman and Southern liberal who once voted liberal on all matters except civil rights and oil depletion. A record of 50 roll-call votes stretching over twenty years from 1940 to 1960, from Congress through the Senate, showed how great, and real, was the change. Over those twenty years, Lyndon Johnson had voted as a Southerner, with the other Southerners, no less than 39 times on matters of civil rights: 6 times against proposals to abolish the poll tax, 6 times against proposals to eliminate discrimination in Federal programs, twice against legislation to prohibit and punish lynching, twice to support segregation in the Armed Forces, once against a Federal Fair Employment Practices Commission, once to maintain segregation in the District of Columbia.[2]

Change had come over Lyndon Johnson slowly; but one can mark the beginning of the change with a specific date—January, 1953, and his election as Minority Leader of the Democrats in the Senate. Johnson and Mr. Sam Rayburn of the House had set out to secure Johnson's election the day after voting in 1952; and they wanted the Democratic vote to be unanimous. A handful of Northern Liberal Senators held out against the choice of Johnson; but Humphrey, a leader of the Democratic liberal caucus, insisted they *must* go with Johnson—and on a no-deal basis. When, indeed, Johnson's choice *was* made unanimous by Senate Democrats (with the vote of freshman Senator John F. Kennedy, too), something hitherto frozen in the structure of the Senate began to melt. First came a basic alteration of the seniority rules, giving each Democratic

---

[2] Campaigning in 1948 for the Senate seat in Texas, he had said: "This civil rights program, about which you have heard so much, is a farce and a sham—an effort to set up a police state in the guise of liberty. I am opposed to that program. I fought it in Congress. It is the province of the state to run its own elections. I am opposed to the anti-lynching bill because the federal government has no more business enacting a law against one kind of murder than another. I am against the FEPC because if a man can tell you whom you must hire, he can tell you whom you cannot employ. I have met this head on" (as cited in *The Texas Observer*, June 10th, 1960).

Senator at least one assignment to a major committee; there followed, then, the beginning of a dialogue between Johnson and Humphrey to define some kind of civil-rights bill which they might propose, and pass, without the threat of filibuster. The Supreme Court decision on school desegregation in 1954 hardened, however, the mood of the Southern Senators which Johnson had believed to be mellowing; and it was left to Eisenhower, in 1957, to propose and then pass the first modern civil-rights legislation of the United States Congress.

Johnson by then was Majority Leader and had been swept far from his Southern anchorages to a sense of the broader flow of the country's politics. Pressed by these national forces, enticed by the Presidential campaign of 1960, in which he hoped to be a candidate, he followed a new direction in his voting. He broke more and more consistently with his old friends, until, in 1960, he defied them on eight identifiable roll calls to vote *for* civil rights.

His service as Vice-President deepened his commitment; as Chairman of the President's Committee on Equal Employment Opportunity, he had taken his work with great seriousness; he had met in this commission some of the ablest Negroes in American government and politics and had been impressed by them—as they were by him. Grappling with the problems of Negro opportunity, he became, so say the Negroes who sat with him, utterly fascinated by the complexity of the matter, almost as if it were a challenge to his ingenuity. Thus, when he rose to speak in Congress on November 27th, five days after John F. Kennedy's death, no man could have been more sincere as he said, "No memorial oration or eulogy could more eloquently honor President Kennedy's memory than the earliest possible passage of the Civil Rights Bill for which he fought so long. We have talked long enough in this country about equal rights. . . . It is time now to write the next chapter—and to write it in the books of law."

As President, now in the spring months, he preached harmony with a hard and stubborn core: he meant to pass the Civil Rights Act before Congress quit for the summer. While Senators Humphrey and Kuchel led the floor fight in the Senate, the President backed them not only privately ("Hold their feet over the fire, Hubert, hold their feet over the fire") but publicly in formal statement, in speech, in press conference and in travel, preaching the new gospel not only in the North but in the South too.

The drumbeat of headlines, the snatch of news film at the end of each day might sound repetitious. But the country was being indoctrinated—and so skillfully, with such apparently artless simplicity, that it was scarcely aware of the process. A White House dinner for 61 executives would be balanced by a speech to the United Automobile Workers; a speech to the U.S. Chamber of Commerce would be balanced by a speech to the Communications Workers; a visit of ten businessmen to the White House would be followed by a visit of precisely ten labor leaders. And almost without notice, as if musing aloud, he slipped into a speech at Atlantic City at a dinner on May 9th, 1964 (attended by the Duke and Duchess of Windsor), the remark that he was going to the country with a program aimed at building "a great society." Two weeks later at Ann Arbor, Michigan, Lyndon Johnson had elevated the phrase to "the GREAT SOCIETY" (in capitals) and was describing (as we shall see later) all the uplands of the new civilization to which America could be guided. Kennedy had demanded sacrifice; Johnson promised happiness.

Even the quaking globe seemed to settle down during spring and summer to permit Johnson to conduct his foreign affairs from what may be called an at-ease position. Vietnam was the only crisis, slowly worsening from week to week—but the President arranged temporarily to sterilize that politically. Kennedy had sent Republican Henry Cabot Lodge to Vietnam ten months earlier to lift that war out of domestic debate; Harry Truman, twenty years earlier, had tried to sterilize the China issue by sending the greatest of our generals, General George Marshall, to Nanking; now Johnson similarly offered the non-political name of General Maxwell Taylor, the scholarly chairman of the Joint Chiefs, to supervise that problem. Beyond that he was cool: De Gaulle would kick America in the shins—the President would not even wince. The MLF, by sleight of hand, was slowly being withdrawn from urgency and gradually pushed back to simmer on a back burner. Africa and Latin America were restless, but none of their rumblings seemed to fluster him. At the rate of two, sometimes three, a week, came the ceremonious arrivals of chiefs of state and foreign ministers at the White House: overnight in Williamsburg or Philadelphia, then by helicopter to the White House lawn, where the Marine Band blew, the honor guard glittered, the twenty-one-gun salute boomed, and the burly President towered down to welcome them first to his office, then to Blair House. The visits came so often, the ceremonies were so repetitious, that a complacent boredom surrounded each of them—comforting by its very dullness.

Months before Barry Goldwater was nominated, the political impact of the "non-political" Johnson could be measured by any inquirer. A survey of the Business Council in May, at its annual gathering at Hot Springs, Virginia, found a majority already leaning to Lyndon Johnson. At the end of May, Henry Ford II announced that, whoever the Republican candidate was in July, *he* was going to vote for Lyndon Johnson. So, too, was Walter Reuther, a fellow Detroiter who, as the greatest of American labor leaders, had initially opposed Lyndon Johnson's nomination as Vice-President in Los Angeles in 1960.

Correspondents on the road with Rockefeller and Goldwater as they fought the Republican primaries would poll the outer fringes of the Republican crowds who had come to hear the aspirants—and over and over again one would hear: "I like that Lyndon Johnson, though," or "Ain't made up my mind yet between Goldwater and Rockefeller —might even vote for that Johnson in November." The irritations and annoyances that one found in Johnson, close up, translated into positive echoes in the country—the penny-pinching darkening of the lights at the White House pleased home penny-pinchers in a nation whose millions remembered fathers or mothers insisting that the light be turned off. Nor could the pollsters miss what was happening: they called it the Johnson phenomenon. There he was in June with 81 percent of the vote against Goldwater (Gallup) or 74 percent of the vote (Harris).

But wait, said everyone, wait until the Republicans choose *their* candidate and the nation gets down to the business of politics.

No, said the White House, no—we aren't in politics yet; no, we aren't campaigning; no, we have no political speeches scheduled.

And yet at some time which can be made no more precise than the turning of March into April, Lyndon Johnson had begun to consider the mechanics of politics and the response he hoped to draw in the first partisan election he had ever directed. It is at this point that one must try to pick up the first strands of the national campaign as woven from the White House.

Lyndon Johnson had never before in his life fought a partisan campaign. He had grown up in the Texas Democratic Party in the days when Texas, like all the rest of the South, was one-party country. Politics was fought, not party against party, but personality against personality in the Democratic family's primary. If a Democrat elected to stay on in Texas state politics, he was involved in the unending feuding of the state legislature at Austin, the guerrilla war of liberals against conservatives from county caucus to railway commission. But if he went on to Washington, as Lyndon Johnson did, he left that kind of politics behind and could run year after year, against zero opposition, confident of re-election as a liberal just so long as he took care of the home folks and did not offend the big oil people.

Lyndon Johnson had never, until 1964, crossed swords with a Republican, and political talk—as party politics are talked about in the

North—was alien to him. His Washington politics was the politics of the Hill and the Senate, at which he was the acknowledged master; his one major effort to translate that kind of politics into national politics in 1960 had been such a hapless effort that the Kennedys had minced him to ground meat everywhere outside the South.

One could now pick up the first technical strand of Johnson's campaign, his own first try at a national contest, when by late March, 1964, he directed his attention to television. A television-station owner himself, he was fascinated by the home set (in the White House he had installed in his bedroom a horizontal three-set deck so that he could watch all three networks at once and control the voices with a flip switch), and instinct as well as reality told him how important television would be in this campaign. Kennedy had settled on the New York advertising firm of Doyle Dane Bernbach (see Chapter Eleven) as his preferred choice to direct the Democratic television effort in the campaign. Now on March 19th, 1964, Johnson confirmed the choice and directed the agency to plan and coordinate with him through Bill Moyers at the White House and Moyers' young friend Lloyd Wright at the National Committee. Thereafter, though the President's personal attention seemed to flag, events began to move. With a sure candidate, good lead time, and the attentive guidance of two young men who held the President's authority, the New York specialists were able to begin on a political television campaign that fully warranted the hackneyed adjective "creative." (See Chapter Eleven.)

A second strand of the Johnsonian thinking could be traced in the gradual growth of his interest in polling from curiosity to fascination. Like any Southern political figure accustomed to a personality contest, Johnson during campaigning had most of all wanted to know "How'm I doing?" and polls in his earlier career had been useful only as measures of personal impact. This attitude began to change slowly in the spring with the record of George Wallace in the Northern primaries (see Chapter Eight), for the President of the United States, so intent on passage of his Civil Rights Bill, could not but be concerned with the meaning of the surprising Wallace vote in the Democratic primary. He had read, like every political figure, the Gallup and the Harris polls, generally published. But he had read them as personality readings.

Now, in the spring, Walter Jenkins commissioned for the President a confidential poll on the Maryland primary—where Wallace had won 43 percent of the Democratic vote—conducted by Oliver Quayle and Company. Out of it, Quayle had produced a fifty-five-page technical report heavy with the terminology of the pollster: themes, voter profiles, issue measurements. Johnson read the poll overnight and was delighted; calling for more, he was supplied with Quayle polls on the Indiana and Wisconsin results which confirmed the original Maryland poll: that *backlash* was a *potential* threat, not yet a *real* threat. The studies fascinated Johnson with their contrasts of Republican and Democratic attitudes, with their measures of voter concern (number one: the frustration of Americans at the endless cold-war vexations) and approval (Johnson was doing a marvelous job on bread-and-butter issues). Backlash and the Negro revolt were indeed chipping away some Democratic strength in the big urban centers—but this was more than overmatched by a contrary drift of Republicans to Lyndon Johnson himself. By June, Johnson had become converted to polls, with the conversion of a man discovering a new science. From Maine, Quayle brought back a survey indicating that Johnson might pull as high as 77 percent of a vote held at the moment; and that Nixon voters of 1960, by a measure of roughly 50 percent, were willing to consider Johnson as their likely choice! From a survey of Wisconsin dairy farmers, traditionally Republican, came remarkable indication of Johnson strength. It was some time later, after Goldwater's nomination, that Quayle offered Johnson the word "frontlash," which the leader then used with a cleverness that transformed the catchword from a descriptive term into a mobilizing force.

But most of all Lyndon Johnson learned from the polls, which became his favorite reading material by June, that he was completely free to choose as Vice-President any running mate he fancied. No name suggested in any Quayle poll as Johnson's partner added to or diminished the President's winning margin more than 2 percent. Theoretically free, as any President always is, to impose his own man as Vice-President, he was politically free, too.

Yet a third strand could be unraveled in May and June at the Democratic National Committee. Here the old Kennedy machinery still functioned under John Bailey. The massive vote-registration drive which Kennedy had ordered ten days before his death had been put in the hands of robust Matthew A. Reese. (In any other campaign but the landslide of 1964 the nearly incredible output and mechanics of the Reese office would have stood out like a Matterhorn. The two million new votes that Reese's office was to register in 1964 would have changed the entire coloration of the 1960 campaign; but in Lyndon Johnson's campaign the two million new votes were only one more ingredient in a coming victory of historic dimensions.) The mobilization of Negro leadership and votes had been confided to Louis Martin of the National Committee, one of the wisest men ever to hold responsibility in high place for his community. Financing was placed in the hands of Richard Maguire, of the old Kennedy team, one of the most silent and most important men in American politics, as inscrutable as he is efficient. A President's Club—membership $1,000—was set up, state by state (and thus below the level of Federal surveillance of expenditures), which was to collect staggering sums of money to finance the campaign. Over the whole presided Chairman John Bailey, busy with the planning of the Atlantic City Convention.

At the White House in May and June one could detect a new sound and tone. Speech writer Richard N. Goodwin of the old Kennedy staff had been conscripted for duty, and as he joined the inner circle, the Presidential speeches began to glow with a new polish. ("The President's concert pianist," someone called him.) So, too, had Horace Busby, an old-fashioned Texas liberal, and author Douglass Cater, assigned to prepare a book for the President's signature. White House meetings began to gather more frequently those whose open, though unannounced, concern was practical politics. Jenkins and Moyers and Valenti and Carter of the President's New Guard joined with O'Donnell, O'Brien, Maguire and Bailey of Kennedy's Old Guard to plan forward.

Yet all who participated in these early meetings or early efforts knew them to be irrelevant—or, at most, merely housekeeping sessions to tidy up structure, organization and management of the approaching Convention at Atlantic City. Why they were irrelevant was simple: None of these meetings and no political strategist could deal with the one question central to the politics of the Democratic Party in the summer of 1964—the attitude of Lyndon B. Johnson to the Kennedy family as represented by its leader, the Attorney General, Robert F. Kennedy.

One would have to probe deep into the nature of two enormously sensitive men to savor the exquisite drama that was being enacted between them. What was at stake was the Vice-Presidency of the United States—and the succession that would almost certainly, within years, carry the man thus chosen to candidacy for the greatest office in the world.

Of Johnson, one of his friends has said that there might well be inscribed on his tombstone the epitaph which, according to legend, the Roman Sulla caused to be inscribed on his: "No man ever gave me favor or did me ill but what I have repaid him in full."

It was not that Johnson had suffered at the hands of Robert F. Kennedy. The brief tempest over Johnson's Vice-Presidential nomination (see Appendix B) he still ascribed to Bobby, although Bobby, that day, had been chiefly attempting to warn Johnson of the fight he faced from

Northern and Eastern liberal and labor forces, and alert him to the possibility of a sharply disputed floor struggle. But Johnson had forgiven other more dedicated enemies than Bobby, and Johnson's genius lay in reconciling enmities.

What Johnson found difficult to forgive was indifference. And the indifference of Robert F. Kennedy to Lyndon Johnson during his brother's Presidency had been embittering. So, too, was Kennedy's abruptness of manner. Kennedy, a man as straightforward as Johnson is complicated, had not so much offended Johnson as ignored him. Where President Kennedy had consulted Vice-President Johnson on all manner of matters and exerted himself to include Johnson both in decision and in counsel, the Attorney General, in all the vast range of his vigorous administration of the Department of Justice, had refrained from any initiative in seeking the advice of the Vice-President, one of the elder statesmen of Congress, on any of the problems of his Department. Had Johnson offered to help the Attorney General in his legislation, the Attorney General would doubtless have been grateful. But only once— on the Civil Rights Bill—had the Vice-President offered to throw his vast legislative skills and resources behind a project of the Attorney General. Two proud men stood cordially, yet suspiciously, apart. Others in the Kennedy group took lead from the younger brother. "Johnson," says one of his present counselors, "sat on the sidelines. He thought they were messing it up in Congress. He was like an old coach watching the boys play and nobody let him explain the game."

It is sometimes perilous to extract great meaning from a clash of personalities. But in the clash of Robert F. Kennedy and Lyndon Johnson there is a meaning which may continue to agitate the Democratic Party for some years to come.

Robert F. Kennedy, who loved his brother more than he loved himself, saw John F. Kennedy, even while alive, as more than a person—as the flag of a cause. His brother was for him not only the occasion of brotherly love but a new departure in American purpose. Unspoken in any conversation with Robert F. Kennedy was the feeling that the old order had passed; that it was the time of youth, the postwar breed, and now was the moment to do things. Impatient, strong-willed, he even more sharply than his brother expressed the single-minded clarity with which young people see things. Always respectful to his elders but excluding them from participation, he saw the leap between what was unpleasant reality and what must be done to set it right as the leap that must be taken at all times—and at once. For him, Lyndon Johnson was all the yesterdays; for him, Lyndon Johnson was his father's generation. And when Lyndon Johnson became President, all the yesterdays were restored. Something had been snatched from his own generation and the high vision of the Kennedy administration by an assassin's bullet.

There are millions of people all across this country who feel as Robert F. Kennedy does; for them the name of Kennedy is magic, as was the name Stuart under the Hanoverian reign of the Georges; and wherever old or young devotants of the Kennedy loyalty gather, the Bonny Prince Charlie of the faith is Robert F. Kennedy. Like the Jacobites, they await the Restoration.

There is only one bitter, inescapable appreciation of American politics which the two men—Kennedy and Johnson—share: the knowledge that Lyndon Johnson was never able before 1964 to become President on his own. No Democratic Convention would nominate him; the Kennedys opened power to him.

From there the emotional departure begins. For Lyndon Johnson, thus, it was essential to prove that he could draw from the American people their own voluntary approval of his national leadership. But for Robert F. Kennedy the title papers of Lyndon Johnson to the Presidency will be forever flawed—flawed by the bullet and flawed by the generation of his age.

There were other aggravations of personality to add to this underlying difference—aggravations of language, of manner, of episode. They had begun within hours of the assassination in Dallas—all in innocence. As the cursed plane approached Washington bearing the body of John F. Kennedy, Johnson had felt that he must, in order to establish continuity, accompany the casket from the plane to the ground in the cargo lift with the family and the blood-stained widow, in the view of the nation. Robert F. Kennedy had felt otherwise—that this was a moment of private grief and that the use of the moment for political purpose was totally unfitting. He had countermanded Johnson, the President. The body of Kennedy and the family had come first from the plane, in grief before the eyes of the nation. Then, moments later, the new President, himself tortured by the occasion, had come down alone from the plane. The matter rankled on both sides. Several other episodes over the mourning weekend had offended the Kennedy family as being in bad taste. But national purpose had bound both new and old in the superb pageantry that marked the passing of power.

Johnson had tried to be kind thereafter. He had in January offered Robert F. Kennedy the therapy of action—commissioned him to travel to Southeast Asia and take the feel of the chieftains of the Malaysian dispersion. The Attorney General had worked hard at the mission, touched down and talked long with Sukarno of Indonesia, Tengku Abdul Rahman of Malaysia, Macapagal of the Philippines, Premier Ikeda of Japan. Kennedy had taken the mission seriously. But when he had returned to tell the President of his findings, he had been received just once in the Oval Office—to find himself required to brief a clutch of Congressmen in the new Presidential presence. Thereafter—no contact with the President nor any solicitation of his findings from the State Department. Johnson had tried to give Kennedy a sense of participation, but Kennedy had participated in nothing after this exertion, and felt he had been used as a decoration to paste the Kennedy name over the politics of another man.

There followed the politics of the spring. The Kennedy name was a property that conniving politicians wanted to make use of. Kennedy-for-Vice-President clubs and statements sprouted everywhere. In New Hampshire the most boorish and vulgar attempt developed, unauthorized, to use the Attorney General's name as a write-in on the ballot in the primary. It upset the President; it upset Bobby Kennedy too.

For Kennedy knew as well as anyone that only a President can choose a Vice-President; no one can force a running mate on him. It seemed incredible to the Attorney General that anyone would think that the Kennedys, with all their know-how and all their imposing connections in New England, had authorized such a clumsy effort to out-write the President of the United States on a New Hampshire ballot. Kennedy's own course for the future was not clear even in his own mind —but at least one part of it was absolutely clear: he would not organize or cause to be organized, or permit to be organized, any move to force himself on the President as a partner in the coming Presidential contest. Nor did he, at any moment. The Kennedy loyalists were a permanent force or element in the politics of America. Robert F. Kennedy would not mobilize them—nor would he repudiate them. The Vice-Presidential nomination must be offered him by the President of the United States—and at that point, if it were offered, they could discuss the terms of their partnership. He would not move. Nor did he move.

Thus Robert F. Kennedy as he waited.

From Lyndon Johnson's point of view—much more obscure[3]—it must have been otherwise. It was not comfortable to be considered a

---

[3] This reporter's analyses of Lyndon B. Johnson are subject to the possibility of involuntary error. Although Lyndon Johnson occupied a major part of my time and travels for over a year, I followed or accompanied him in the group of reporters assigned to the White House. I am grateful for the courtesy and confidences he permitted so many members of his staff to offer me; but I must record that the President never discussed his planning or motivations with me personally.

usurper. A President must be President in his own right. An astute
politician, Johnson knew how resoundingly he had been defeated in 1960
by the Kennedys at Los Angeles. But now he must govern as President
on his own. To attach a Kennedy name to his own name would mean
forever sharing the title of the Presidency with a ghost of the past. Moreover—a President must be free. The New Hampshire write-in
for Kennedy had annoyed him; though Kennedy himself deplored the
write-in, nonetheless Johnson felt he had been crowded. Further, the
press unendingly murmured with nostalgic comparisons of Johnson's administration and the Kennedy administration. If Johnson were to prove
himself in a campaign for the Presidency, he must prove himself *alone*.

By June, certainly, the decision had firmed in his mind. His polls
had now reported again and again that no choice of Vice-Presidential
candidate would add or subtract more than a few percentiles from the
measure of public opinion.

Yet how would he go about it?

The answer to the question (had it been answerable at the
time) would have been found only within the innermost, undefinable
command cell of the Johnson campaign. For almost all the men who
joined Johnson in his campaign were either men of his personal staff
or political technicians—administrators, specialists, operators, servants;
but few confidants. Those he could trust with his own inner ruminative
thinking were, indeed, only three: Abe Fortas, first, like himself a New
Deal graduate; Clark Clifford, who, along with Dean Acheson, had been
one of Harry Truman's two great lieutenants of policy; and James Rowe,
the President's friend since Roosevelt days and the chief Johnson lieutenant in the preconvention campaign of 1960. From beginning to end,
only this trio of Fortas, Rowe and Clifford understood or was consulted
by Lyndon Johnson on the true and inner problems of his own decision.

It was with these men that, in late June and early July, Johnson
had first discussed and then decided on the necessity of dumping Bobby
Kennedy. But how—and when?

What triggered the Johnson action no one can tell. Early in June,
planning for the Democratic Convention had begun. It was to be entertained by various films which might amuse delegates otherwise
bored. One film was to be the story of Lyndon Johnson. Another would
be a tribute to John F. Kennedy—and to the makers of this film, instructions came from the White House: no clips of Bobby Kennedy in
the film. By late July another change had come about: the memorial
film to John F. Kennedy would be shown *after* the nomination for Vice-
President, not before. But then a counter-jolt: the press speculated that
Jacqueline Kennedy herself might come before the Convention to

plead the cause of her brother-in-law, the Attorney General, for Vice-President.

On Monday, July 27th, the President telephoned the Attorney General—would the Attorney General come and see him?

Of course the Attorney General would come and see him, was the reply—he had planned to fly to New York the next day for a meeting about the Kennedy Memorial Library, but that could be canceled. No, said the President, it wasn't that important; let the Attorney General keep his date in New York on Tuesday; Wednesday would do just as well.

There are several versions of the events that stretched over the days of Wednesday and Thursday, July 29th and July 30th, 1960—for at 1:00 P.M. on Wednesday the President of the United States eliminated Robert F. Kennedy from consideration for the Vice-Presidency in a direct and secret confrontation, and on Thursday evening announced to the nation that he had eliminated his entire Cabinet.

The first of the versions came from the President of the United States. On Friday, still unwinding from the tension of the decision, the President spontaneously decided to invite three eminent Washington correspondents to lunch with him. Now, any of the senior correspondents in Washington is as trustworthy as a Justice of the Supreme Court. Perhaps two together are equally trustworthy. But if three or more correspondents are together in a room, then no confidence can reasonably be expected to bind them; no off-the-record agreement can hold other correspondents, excluded from the original confidence, who will inevitably learn of it through the conversation of their fellows.

The President invited the three correspondents upstairs to his private chambers and served excellently—a few glasses of sherry, broiled half-lobster, tomato salad, watermelon, iced tea. The lunch lasted from one o'clock to five in the afternoon, and as it proceeded, the President relaxed. What impressed his guests most was how shaken he must have been by the ordeal, and how deeply he must have felt what he had assumed to be a challenge to his power over his Party. The President showed them a sheaf of fifty overnight telegrams—one from every Democratic state chairman in the Union—pledging full support and loyalty to his previous night's decision. It was as if a coup had been accomplished, as if a tenuous control had been reinforced and confirmed by a stroke of action. Then he began to talk about Bobby Kennedy:

When he'd called Bobby on Monday—the President told his three visitors—he could sense that Bobby's voice was kind of funny; he could tell it from the sound because he'd once been in the same situation as Bobby. Then, on Wednesday, when Bobby had come at one o'clock,

Bobby had sat there in the chair to the right of his desk and he'd sat behind the desk. He'd told Bobby he approved of his desire to run the
country some day, but after giving the matter a lot of thought he had decided that he wasn't going to ask him to run for Vice-President with him
this time. The President had watched Bobby when he said that, and he
remembered that Bobby had said nothing, just gulped. The President, at
this point in his recounting of events, gulped himself to show how Bobby
had done it. But Bobby had taken it quite well, the President acknowledged—he had said he was still supporting Johnson for President and
wanted to help. The President had asked Bobby to run the campaign for
him, the way he had done for his brother. But Bobby had said that
he would have to resign the Attorney General post; the President had
urged him to go ahead, but Bobby had wanted a commitment that Nicholas Katzenbach would replace him.

They had then, according to the President, discussed how the announcement should be made, and Bobby had said he wanted time to
think it over. So the President had waited all afternoon Wednesday and
all day Thursday for some way of getting the news out. But no response
had come. He'd tried to discuss the matter with Kenny O'Donnell and
Larry O'Brien, the two men in the White House in his own service who
also held power of attorney for Robert F. Kennedy. O'Donnell had said
that Bobby felt it would be arrogant for him, Bobby, to make the statement; it must come from the White House. The news had now begun to
leak. The President, who has a raconteur's relish of, as well as the art for,
a good anecdote, told how on Thursday afternoon one of those things had
happened that shouldn't happen. Earlier in the day Moyers had called
him on the direct talk-box in the President's office to say why not Sargent Shriver, Bobby's brother-in-law, as Vice-President? The President
had replied that such a choice would further offend the Kennedys. In
the afternoon, as the President had sat with Kenny O'Donnell discussing the problem, the direct talk-box on his desk had opened up once
more with the voice of Moyers calling in. Moyers had proceeded directly to the point, not knowing that O'Donnell was also listening. Moyers had reported that he had checked with Sarge Shriver directly and
Sarge Shriver had replied that the family would not object to his being
named Vice-Presidential running mate and that Bobby would not mind.
At which point O'Donnell, a man of absolutely unalloyed and perfect
loyalties, had reached directly over the President's desk, flicked down
the "talk" switch and snapped to Moyers, "The hell he wouldn't!"

The conversation of Lyndon Johnson is so fragrant, his phrases so
memorable, that it is difficult for any man not to delight friends with
their repetition. However it was, within days a full report of the four-

hour Presidential unburdening had been carried to the Attorney General. The Attorney General had believed that their private conversation and his elimination would be a matter of secrecy. He now learned that it was the talk of the town. Several days later the Attorney General met the President again and protested this breach of confidence. The President assured the Attorney General that he hadn't told *anyone* about their conversation. The Attorney General observed directly to the President that the President was not telling the truth. The President said he would check his records and calendar to see whether he had forgotten some conversation he might have had.

Thereafter, since the story was out, friends of the Attorney General began to make available Robert F. Kennedy's version of the story.

He had indeed come at one o'clock to the Oval Office and the President had sat behind his desk. The business part of the conversation had taken only a few minutes of the forty-five-minute session. The President had looked at the wall, then looked at the floor, then said that he'd been thinking about the Vice-Presidency in terms of who'd be the biggest help to the country and the Party—and of help to him, personally. And that person wasn't Bobby.

The Attorney General had said fine, and offered to help and support him. The Attorney General had been restrained during the entire conversation—he knew that the President had taken to the habit of recording conversations in his office on tape, and he could see that the buttons were down and the tape recorder was on. The President had said that the Attorney General was fully qualified to become President some day, and, though he couldn't commit himself at the moment, he wanted to help all he could. He had offered the Attorney General his choice of jobs—the embassy in London, Paris or Rome, or a Cabinet post as soon as any was made available by present incumbents. Kennedy had said he liked his own position best—Attorney General. The President had complimented him on his outstanding staff and, briefly, regretted the low quality of his own staff. Which had taken the President to a discussion of the approaching campaign, and who would run it, and some of the problems. The longest portion of the conversation had been devoted to a discussion of the impact on the campaign of the Bobby Baker case, which the Attorney General had brought up. The President expressed a low opinion of Bobby Baker, who, he felt, went where power was and had latched on to other people after 1960 when Johnson left the Senate. The President had felt that the Republicans wouldn't dare open the Baker case up because there was so much to be opened on their side, too. The President had mentioned several Republican luminaries whose records were not such as to bear a moralist's judgment. They had let the Bobby Baker case drop. And the meeting had ended.

Both the President's and the Attorney General's versions of the conversation agreed on the opening and the closing. The President remembered Bobby saying, as he left, "I could have helped you, Mr. President." As the President told it, it sounded wistful. As Bobby's friends told it, it had the smack of sardonic humor.

There is no essential difference in accounts of the events after Lyndon Johnson told Robert Kennedy he must make his way on his own. The President tried to have McGeorge Bundy arrange that Robert Kennedy announce his own withdrawal from consideration. But Kennedy insisted to Bundy that the President alone could make public the dismissal. At one point, on Thursday, the President relayed word that Bobby could draft any statement at all that would put the matter in the best light; and, at another point, passed word through O'Donnell that he would let Bobby name the Vice-President. Finally, O'Donnell, on his own authority, approved for Bobby a suggestion of the President's for a way to handle the matter.

And so on Thursday evening the President appeared before the television cameras and told the nation that he had excluded from consideration any Cabinet member or any official who sat regularly with the Cabinet. Wiped out at one stroke, therefore, were Robert McNamara of Defense, Adlai Stevenson of the United Nations, Sargent Shriver of the Peace Corps, Orville Freeman of Agriculture, Dean Rusk of State and several others. It was like rearranging books on a shelf not by title, author or subject, but by the color of their bindings; and all blue volumes were to be excluded. "I'm sorry," quipped Bobby Kennedy a few days later when he had regained his sense of humor, "I took so many nice fellows over the side with me."

The President, observant of the niceties, made sure that all those who might have felt themselves under consideration were informed in advance of the forthcoming statement. Stevenson, boating off the coast of Maine, was actually informed by *two* Presidential emissaries: first by Luther Hodges, Secretary of Commerce, and next by James Rowe of the President's inner circle. Stevenson, so often maliciously compared to Hamlet, knows that Shakespearean play well—well enough to tell a hawk from a handsaw. On learning the news, he simply telephoned an old friend in Washington and said, congratulating, "Hubert, it's you."

It was indeed to be Hubert Humphrey.

But Adlai Stevenson was premature. No one—not Hubert Humphrey nor James Rowe nor Abe Fortas nor Clark Clifford—could tell exactly what was going on in the story of the Vice-Presidency; for the choice of a Vice-President is an entirely personal decision on the part of the man who stands for the Presidency.

Nothing in the American Constitution is more imperfect than the mechanics of the choice of the Vice-President. Every stipulation of the original Constitution on the Vice-Presidency is obsolete; the thought of the Constitution makers that the electors should freely choose the second most highly qualified man in the nation for the second most important post in the Republic has proven entirely unworkable.

The "why" of this development is not difficult to explain. There is too much potential power in the Vice-Presidency. A Vice-President who chose to set himself up as dissident from the President himself could wreck the entire American system; the President is ill frequently, wrong occasionally, in controversy constantly. A scheming rival next in line of succession could find a dozen occasions in any term to shatter the unity of leadership or perhaps even warp American politics into the terrible intrigue and cruelties of the Kremlin succession or the medieval courts of Europe.

Only a President can choose a Vice-President; like the Roman Emperors, he can best provide for stable succession by exercising his own choice. All political leaders in America acknowledge this necessary freedom. Yet, nonetheless, for all his freedom, a Presidential candidate has a tormenting choice: he can choose the man who will best help him *win* an election or the man who could best govern the country if he himself should die.

In the middle years of the Republic the dilemma could be avoided by choosing an amiable incompetent who, whether elected or defeated, would vanish in the pages of history. In our generations, as the Presidency expands its power, as the President requires more help and support, the choice becomes more difficult. As a candidate, Franklin Roosevelt might find a John Nance Garner forced on him by the need to acquire enough delegates for his nomination. But free, finally, as an elected President, Roosevelt chose his Vice-Presidents like administrative deputies—first Henry A. Wallace and then, either with incredible luck or incredible wisdom, Harry Truman.

Rarely does a Presidential candidate have the good fortune of an Adlai Stevenson, who in 1956 could let the Democratic Convention choose freely on the floor between two unquestionably superior men, Senators Estes Kefauver and John F. Kennedy. For the last twenty years of American history, from 1944 to 1964, every election but one has included a Vice-Presidential candidate who, in turn, succeeded to a full try at the Presidency on his own. What is involved, therefore, in the choice of a Vice-President is the full and serious consideration of succession to power—a consideration that goes on in the mind of one man alone; and by that man's decision one other man is presented to the nation as a candidate to whom power may be transferred.

It is to Lyndon Johnson's credit that the choice of succession had been, from February, 1964, on, considered so seriously. And thus, as through spring and early summer Lyndon Johnson floated the names of personalities under his consideration, one could see all the forces in American politics passing over the stage of the President's attention.

Speculation on a Vice-Presidential choice had begun within two weeks of the assassination. It built slowly. By May, Johnson had begun to discuss it openly with visitors; by June and July it was obvious that he was conducting at once a survey and a teasing contest in which were involved and consulted governors, mayors, Senators, Congressmen, publishers, labor leaders and anyone else from whom Johnson could draw an echo of opinion.

The names that rose in the headlines and columns came up like bubbles, one after the other, in the wake of every Johnsonian conversation—and burst, sooner or later, in the extraordinary candor of other Johnsonian conversations. There were the two Kennedys, Robert F. and in-law Sargent Shriver. There was Adlai Stevenson. There was California's Governor Edmund G. Brown; New York's Mayor Robert F. Wagner; Maine's Senator Edmund S. Muskie. There was the Secretary of Defense, Robert S. McNamara. There was the senior Senator from Minnesota, Hubert H. Humphrey—and the attractive junior Senator, Eugene J. McCarthy.

As the names rippled out of the President's conversation or faded after some pungent Presidential dismissal, one could begin to see the dilemma in terms of the structure of his Party or the geography of the country. Johnson himself had been chosen by Kennedy for two prime reasons: extraordinary ability and political balance. The ability had been proven; and the balance was obvious—Kennedy, a Catholic from the Northeast, needed a Protestant from the South or West. Thus the ticket.

Would Johnson see the country in similar terms? His weakness in 1960 had been precisely in that great Northeastern region of the nation where the Democrats normally win their heaviest electoral majorities— where the workingmen, the unions, the Catholics, the Negroes, the Jews, the Irish provide the mass base of Democratic victories. All these had opposed Johnson in the nominating politics of 1960—but in November he would be their champion against the Republicans. How should he cement them to him?

There were a number of dominant considerations—and they all overlapped.

There was, first, the regional balance of the ticket. If the Republicans chose a Rockefeller, a Romney, a Scranton, Johnson would be exposed on his Northeastern flank—in the big cities where the workingman's vote and the Negro vote are so important. Yet if he chose a Northeasterner—which one? Wagner of New York? Muskie of Maine? Ribicoff of Connecticut?

But the regional problem of the Northeast was complicated by the Catholic problem. Kennedy had shown how powerful an appeal a Catholic makes to his coreligionists at the polls—and, simultaneously, shown that a Catholic of distinction was not nearly as abrasive to Protestant voters as had been feared. If Johnson were to opt for a Catholic, there were other names to be considered—like those of Brown of California and McCarthy of Minnesota.

Yet another matter was involved: should he choose a Catholic at all? Was the Presidential harness of the United States to be so fixed that always a Protestant and Catholic must balance each other, chosen on religious grounds, ultimately to make the leadership of the country as mechanically contrived as the balanced tickets of the big cities or the Hindu-Muslim ratios of the British Empire in India? If Johnson accepted Robert F. Kennedy as his running mate, he would be reaching for something besides a Catholic partner, for the Kennedy appeal was so broad, so romantic, so well earned that it reached millions beyond the Catholic faith. But if he rejected Kennedy and chose another Catholic, would it not be obvious that he was choosing a lesser man simply because he was a Catholic?

And yet again there was another consideration—executive ability. Should not the President choose simply the ablest executive of his knowledge, in case a second tragedy visited the Presidency in his term? The ablest executive of his acquaintance—indeed, the ablest executive in Washington—was Robert McNamara of Defense. But McNamara, former President of the Ford Motor Company, had been a registered Republican until 1960, when he split his ticket to vote for Kennedy as President and a Republican as Governor of Michigan. Johnson raised McNamara's name speculatively with Richard Daley, Mayor of Chicago, archetype of big-city politicians and a Kennedy loyalist. The President explained that he was mortal, might have another heart attack, and must think about the ablest man to lead the country. How about McNamara? Daley let the President know that the President was free to

pick any man he wanted—but since indeed he was mortal, if an accident were to happen, did he actually mean to turn the nation and the
Party over to a man who only four years before had been a registered
Republican? Similar soundings brought similar reactions from labor
leaders; and a crescendo of even more intense protest from other political leaders.

Slowly through June the President realized that some eliminations
had soon to take place on his list. He dined with a group of journalists
and ran through the names with them: Shriver—what did Shriver know
of the common people and of poverty? Bobby—he hadn't seen enough of
Bobby lately. McNamara—the labor people didn't like him. He lingered
over McNamara, though. Wagner—a wrinkle of indifference when he
pronounced Wagner's name. Brown—no, not Brown.

The Goldwater nomination of July clarified matters; Lyndon Johnson would not now have to protect his Northeastern flank with a Northeasterner—no man could be weaker in the Northeast than Goldwater. If
any Goldwater victory were possible—and highly unlikely it seemed—it
would have to come by adding the Midwest to the South. Goldwater,
who had advocated abolishing farm subsidies and the elimination of the
REA, had a natural weakness all through the traditionally Republican
agricultural Midwest.

Slowly the same force of political gravitation that drew the nation's
attention to the two Senators from Minnesota compelled the President
also to examine the two Midwestern Senators: Hubert Humphrey and
Eugene McCarthy.

There they were—McCarthy, a Catholic intellectual with a solid
appeal not only to his coreligionists but to a vast spectrum of intellectuals, opinion-makers, labor leaders—and Southerners, too. And Humphrey—a man whose dazzling talents spanned the range from foreign
affairs to agrarian economics, a glittering phrase-maker, a lusty campaigner, the favorite of all the labor unions, beloved by the Jews second
only to Adlai Stevenson, hero of the Negroes, and touched with affection
even by the Kennedy people who had fought him so long and crushed
him so completely in 1960. Yet Humphrey was the most eloquent and
ardent spokesman for civil rights in America—and, further, a man of so
positive a character that in the Vice-Presidency he might become a power
center in his own right.

What inner drives and what strange adventures brought Lyndon
Johnson and Hubert Humphrey to Washington and the climax of their
careers at the same time will, I hope, some day be examined and told by

an American novelist who can see the wondrous workings of American politics in terms of people.

For a full half-century the leadership of America had been dominated by men of the Eastern Seaboard and its grand academies of American culture and learning. But both Johnson and Humphrey were men of another culture—and the culture of Lyndon Johnson, who grew up in Johnson City, Texas (population 611), was far closer to that of Hubert Humphrey, who had grown up in Doland, South Dakota (population 481), than either was to the culture of Dallas or Minneapolis—or Washington, D.C., or New York. In this century America has known two Presidents who came of the yeomanry—Warren Harding and Harry Truman; but neither of these had roots so deeply entwined in the life of common people, so close to the sorrows and agonies and silent miseries of those who must work for a living. Indeed even today, no other great nation except China has leadership which comes so much from the raw stuff of simple people as these two.

They had entered the Senate at the same time in 1949—Lyndon Johnson from the right as a segregationist, and Hubert Humphrey from the left as the most vocal champion of civil rights to enter the Senate since Reconstruction. Too much in the past bound them together to let them be enemies—poor boys both, their families bowed by the depression, they had wandered the country and then become classroom teachers before entering politics. They had been hungry; and they had made good. They liked each other—yet they bristled at each other. Johnson, the darling of the Southern leadership of the Senate, was to be made Democratic leader in four years (1953); Humphrey was to enter the leadership "club" shortly thereafter, by Lyndon Johnson's intercession, as the first Northern Democratic liberal of the Bourbon inner circle. Johnson would plead Hubert's cause to the Senate elders. ("My link with the bomb throwers," he called him.) Humphrey, on the other hand, would defend Johnson to the liberals at a time when defense of Johnson was equivalent to liberal heresy. They might clash occasionally. Once when Johnson was Majority Leader, he and Humphrey had burst into a bellowing session in which Humphrey had snapped that Johnson might be a big man from a big state, but that *big* wasn't *great*—and that he had a hell of a lot to learn before he became a *great* man. But they respected each other. In the 1960 campaign in West Virginia, Humphrey, foretasting defeat by John F. Kennedy, had said to this correspondent that if he couldn't make it himself, he thought the Presidential nomination ought to go to Johnson, the best man to run the country.

Now Johnson was President. He needed no Hubert Humphrey to help him to victory. It was true that Humphrey would add to the vote-getting appeal of the ticket in the big cities of the East and in the farming states of the prairies—but this appeal would almost certainly be balanced by a loss of votes to Goldwater in the South. Yet if one were to choose a successor to the President—and one had already eliminated Robert F. Kennedy and Robert S. McNamara—to whom else could one safely entrust the government of the United States? This was a question for Johnson alone—but a question to which Humphrey and his friends had been providing, softly, subtly, then insistently, but one answer all through the spring and summer.

All campaigns in politics are unique. But none approached in delicacy the campaign of Hubert Humphrey for the Vice-Presidency of the United States. It had as its audience only one man—and that man no stranger but an old friend who knew all the candidate's strengths and weaknesses.

There were two Humphrey campaigns, actually. There was a large, external and active campaign of Hubert Humphrey's Minnesota loyalists and Northeastern friends, carefully and energetically directed by a team whose two leaders were Humphrey's longtime friend, lawyer Max Kampelman, and William Connell, Humphrey's chief executive assistant. These two did their best at once to inspirit yet control the surge of Humphrey enthusiasts—the sincere and calculating alike (for the obvious perspectives of a Vice-Presidential nomination led on to the Presidency itself and all the power which participants might share). Just as in the non-campaign of Robert F. Kennedy, crackpots, zealots and hitchhikers tried to seize the lead of this external campaign, while Connell and Kampelman strove mightily to keep them under control.

The inner campaign was directed by only two men—Hubert Humphrey and James Rowe, who could not publicly be attached to Humphrey. But Rowe and Kampelman were next-door neighbors on Highland Place.

Humphrey had spoken directly to the President about his own nomination but once, in March. The President had brought the subject up: if, said the President, there were no other political considerations later in the year, Hubert Humphrey was the kind of man the President would like to run with. But the President had made it clear that this was no promise. Humphrey had replied that he would not, of his own initiative, again speak of the matter. He would never burden the President with it, nor would his friends if he could control them—and if any did, it would be not of his knowledge or initiative. "He knows me," said Hubert Humphrey in April. "I'm here."

James Rowe conducted the other branch of the inner campaign. Rowe, who comes of that rare strain of men who love American politics without profiting from it, had had the distinction in 1960 of successively serving the three major candidates of the Democratic Party: he had directed the Humphrey campaign until the West Virginia defeat; conducted the Johnson campaign until the Convention; and participated in the Kennedy campaign thereafter. The new year presented Rowe with the exquisite opportunity of running his favorite 1960 candidates back to back as President and Vice-President. Trusted both by Johnson, who had summoned him as campaign counsel in May, and by Humphrey, with whom his friendship is deep and intimate, Rowe had been present at the sessions which brought about the dismissal of Robert Kennedy. He waited, thus, delicately, in the President's inner circle quietly urging Humphrey on Johnson, yet insistently urging Humphrey to restraint.

All through the spring Humphrey too, by his own instinct, had chosen the course of restraint, stilling his most important supporters. In the Senate, one by one, his colleagues had filed up and offered support—Inouye of Hawaii, Brewster of Maryland, Smathers of Florida, McGovern of South Dakota, Nelson of Wisconsin. At the height of the civil-rights filibuster even two Southern Senators—Robertson of Virginia and Olin Johnston of South Carolina—had offered support. To all Humphrey had said: Wait—speak only if the President asks you.

Meanwhile, at their meetings the President would occasionally tease Humphrey: once at a dinner party Johnson leaned across the hostess and asked Humphrey out loud what he thought of So-and-So for Vice-President. Again, Johnson would read the polls and tease Humphrey on his standing. Once Humphrey clipped the cover of *Life* Magazine and, at a legislative breakfast, placed under the President's orange juice a facsimile of a campaign button showing Johnson and Humphrey as running mates.

And then, on July 29th-30th, came the elimination of Robert F. Kennedy.

The night of the 30th, James Rowe came to Humphrey's Senate office bearing the advice of a friend and an amplification of the President's thinking: that if Hubert Humphrey wanted the Vice-Presidential nomination he must make it crystal-clear, come hell or high water, directly to the President that he would be a *Johnson* man as Vice-President of the United States. For the first time since March, Humphrey then spoke to the President directly about the matter. Lifting the telephone, he called the President and explained that he felt he knew the sense of the public announcement; and he wanted the President to know that he, Hubert Humphrey, understood the rules of the game. If the choice were to fall on him, he would offer with full heart, and no reservation, all the loyalty there was in him as Vice-President to the President.

Rowe offered another observation that evening: it was time for restraint to end. If Humphrey did have broad public support—now was the
time to let it be shown. For weeks Humphrey had been restraining the
Kampelman-Connell operation. He himself had been quietly active: he
had already let it be known to the old Kennedy team, O'Brien and
O'Donnell, that he expected them to be for Bobby and life was too short
for another Humphrey-Kennedy clash; but that if the Presidential decision went against Bobby, he hoped they would be for him. Others, too,
had been contacted. Now they surfaced. Governor Hughes of Iowa said
that Humphrey would be the strongest Democratic Vice-Presidential
candidate in the farm belt. Governor Brown indicated that if it could not
be he, he would prefer Hubert Humphrey. Governor Hughes of New
Jersey had already reported to Johnson his strongest support of Humphrey's candidacy. So had Governor Breathitt of Kentucky. Polls were
fed to the President showing Humphrey strongest of his potential running
mates. Governor Reynolds of Wisconsin spoke for Humphrey. Labor
leaders—chiefly Reuther, Dubinsky and Meany—urged the senior Minnesota Senator on the President. (Once Johnson asked George Meany,
President of the AFL/CIO, who his three choices were for Vice-President; Meany replied he had only one—Humphrey.) Every living former
chairman of the Democratic Party—Farley, McGrath, Mitchell, Mc-
Kinney—was led to urge Humphrey directly on Johnson. Two eminent
publishers, John Cowles and Palmer Hoyt, spoke to the President.

The President probed further. A week after the elimination of Robert F. Kennedy, Rowe was deputized to an embarrassing mission: to
examine Humphrey on all those matters in which the President must
know the record of a man who might run as his partner and become his
eventual successor. The conversation took place on a weekend at Rowe's
home—the kind of interrogation that the CIA might give before assign-
ing a man to a pressure-sensitive post: primitive, cold, fact-establishing
and concerned with the quality and record of a man who might be President. What were Humphrey's debts and assets—did any contributors'
liens exist on his future politics? Answer: None. Humphrey's record in
the war? Answer: Humphrey had applied for a Navy commission and
been rejected for color blindness, tried unsuccessfully to enlist, later
was drafted and sent home because of a double hernia. Possible romantic
escapades? Answer: No blemish.

From this conversation on, Hubert Humphrey was the leading candidate for nomination as Vice-President of the United States. Yet it
could not be certain—it could not be certain because the President of the
United States was operating in the moving pressures of the life of America.

On the one hand, July had been a month of majestic achievement in law-making. One by one, the bills Lyndon Johnson sought to make into law were becoming Acts of the Republic. On July 2nd, 1964, the Civil Rights Act had become law of the land. On July 6th the Mass Transit Act became law by the President's signature. The Civilian Pay Raise Act, the first down payment on the great debt owed by the government to the underpaid civil servants who keep it running, had been passed. Extension of the Hill-Burton Act for Hospital Construction, and the Anti-Poverty Bill were already certain of enactment when Congress should reconvene after the recess for the Republican Convention. The deft response of American planes to the jabbing of North Vietnam's torpedo boats in the Gulf of Tonkin had been carried off with the nicest balance between boldness and precision. Of the 52 major proposals before the second session of the Eighty-eighth Congress, Johnson could already look proudly forward to the final enactment of 45 measures, or 87 percent—the highest score of all the four years of the Kennedy-Johnson administration.

Yet, on the other hand, this prodigious achievement had been obscured in public attention, first by the Goldwater Convention (see Chapter Seven) and then by the detonation of Negro riots in the big cities (see Chapter Eight). If the violence of the Negroes could not be contained—if it rose from crest to crest to the final validation of Goldwater's incantation about law and order—then possibly it might boil over into an anarchy which would lead the American people to vote for the Prophet from Arizona, preferring the domestic tranquillity he promised to any risk of war abroad.

Robert F. Kennedy had, indeed, been eliminated for personal reasons—but also, secondarily, because to both North and South he was the figure and symbol of the upheaval opening equality to Negroes at every level. Yet Humphrey believed as ardently in civil rights as did Kennedy, and was almost as prominently tagged with the political burden. In the domestic turbulence of July, with no man able to say how the turbulence would crest, Humphrey, too, might prove too extreme a choice for frightened white Americans to accept. The private support of so many Southern Senators was an offsetting factor—yet the President had to balance this against the public reaction to continuing violence.

Thus, through August, as the people of the big cities where the Democrats are concentrated considered the violence of Negro communities, the President had to consider his approaching Convention, where the Negroes of Mississippi proposed to present a flat and open challenge to the custom of the community, Party and Convention. His choice of Humphrey had probably been 90 percent firm at the end of July. (During Convention week the President said to Humphrey, "If you didn't know you were Vice-President thirty days ago, maybe you're too stupid to be Vice-President.") The remaining 10 percent of indecision in the Presidential mind was a necessary margin for reconsideration; yet, as August wore on and the Negro riots came under control (see Chapter Eight), as the measures of White House and Negro leaders began to take effect in the ghettos, the margin narrowed and narrowed again. As convention week opened, what little indecision appeared to remain in the President's mind could probably be estimated as one part real and nine parts stage management.

The problem of stage management for the Democratic National Convention was at all times perplexing. The Convention offered two perspectives as it drew near. Either it would boil with authentic and violent drama as the Negroes of Mississippi presented their case to the nation—and if this happened, and if, further, the millions of black Americans who lived in the close-by ghettos of Washington, Baltimore, Philadelphia and New York joined in the drama, it might be frightful. Or else, if this drama could be avoided, then it would be the dullest gathering of delegates since San Francisco in 1956 and a public tedium.

In either case the stage was to be Atlantic City.

Of Atlantic City it may be written: better it shouldn't have happened. On a six-and-a-half-mile boardwalk, once the most fashionable resort of the Atlantic Coast, its old hotels rear themselves like monuments of another era, a strung-out Angkor Wat entwined in salt-water taffy, along a long, curving strand which is still one of the best beaches in the world. Time has overtaken it, and it has now become one of those sad gray places of entertainment which one can find across America, from Coney Island in New York to Knott's Berry Farm in California, where the poor and lower middle class grasp so hungrily for the first taste of pleasure that the affluent society begins to offer them—and find shoddy instead. Frequented now by old people on budget, by teen-agers who come for a sporting weekend, by families of limited means trying to squeeze into cramped motel rooms, it is run down and glamourless. One of the saddest minor notes of the Convention was the conversation of middle-aged and eminent delegates whose fathers, forty years before in the days of the resort's glitter, had brought them here as children on fashionable family vacations. The buildings were the same, but all else had changed.

Atlantic City is, of course, trying to recapture some of its lost glamour, and the Democratic National Convention of 1964 was to be a marker on the road back. The Convention had come to it by a process of elimination. In 1963 John F. Kennedy had hoped to hold it in San Francisco—and thereafter part of his thinking was that he and Jacqueline and the children would vacation on some mountain ranch or on the Pacific, to enjoy the Western country. But the time gap between the Republican and Democratic conventions was too long. He would have preferred Chicago next—but the city fathers of Chicago could not raise the money in the spirited bidding for such a national convention. Miami and Atlantic City were the chief rivals when it came down to final choice. But Miami was full of Cuban refugees; demonstrations against Castro might ring the Convention before the television eye and press home on the nation a point that the Republicans were only too eager to have pressed. While New Jersey Democrats promised to raise $625,000 for the event. Thus by a process of elimination the prize fell to Atlantic City. And for the leading citizens of Atlantic City, it may be said, prizes like this should be wished on their enemies.

Fifty-five hundred newsmen and technicians outnumbered the 5,260 delegates and alternates; but the newspapermen, who had come but recently from the sparkle and elegance of San Francisco, suffered from what can only be called "culture shock"—which they immediately transmitted to the nation. The bulbous souvenir busts of John F. Kennedy and Lyndon Johnson; the hawkers; the knockdown discount auctions; the hotels where room service did not function and where prices soared; the restaurants where one could not be seated and the food was bad; the honky-tonk and the tawdriness combined to produce an immediate aphorism: "This is the original Bay of Pigs." One or two of the hotels—notably such hostelries as the Shelburne and Haddon Hall—rose to the occasion with grace and efficiency. But most, accustomed to smaller conventions of Elks, Shriners, dentists and chiropractors, broke down under the demands of a political convention; by mid-morning, switchboards would collapse and the flustered operators refuse to take messages for political guests whose message mating and communication are of the essence; promised television sets did not work, and promised air conditioning proved nonexistent.

The California delegation put out a daily bulletin—sample items: "Jack Woolfe of KNXT-Los Angeles left home with instructions to send home a souvenir. He will send the piece that came off in his hand when he tried to open a window"; "Sandy Kahn of San Diego leaned on the footboard of his bed and when he straightened up, the ornament stuck to his hand"; "Ron Rieder of the Van Nuys *News* tugged on the handle of his door. It fell off. He reached for the shower faucet. You guessed it." A sample conversation overheard in a hotel corridor ran like this: First guest: "My shower doesn't work." Second guest: "Mine works except it doesn't spray into the tub." First guest: "So what do you do?" Second guest: "I just stand anywhere in the bathroom where it happens to be spraying." Never have a town and a chamber of commerce made a greater effort only to end by exposing themselves to ridicule. In fairness, however, one compliment must be paid: the kosher delicatessens of Atlantic City proved themselves outstanding, perhaps the best in the nation, spreading the reputation of short-order Jewish cuisine from the boardwalk all across the nation in that cultural cross-fertilization of American life which continues forever.

The true drama of the Convention was short-lived—or perhaps what was true in the drama was prelude to other and future conventions.

It was the Mississippi Freedom Democratic Party that provided the drama; for their delegation was the final triumph of Robert Moses' heroic summer project in the state of Mississippi (see Chapter Six). They had performed as conscience told them to perform; they had created in Mississippi an entirely new organization of Negroes called the Freedom Democratic Party; they had paid with sacrifice of life; they had now sent on to Atlantic City sixty-eight delegates and alternates (including four white civil-rights Mississipians) to demand seats as full delegates of the full Convention.

It was known that they were coming. On August 14th, ten days before the Convention, five Southern governors had gathered to consider the problem; four days after that, on August 18th, the President himself, under Southern pressure, had gathered a number of his advisers at the White House to consider the problem. He had with difficulty just stilled the Negro riots in the cities—yet he was, with equal difficulty, negotiating with and trying to get the cooperation of Mississippi's Governor Paul B. Johnson, Jr., to discover the murderers of the civil-rights trio. He felt that most of the Southern states would accept any compromise he might bring—so long as it moved in the channel of Convention law and order. To work out this compromise he appointed David Lawrence of Pittsburgh, former Pennsylvania Governor, to head the Credentials Committee; and Hubert Humphrey to preside over compromise.

It is difficult to compress the emotion that the Freedom Democratic Party aroused at Atlantic City into the narrow proportions of importance it holds in the story of the Convention. There was no moment when the Convention machinery of Johnson, Lawrence and Humphrey might not have imposed a solution. But the intensity of the emotion was so deep, and all other proceedings were so dull, that for three days the Convention paused to consider its only excitement.

One gets the flavor best, not by considering the issues raised, but by considering and listening to a voice. On Saturday afternoon, as the Convention Credentials Committee moved to consider the situation in Mississippi, a robust Negress rose to testify. She gave her name as Mrs. Fannie Lou Hamer, of 626 East Lafayette Street, Ruleville, Mississippi, Sunflower County, "the home of Senator James O. Eastland and Senator Stennis." Then she proceeded to tell her effort to register to vote, legally, going back as far as 1962; and as her fine, mellow voice rose, it began to chant with the grief and the sobbing that are the source of all the blues in the world. The hot, muggy room was electrified as she concluded her narrative of a Mississippi Negro's life when one attempts to register:

> I was carried to the county jail and put in the booking room. They left some of the people in the booking room and began to place us in cells. I was placed in a cell with . . . Miss Ivesta Simpson. After I was placed in the cell I began to hear sounds of licks and screams. I could hear the sounds of licks and horrible screams, and I could hear somebody say, "Can you say 'Yes sir,' nigger? Can you say 'Yessir'?"
>
> And they would say other horrible names.
>
> She would say, "Yes, I can say 'yes sir.' "
>
> "So say it."
>
> She says, "I don't know you well enough."
>
> They beat her, I don't know how long, and after awhile she began to pray and asked God to have mercy on these people.
>
> And it wasn't too long before three white men came to my cell. . . .
>
> I was carried out of the cell into another cell where they had two Negro prisoners. The State Highway Patrolman ordered the first Negro to take the blackjack.
>
> The first Negro prisoner ordered me, by orders from the State Highway Patrolman for me, to lay down on a bunk bed on my face, and I laid on my face.
>
> The first Negro began to beat, and I was beat until he was exhausted. . . . After the first Negro . . . was exhausted, the State Highway Patrolman ordered the second Negro to take the blackjack. The second Negro began to beat and I began to work my feet, and the State Highway Patrolman ordered the first Negro who had beat to set on my feet and keep me from working my feet. I began to scream, and one white man got up and began to beat me on my head and tell me to "hush."
>
> One white man—my dress had worked up high—he

walked over and pulled my dress down and he pulled my dress
back, back up. I was in jail when Medgar Evers was murdered.
All of this is on account we want to register, to become first-
class citizens, and if the Freedom Democratic Party is not
seated now, I question America. . . .

There was more of the same—and slowly the civilized delegates of
other states—of North, West and New South—realized how difficult
compromise really was. For there were two absurdities face to face before the Credentials Committee. One was the absurdity of the white Mississippi delegation—though the white delegation was legal, it was morally
absurd; it had been elected under laws administered in sin. The other was
the black Mississippi delegation—though its legal case was absurd (for
it had been chosen outside any publicly administered law and now insisted on a legal delegate voice in a lawful assembly), its moral case was
impeccable; these were citizens denied the founding rights of Americans.

For three days the compromise committee wrestled with the problem as moral fervor swept delegations from the big-city states while
counter-indignation roused Southerners who insisted on legality. The dialogue in itself was interesting, for it dealt with deep moral and political
phenomena. Led by the saintly Robert Moses, the Freedom delegates
would agree to no compromise which was not based on a moral principle
—and Moses' moral principle led him to insist that this meant recognition of Mississippi Negroes' right to vote by giving them votes here on the
Convention floor; he would not bargain or accept either fraternal seating
at the back of the hall or a deal of two actual votes. The Southern delegates, and many Northern delegates as well, pointed out that if any unofficial group could pressure or panic a convention into seating new delegates with votes, with legal participation in choice of a Presidential
candidate, a precedent would be set up whereby other pressures, other
demonstrations could force a convention to seat delegates similarly
chosen outside the law. If force could impose outside votes on a convention, the most delicate conjunction of the American political process,
where and how would it end?

This dialogue went on all around and outside the Convention hall,
in midnight sessions at Negro headquarters and Negro churches, at hotels where a mesh of overlapping committees worked until two or three
or four in the morning. Leadership was Humphrey's—exercised through
the lieutenants of his Minnesota delegation, Mondale and Joseph;
through his old friends Walter Reuther and Joseph Rauh; through all the
links he had built to the civil-rights movement since the day when, in
1948, at his first national convention, he had first made civil rights an

issue on the floor. Negro leaders themselves were divided: Martin Luther King and Roy Wilkins sought compromise; Moses sought victory. But when all Humphrey's loyalties had been claimed, and all Johnson's pressures exerted, what was worked out was, in all historical aspects, a true triumph for Robert Moses' heroes of the summer project. The compromise read that no Mississippi regular delegate could sit unless he pledged allegiance to the ticket; that two of the Freedom Party leaders would sit as delegates at large with full right of vote; and that *at the Convention of 1968, and thereafter, no delegations would be seated from states where the Party process deprived citizens of the right to vote by reason of their race or color.*

Much overnight pressure and persuasion was exerted on delegates both North and South. By 9:27 on Tuesday evening, when temporary Chairman John O. Pastore read the compromise to the Convention floor, a single bang of the gavel over contending voices from the floor caused the compromise to be accepted in thirty seconds, and a victory had been achieved.

Whereafter, as in the process of many revolutions, the triumph was misread by those who had triumphed. For a turbulent two hours the Convention saw its floor invaded by Negroes of the Freedom Party, entering on purloined passes with purloined badges and seating themselves illegally in the Mississippi seats, insisting by physical presence that they repudiated the triumph which morality had wrung for them out of the law of the Convention. Stolidly, arms linked, defying the sergeants-at-arms of a national convention that had gone as far as it could to accommodate morality and law, they defied law. It was as if the illegality of Mississippi had authorized them to commit a counter-illegality on a national body. What seeds of danger lay in their intrusion only future conventions can tell. But when finally, after three hours, order had been restored on the floor by direction of Lyndon Johnson, the Freedom Democratic Party had stained the honor that so much courage and suffering had won it.

Again, only as one draws away from the dramatic episode to reach for perspectives can one read events clearly.

The episode marked the end of a chapter of politics. What the next chapter would read no one could tell. But the chapter that had begun with Hubert Humphrey's advocacy of a strong civil-rights plank in the Party platform in 1948 and a walkout from the Democratic Party of four Southern states was over. The South, reduced over the years from its one-time absolute veto on Democratic nominees, had now been forced to accept the best compromise it could get: a four-year delayed sentence. As Negro demonstrators chanted for integration outside the hall and demanded "Freedom Now," the end of the long process was so obvious inside as to be ignored. In the delegations of California and Illinois, Ohio and Michigan, Pennsylvania and New York, Negro delegates sat so casually, in such old-shoe companionship with white politicians with whom they had worked for years, that their presence was too commonplace for any commentator to note or record. As they jostled and paraded and shoved, along with all the white jostlers, paraders and shovers in Convention demonstrations, not even Southern delegates protested; already at least three of the Southern Democratic delegations (Tennessee, North Carolina and Georgia) had brought with them Negro delegates of their own.[4] Mississippi and Alabama, as they walked out, were the last of a rear-guard action that had for sixteen years been in retreat; not even South Carolina or Virginia or Arkansas followed them.

Drawing yet farther away from the turbulence of the Credentials Committee, one could see an even broader outline. Completely overshadowed by the Mississippi episode was the work of the Platform Committee. Yet the work of the Platform Committee was what Lyndon Johnson purposed this Convention to be. No controversy surrounded the platform hearings—the platform had been prepared at Lyndon Johnson's direction by Willard Wirtz in Washington; the labors of its members in Atlantic City were almost entirely editorial, adjusting minor points to interventions made before it.

Its theme was clear: One Nation—One Party. All Lyndon Johnson's life and political art had been spent in trying to reconcile differences, to bind and hold them together and, out of them, to make law. This had been the underlying theme of all his spring travels and spring speeches. Now, in the platform, they were to become the campaign program, the campaign theme. In this sense the Convention was one of Johnson's greatest triumphs—a triumph largely ignored because the emotions of the Mississippi division hid the great significance of the work of the Platform Committee under the skillful direction of Congressman Carl Albert, a one-time Rhodes scholar from Oklahoma. In the Platform Committee the great majority of white Southern delegates had at last freely accepted a revolutionary civil-rights program and platform that committed them, their state parties and their states to the forward movement of the rest of America.

[4] On the Sunday before the Convention the outstanding Georgia delegation had what was perhaps the first integrated social reception in its history. Georgia had one Negro delegate and two Negro alternates. As they arrived with their wives and children at the reception, where all ate easily together from the long board, one astounded Southern white delegate said to a friend in self-amazement, "Whoever thought—we got fifteen Nigrahs here and seven Jews and everybody's having a good time."

In the flow of history, this, certainly, was the largest achievement of the Democratic National Convention of 1964. But in the flicker of public attention it passed almost unnoticed; and when the real drama of the Mississippi delegation had been settled on Tuesday evening, there was little else to occupy public attention except the choice of the Vice-President. To summon public attention back to his Convention, the President now proceeded to direct a melodrama.

The melodrama of August 26th, 1964, was staged and produced entirely by Lyndon Johnson.

In its mixture of comedy, tension and teasing, it was a work of art; it was as if, said someone, Caligula were directing *I've Got a Secret*. It is difficult to see it as any appropriate way for handing on succession to the mightiest office on earth; but as excitement and a study of personality in power, it was unmatched.

Lyndon Johnson had awakened well on Wednesday morning, August 26th. For days his sense of euphoria and goodwill had been building. Master of all he surveyed, stage manager of his own Convention, he faced the appetizing prospect of a campaign against Barry Goldwater, and only the Mississippi episode had disturbed him. His man, John Pastore, had been the keynoter of the Convention, placing his seal on the Convention. His platform had been approved. Seats and schedules and decorations and arrangements and cameras had danced at his command. There were perhaps 30,000 people assembled at Atlantic City, but all of them danced on the strings that ran from the White House to the Pageant Motel, where his personal staff edited, approved, disapproved, arranged matters to suit his fancy.

On Monday the President had permitted himself an initial burst of exuberance as he led the White House press corps around the White House lawn nine times and then invited them to his office for further briefing. He had parried them; no, he had not made up his mind about the Vice-Presidency; no, he was still in contact with leaders all around the country, seeking their opinions. He could, indeed, offer some of his thinking on what a Vice-President should be: a man capable of taking over his own job; a man who would be a full working partner; a man who would be responsible for space and foreign affairs and civil rights; a man who would be an active force in dealing with both Houses of Congress; a man who would be given much more eminence than Vice-Presidents had ever been given before, who would have his own official residence, his own staff, his own responsibilities. "I expect to work hell out of him," said the President about the successor candidate, whoever he might be. But as to who it might be he was silent.

Tuesday the President was, apparently, busy at his work behind the screen of the White House lobby. By Tuesday night the drama of Mississippi was over, the compromise settled, the stage cleared for the last act.

On Wednesday the President's euphoria built steadily, at first slowly, then intensely to a crest. He presided in the Cabinet Room over the signing of a bill which let the Atomic Energy Commission sell atomic fuels to private industry. He handed out pens. He retreated to his Oval Office. And then, shortly after noon, the White House correspondents, as befuddled as the nation by the President's secrecy, were invited out to the south lawn for another walk.

It was 89° on the White House lawn.[5] Sweat rolled from the correspondents; the birds sang in the trees; gardeners mowed the lawn at its distant edges; the dogs yipped and barked; and at the head of his tagging procession moved the President of the United States, rolling his secret about mischievously in his head. It is a quarter mile around the White House south-lawn drive; fifteen laps make a hike of almost four miles. And round and round the President jovially led his procession for an hour and a half as he talked on and on and on, expansive, friendly, happy. One by one the women correspondents dropped out of the tagging, perspiring procession, to pant under the shade trees in the grass; so did the weak, the halt and the simply tired. But the President was examining the Presidency and the Vice-Presidency—and Lyndon Johnson—in a public circus that perhaps has never been equaled before anywhere.

He was giving very serious consideration to the problem of the Vice-Presidency, he said. Why, just that day, just that very day, he had been on the phone discussing it with everyone: with Daley of Chicago and Wagner of New York, with Connally of Texas and Shelley of San Francisco, with Lawrence of Pennsylvania and Bobby Kennedy too; with Smathers of Florida, and McCormack of Massachusetts; with Pastore of Rhode Island. (Back at the Convention, the telephone ring of the President seemed to be the equivalent, for the recipient, of honors and knighthood at the Court of St. James; to have received a call from the White House meant that one was really "in.")[6] The President was enjoying

---

[5] I permit myself here, with his permission, to rely on the recollections of Hugh Sidey, of *Time* Magazine, whose warm and perceptive reporting is that of an artist. I was in Atlantic City on the day of the Washington lawn walk, and Hugh Sidey, with his usual generosity, has shared his reporting with me.

[6] By the end of the hectic day the Presidential call had become one of the prodigies of history in itself. He estimated for one newspaperman that he had, in the midst of all other matters on his schedule, reached 25 to 30 governors, 35 Senators, at least 35 Congressmen—and labor leaders and businessmen in numbers he could not recall. It was as if some compulsion made him reach back and forth

himself. No, he had not decided on a Vice-President yet; he was still talking to people and discussing it. There was still the time until tomorrow morning to choose one. But maybe he'd do it this afternoon, so as to let the man have time to write his speech—he hadn't had much time himself to decide at Los Angeles in 1960. About Humphrey? Well, he was asking Humphrey to come down and see him this afternoon, just to get his ideas on the Vice-Presidency and discuss it with him—that is, if the visit didn't conflict with Humphrey's plans. "If he can come down," said the President, "I'll let you know."

Their tongues hanging out, sweat plastering their shirts to their bodies, the correspondents trotted after the President as he strode around the walk, coins jingling in his pocket. There was not a cloud on Lyndon Johnson's world, and sunnily he responded to any question. The campaign? He was going to sit right in that house (pointing to the Oval Office) and do his job week by week. Goldwater? Goldwater was frightening people. Polls—on polls the President held forth at enormous length, the Gallup, the Indiana poll, the Texas poll, all overwhelmingly in his favor. He chided the press for its talk of backlash and hammered instead at the frontlash. He discussed his health—the doctors had just checked him up, he was just fine, they could have the medical report.[7]

He discussed the Convention—what a fine affair it had been; how marvelous John Pastore's speech had been; what a good job John McCormack had done; and what a good job Carl Albert had done. Interspersed with the banter were gems of insight. He was going to campaign on peace, and then, shrewdly, "I don't want the Russians to get the idea I'm a-goin' to start a nuclear war"; and then again on his own party, "The Democratic Party has been the House of Protest since it was born."

It was close to three o'clock when the panting group trudged back with the exhilarated and untired President to the Oval Office, where he

---

across the country, touching, touching, touching again and again, feeling the response to the touch.

[7] The medical report read:

> President Lyndon B. Johnson has no symptoms. His exercise tolerance continues to be superb. Physical examination, including the examination of the eyes, lung, abdomen, lower intestinal tract and reflexes, is normal. His blood pressure is normal.
>
> The electrocardiogram is normal. The examination of his blood, including blood counts and serum cholesterol, is normal. The urinalysis is normal.
>
> There is no health reason why he could not continue an active vigorous life.

/s J. C. Cain, M.D.   /s George G. Burkley, M.D.
/s J. Willis Hurst, M.D.   /s James M. Young, M.D.
                                      August 24, 1964

sipped orange drink and offered to all of them his invitation: to come
on up to the living quarters that evening for some hors d'oeuvres and
drinks and watch the Convention and nomination with him on TV.

By now the wires had flashed the bulletin from the lawn parade that
the President had asked Hubert Humphrey to come down from Atlantic
City; before the walk was over, other wires of the news media had responded that Hubert Humphrey was already en route. The correspondents left the President to his privacy and spun madly to their notes, their
typewriters, their telephones, and made ready for the ceremonial arrival
of the man who would be next in line of succession.

We must shift the scene of events now to Atlantic City; for two
separate and independent personalities were involved: the President,
in Washington, and the next Vice-President, at the Convention
in New Jersey, 140 miles away.

The flash from the White House lawn had interrupted Hubert
Humphrey in a homely afternoon moment. He was letting himself
be interviewed by Radio Station WCCO of Minneapolis by telephone
in a program which let the home folks ask their favorite Senator direct
questions ("How are you, Hubert?" "How is Muriel?" "How are you
enjoying Atlantic City?"). The host of the program broke in to say that
CBS was proclaiming a net alert. Then followed a bulletin from Washington; and Humphrey heard over the telephone the voice of the an-
nouncer in Minneapolis saying that President Johnson had just invited
Hubert Humphrey down to Washington. At this point Humphrey decided
to close the interview, saying, "Somebody's probably trying to reach me
on this telephone; I'd better hang up." And, indeed, within a few minutes, someone *was* trying to reach him—Walter Jenkins with a personal
message from the President to come at once.

The Jenkins call did not take Humphrey by surprise at all—for
Humphrey was already as sure as anyone could be of the Presidential
mind, and had been for almost twenty-four hours. Here the reader
must backtrack from the Wednesday of public fascination to Tuesday,
the day before, in order to find the window on the choice of the Vice-
President of the United States.

On Tuesday the turmoil over the Mississippi Freedom Democratic
Party was nearing climax. Yet by mid-morning of Tuesday the President had finally made up his mind. At 2:30 that afternoon Rowe, sitting in one of his rooms in Atlantic City, heard the beeper of his pocket
radio telephone go off—a signal that he must call Walter Jenkins, established in his hideaway suite at the Deauville Hotel. He called, then
proceeded to Jenkins' suite at the Deauville to hear the message of the

maestro from Washington. The master wanted Hubert Humphrey to read the interview he had given correspondents Monday on qualifications for the Vice-Presidency; he preferred that Humphrey read the Washington *Star* text, not the Washington *Post* text. He wanted—so read Jenkins from his notes—Humphrey to understand certain things about the Vice-Presidency: that the Vice-President must have no public disputes with the President; that there must be no lobbying for special interests; that any public speech of the Vice-President must be cleared at the White House for national policy; that any amount of internal debate was permitted before decision, but once decision was taken the Vice-President must support the President; and that, in addition to all this, sometimes the President must share secrets with a Vice-President that the Vice-President could not even share with his wife. Late in the afternoon of Tuesday, Rowe, in a charade of hidden rooms and mysterious entrance knocks, was able to find Humphrey at the Hotel Traymore, where Humphrey and Walter Reuther and a number of others were trying to work out a compromise for the Mississippi problem. There, in the presence of the others, Rowe was able to deliver his message. As remembered by one of those present, the burden of the message was very clear: Humphrey was to be Vice-President, the seconding speech was to come from Georgia, other seconders were named.

The first message having been relayed to Humphrey, there came a second Presidential message to Jenkins in Atlantic City. The President wanted Hubert Humphrey *and* his wife, Muriel, to stand by for a plane trip to Washington. With the delivery of this message by Rowe, Humphrey breathed relief—an invitation to come *with* his wife meant that the President was about to say yes, not no. The inner group—Hubert and Muriel Humphrey, Kampelman, Rowe and Connell—gathered now at a room at the Colony Motel in Atlantic City to see whether the Mississippi struggle might still boil over in a floor fight requiring Humphrey's presence—or whether he could fly to Washington directly. They watched the floor proceedings on television; and the Mississippi compromise was accepted; and they were ready to go. But by now fog had settled in and no plane could fly. Max Kampelman suggested that they *drive* down to Washington so that Humphrey and Muriel could sleep peacefully in their own beds. The idea was vetoed; the President had said *fly*, and the President was writing the script. For a moment Humphrey, his nerves under strain for months, exploded, and Rowe calmed him. Tomorrow he would be the nominee for Vice-President, said Rowe; tonight he was just Hubert Humphrey of Minnesota, and he must live with this condition for one night more. The group had finally ordered steak and potatoes, and as they relaxed,

Humphrey began to see himself from the outside, as only he can, and let himself enjoy the moment.

Thus, the telephone call from Walter Jenkins, which followed as soon as Humphrey had freed himself from his broadcast interview with WCCO on Wednesday, had come as no surprise. Humphrey had been quite relaxed that morning. He had slept late after the exhausting and triumphant conclusion of the Mississippi struggle and been wakened by his press aide, Robert Jensen, with good news: that Senator Eugene McCarthy, his old friend from Minnesota, had voluntarily and gracefully withdrawn his name from consideration for the Vice-Presidency in a telegram to the President. Humphrey breakfasted with his family, asked one of the boys how he would like his Daddy to be Vice-President, then gazed longingly out on the sunny beach. He wanted to go out and walk on the sand and get some exercise and fresh air; his aides dissauded him —he would be mobbed, they assured him, if he showed himself in public. Thus, he had waited in his room to find out how the President, his stage-master, wanted the script scored for the day; and when the Jenkins call came he was neither discouraged nor depressed by the news that Muriel was not, after all, to fly down with him to Washington; he felt 99.5 per-cent sure that he was now, at last, the President's real choice for running-mate. Nor was he upset when Jenkins said that some other undisclosed Senator would join him on the plane flight to the capital. Humphrey, too numb for protest, assented to the script however the President wanted to stage it.[8] Exhausted, Humphrey now let himself sleep on the plane. The plane arrived in Washington at about 4:30, to be met at the airport by Jack Valenti in a big black White House Cadillac.

The Vice-President-putative was now ready to face fate. Yet the Presidential script read otherwise. At that moment the nation's tele-

[8] Not so, however, for the few Convention delegates who could not guess the outcome of the script. The news that Senator Thomas J. Dodd was accompanying Humphrey produced exactly the effect that Johnson had calculated—consternation! Was the President shifting signals? Was Dodd a real candidate? Immediately all news and broadcast agencies began to assemble their experts on the credentials, ca-reer and capacities of Tom Dodd. So deep was the surprise, so skillfully had the President managed his cliff-hanger, that some of the most adroit politicians at the Convention were duped into folly. One group immediately attempted to organize a floor revolt—the President could not play thus with a national convention; a pledge of $50,000 in cash to back an overnight floor fight was raised within half an hour; three state delegations were sounded out and agreed to participate in a floor fight for Humphrey against Dodd; the entire stillborn enterprise petered out in forty-five minutes when one of its members reached Robert Wagner, boss of the New York delegation, to find out whether New York would join. Before the Mayor could be sounded out, he reported that he had just received one of the Johnson calls from the White House—and Johnson had told Wagner that the choice was indeed and actually Humphrey.

vision cameras were focused on the helicopter arrival of Lady Bird Johnson in Atlantic City. It was imperative not to interfere with this entr'acte scene, and for almost an hour Humphrey, Dodd and Valenti made the tourist circuit of Washington's monuments while the nation waited for denoument. Finally the Cadillac entered the White House grounds by the South Gate, then rested by the helicopter pad and waited —for almost twenty more minutes. Exhausted, Humphrey fell asleep in the back seat—to be awakened by a gentle but persistent knocking on the window pane of the car. Coming awake, he saw it was the President knocking.

"Hubert, come on in," said the President.

Humphrey, Dodd, the President all walked across the lawn to the White House as a bedlam of reporters surrounded them. It was Dodd who entered the Presidential presence first, while Humphrey sought escape from reporters and cameras in the Cabinet Room. Then, at six, the President, having released Dodd to the press, sought Humphrey in the Cabinet Room and asked him in to the Oval Office.

They walked in together.

The President put his arm around Hubert Humphrey and said, simply, "Hubert, how would you like to be my Vice-President?" Humphrey said that he would love it and that it would be a great honor, Mr. President. The President then said that that was what he'd been thinking about—let's talk about it a little bit.

In discussing any phase of Lyndon Johnson's behavior, one must always be prepared for the most extreme swings between tenderness and cruelty, between dedication and cynicism, between comedy and high purpose. Thus, in late afternoon two men sat down in the Oval Office of the Presidency, the President in his rocking chair and the other on the sofa, for the highest and most somber conversation about the Presidency and the power of America.

The President discussed the intricate, delicate relationship between a President and a Vice-President, and all the people who would try to come between. He recalled with tenderness a moment in his own unhappy Vice-Presidency when the President had stood up to support him, Johnson, against the Presidential staff. One of the happiest days in his life, as the new President recalled it. Johnson had had research done on the Vice-Presidency during his own incumbency; he could not remember a single President who had gotten along with his Vice-President.

Yes, sir, I know that, Mr. President, said Humphrey.

The President said that he knew no two men who got along better than himself and Humphrey. This, too, was true. For if it had been

John F. Kennedy who had given Johnson his opening to power in 1960, it was Johnson who for years had opened the way to Humphrey's power in the Senate. Both of them, country boys discussing the government of America, knew which one in this situation—and for years—had been the leader.

Johnson made it clear: The man who would be Vice-President must know that his duties and responsibilities were very much in the hands of the President.

Humphrey wanted the President to know that if indeed he had the opportunity to be Vice-President, he would be loyal. There would be mistakes; he knew he might cause the President concern—but if such things happened, the President should know it was not *because* Hubert Humphrey wanted them to happen, it would be because they were mistakes.

They discussed in some detail the new and developing concept of the Vice-Presidency and its duties. The President would want Humphrey to keep an eye on agriculture (on which subject Humphrey is one of America's great experts) and on health, education, welfare. There were also the duties Kennedy had previously assigned to him, Johnson: space, equal employment opportunity for Negroes. There would be political chores—speaking around the country for the Party. And then there would be the matter of foreign affairs and watching that.

Humphrey observed that the Vice-Presidency was what the President wanted to make of it; the President alone would decide what to share with his Vice-President.

The sunny day had turned gray outside, and rain was intermittently falling on Washington. It was darkening. Both agreed that it was up to them to make the relationship work. Johnson added: Any time you want to see me, come and see me; don't let anyone stop you from coming in.

Thus they discussed the power between them. Softly and seriously.

And then suddenly it was the other Johnson again, exuberant, gay —Let's call up some people, said the President. And so they called Dick Daley and George Smathers and half a dozen other Democratic leaders. And then the President said, Come on in and meet some other people—and led the Vice-President-designate out of the Oval Office into the long, dark Cabinet Room.

You sit right down there, said the President, motioning his choice to the Vice-President's chair opposite the President's place. And the Vice-President-designate sat down with Secretary of State Dean Rusk, Secretary of Defense Robert S. McNamara and National Security Special Assistant McGeorge Bundy.

Hubert, said the President, I want you to know how these men here have been pushing you. Right in my own Cabinet you've had your campaign going. You know what I think of Bob McNamara —he's for you. But then you've got that Georgian, Rusk—he's been carrying on a one-man campaign for you. Why, I've been asking everyone, right here in the White House, and everyone's been for you right down to the cleaning women who work here.

Humphrey murmured his thanks for their support, and then the President, in a burst of warmth, described to the three men responsible for America's outer security how great and fine a person was Hubert Humphrey. It was, by report of most of those present, a moving moment.

Then the President, in another cycle of emotion, decided that the good news must be shared. Let's call Muriel, he said, and the White House operators found Muriel Humphrey immediately in Atlantic City. Muriel, said the President, how would you like to have your boy be my Vice-President? The senior members of the Cabinet watched and the President smiled and Muriel agreed.

After which, again, the mood changed and it was Dusk at Dogpatch.

A brief press conference, gay and incredible (at which Mr. Humphrey was addressed as "Mr. Vice-President" and made angel-eyes at the sky as if not hearing), was followed by a brisk two-lap walk around the south lawn. Then the President, Humphrey and the White House correspondents who had been invited to hors d'oeuvres trooped upstairs to the private quarters where the President lives. By now Humphrey had been informed that he was to fly with the President back to Atlantic City in his rumpled suit and baggy trousers, and, buffeted by events, he followed along with the throng. Upstairs in the private quarters of the President, to which so few have admittance, the correspondents stood by and ate and assisted, as if they had come early to a party for which neither host nor hostess was quite prepared. Those who had not eaten or drunk that hectic day now filled themselves on excellent caviar sandwiches and good cheese; and those who needed stimulant drank heavily as the White House waiters rushed about to make sure all were fed. On the lawn the Armed Forces of the United States scrambled to assemble enough helicopters to transfer the President, his party and all the correspondents he had invited to Atlantic City. Aides rushed about, bringing briefcases full of speech material for the convention; Humphrey, dazzled, eating sandwiches, finally accepted numbly the congratulations of correspondents and reporters.

Serene, delighted, happy as only a President of the United States

can be when the whole country stands by for certain applause, Lyndon
Johnson was at his ease. His bedroom door was flung open as he prepared for his descent on his Convention, and reporters could watch him
as he prepared. Inside the bedroom the President's three television
sets (tuned to the three networks) showed him what the nation was
seeing; every now and then he would pause to make another telephone call to another dignitary to announce the good news. Meanwhile
the President dressed. A lady reporter, gawking through the open door
of the bedroom, recoiled as the President stalked by in his undershorts.
A few members of his staff had gathered contributions to give the
President a birthday gift (it was to be his birthday the next day), a
wooden contraption designed to hold papers for reading in bed. The President accepted it gracefully and then said, "What this is for is for her
to load me up with so much reading that if I read as fast as I can, I can
get finished by two A.M."

Then off—dignitaries, staff, and reporters—by helicopter from
the lawn to the field at Andrews Air Force Base. Then to Atlantic City.
On the plane the jumping and jittering television set managed to pick up
the Convention scene as Alabama yielded to Texas for his nomination.
The flickering image of Governor John Connally of Texas nominating
Lyndon Johnson for President of the United States appeared next. Hubert Humphrey munched cashew nuts, hoping a clean shirt would be
ready for him on arrival at Atlantic City. But the President was entranced
as he sipped Sanka from a paper cup. "All right," he called above the
plane's engines, "we've got this show on the road!"

But the President had entangled himself. John Connally's nominating speech was coupled with a speech from Governor Brown of
California, coequal nominator. And thus when the speeches had
ended and the time had come for climactic demonstration on the floor,
with governors, Senators, Cabinet members and sub-Cabinet members stomping around in parade, the President himself arrived at the
Atlantic City airport. He was competing with himself for the nation's
attention. Thus the TV directors were posed with a choice of what
to offer the nation: either the President himself arriving to command
his Convention, or the demonstration on the floor in the President's
honor. They chose correctly and hit their switches to offer the sight of
the live President.

Thence to the Pageant Motel; thence to the Convention floor;
thence to the rostrum, where the President took the gavel from the
hands of Speaker John McCormack, pounded the delegates to order,
and suggested to the Convention the name of Hubert Humphrey as
Vice-President. Which the Convention accepted. Whereafter tumult.

And then, in complete disregard of the tumult, the President came down from his imperial box and made his way through the overchoked floor, shaking hands, while the terrified Secret Service and New Jersey state troopers did their best to protect the only living President they had. And the President enjoyed himself mightily.

Of the last day of the Convention there remains in this reporter's memory only a succession of scenes without any real connection—as if they were all patched together from fragments of a broken stained-glass window:

§ The first abashed bow of the convention to the memory of John F. Kennedy. Robert F. Kennedy, standing before the gathering, trying to speak, unable to speak, the applause building and building and building (one Chicago delegate to another: "Let's not let them stop it —first our row will clap, and then when we get tired we'll get the row behind us to clap") and going on for twenty-two minutes until finally Bobby could deliver an evocation of his brother's memory which he climaxed with a passage from _Romeo and Juliet:_

> . . . when he shall die
> Take him and cut him out in little stars,
> And he will make the face of heav'n so fine
> That all the world will be in love with Night
> And pay no worship to the garish Sun.

(Whereupon one of the Chicago delegates growled, "See? I bet nobody else could quote Shakespeare to a Democratic Convention and get away with it. I feel sorry for this guy Keating.") Then a film of memorial, showing not only the youth but the strength of the President who had been murdered, and closing on the sound of song and the sense of the joy of life and youth and the dead President teaching his baby boy how to tickle his chin with a buttercup. And we all wept.

§ Adlai Stevenson's eulogy to the memory of Eleanor Roosevelt. (Someone said that three conventions were in session at the same time: the convention of Franklin Roosevelt and the men whom he had made; and the convention of John F. Kennedy and the people he had brought there; and Lyndon Johnson's convention—and that was the real Convention. But Adlai Stevenson's speech covered all three.)

§ There were the stacked banners of the civil-rights people who had finally ceased from demonstration now that all was in the hands of Lyndon Johnson—banners and placards in a neat mound under the bust of John F. Kennedy.

§ There were the acceptance speeches: Lyndon Johnson's speech, the poorest he made in the campaign, not half so good as the speeches he could and later *did* deliver of his own composition. His speech was a consensus of the worst thinking of the best thinkers attached to the White House. Willard Wirtz, Richard Goodwin, McGeorge Bundy, Horace Busby and Bill Moyers together proved that no great speech can be written by a committee. Johnson's speech was dull. Hubert Humphrey's speech was better, and one remembers the belting out of the theme that was to become so familiar during the campaign: ". . . *but not Senator Goldwater!*"

§ There is the memory of walking up the boardwalk and pacing by the huge illuminated sign which presented Barry Goldwater, jutjawed in electric lights proclaiming for the first time, IN YOUR HEART, YOU KNOW HE'S RIGHT—and then, minutes later, coming across a familiar figure and, accosting him, finding Senator Pierre Salinger of California, most populous state of the Union, red-eyed with weeping and almost incoherent. He had watched the proceedings in a hotel room with other comrades of the 1960 Kennedy campaign—O'Donnell, O'Brien and Powers—and had not recovered, could not recover, his balance. He mourned for John F. Kennedy and for another Convention four years before on the banks of the other ocean; he could not erase past from present.

Finally, it was all over but the dancing. One sat by the boardwalk in the evening as the ocean rolled in at its powerful leisure, and the sky was clear, the weather balmy. And it was all over except for the celebration.

It was Lyndon Johnson's birthday. Fifty-six years before, on August 27th, 1908, he had been born in a frame house by the Pedernales; and after that he had known hardship, and understood evil, and done good. And now they were going to give him a birthday party. The barefoot boy from the harsh caliche soil was going to have the largest birthday party, with the biggest birthday cake, and the loudest birthday bang in the world.

The crush on the upper floor of the Convention hall was so great that even those who had tickets ($1,000 each) for the President's birthday party could not get in. On television one could see the President cut the huge cake and take a giant slice; and he ate it all, apparently enjoying every crumb as he munched it down.

But it was even better to be just a bit removed: the fireworks were the show. They were the greatest fireworks ever seen—three tons of them. They went up delicately first, with finesse—blue and white and green and red, streaking in arcs over the night sky.

Then they became fancier; the colors in the sky began to criss-cross each other and to tumble in starry waterfalls; cascades and inverted fountains of color, like the tracer bullets of anti-aircraft that used to fall over cities during the war; and with each volley the smoke clouds over the boardwalk grew bigger and thicker, so that one could see only the stab of light gash through the smoke cloud. And now that light and delicacy were obscured, the noise became the thing. The bangs were bigger and bigger. They beat on the ears with deafening thunder. And from where I watched a quarter of a mile away, I became aware of the smell of burnt powder in the nostrils and, simultaneously, of tiny flakes of shattered cardboard that were fluttering down from the bursting rockets far away. Big Daddy had had the biggest birthday party in the biggest and best country in the world.

He took off at noon the next day for home and the ranch, carrying Hubert Humphrey with him. There remained only a mopping-up. And so we proceed to the mopping-up which was the election itself.

CHAPTER TEN

THE ISSUES: CRY BABYLON!

THE campaign of 1964 was that rare thing in American political history, a campaign based on issues.

War and peace; the nature and role of government; the morality and mercy of society; the quality of life—all were discussed in a campaign that will leave its mark behind in American life for a generation. A myriad cross-currents of conscience, judgment and tradition were engaged in the voter choice of 1964—but they were cross-currents resolved in individual decisions in solitary voting booths, and the massive totals of the late fall tally gave little indication of how much and how deeply Americans had been stirred to think.

In retrospect, the clear and crushing margin of November's decision was as distorted a reflection of the confusion in American thinking as the thin and ephemeral decision of 1960 was a distortion of American response to John F. Kennedy's clear call.

Up and down the nation, at prop stop and whistle stop, by dawn and dusk and at high noon, Barry Goldwater first challenged and then, as November approached, grieved: "What kind of country do we want to have?"—and to this challenge Lyndon Johnson replied in as masterful a campaign as the Democrats have ever conducted. A hundred million Americans were asked to hear the two leaders give their vision of America, as if they were citizens of a Greek *polis* two thousand years before; and rarely, when questions were asked about the nature of fate and country, have citizens heard their leaders give answers so violently and unequivocally opposed.

The puzzling complexity of the campaign-about-to-be was apparent from the moment at the Democratic National Convention when the boardwalk first flared with the huge illuminated sign that showed Barry

Goldwater, golden profile against dark blue background. IN YOUR HEART, read the giant sign, YOU KNOW HE'S RIGHT. This was the challenger. But what lay in *his* heart? What did he want? What did he seek?

No Presidential candidate since Franklin Roosevelt in 1932 had so completely separated himself from the then-reigning dogmas of American life and society as did Barry Goldwater in 1964.

For thirty years Republicans and Democrats had fought each other in the arena of the center, where their differences, though real, were small, narrow or administrative—differences of pace, posture and management in a direction that both parties alike pursued. Goldwater, however, proposed to give the nation a choice, not an echo—and not just one choice but a whole series of choices, a whole system of ideas which clashed with the governing ideas that had ruled America for a generation: a choice on nuclear weapons, a choice on defense posture, a choice on the treatment of Negroes, a choice on dealing with Communism, a choice on the nature of central government. There were many choices, but the rub was that no voter could pick at will among those choices. A voter must buy the entire Goldwater package of ideas—or reject them entirely. He could choose America as it was moving, in the direction it was moving, in full tide of power and prosperity—or he must choose to reverse the direction entirely. In Goldwater, one had to take all or nothing.

Over and over again, the Republican candidate and his staff members complained, with always increasing bitterness, that they had hoped to make this a historic campaign of issues, to make of it the great dialogue between the two philosophies, the conservative and the liberal—but that Lyndon Johnson and the Democrats refused to meet them issue by issue on the high ground of public debate. They were right in their exasperation; but so was the Democrats' response. With firm decision, the Democrats insisted that the nation must choose the entire Goldwater package, the entire man, the entire attitude and bundle of emotions of the Republican candidate. No Democrat chose to debate issues at retail —only in the large. If Goldwater challenged, "What kind of country do you want to have?" the Democrats responded: "The kind of country we've made of it over this past generation."

The response of the people to the personalities of the two candidates must be examined later, for this is a book of politics, and politics is a study in leadership and response.

Yet in 1964 the issues acted so powerfully on voters' emotions that one must first examine all the subtleties involved to see how the final answer came to be as it was.

What exactly were the issues? What precisely were the choices offered in 1964? How valid were they? How did the craftsmen of politics convert these ideas technically and politically into the emotions which move American voters to vote? How ephemeral or how permanent will they prove in the future to be?

* * *

There is no doubt in the mind of any candidate, any political observer or any man who trooped the long marches from coast to coast with both candidates that one primordial issue overshadowed all others: the problem of War and Peace, pegged on America's use of its arsenal of nuclear weapons.

How, when, by whose authority, where should such weapons be used?

The issue entered the campaign almost obliquely, obscure from the very moment of its posing. On October 24th, 1963, a year before the election, Barry Goldwater journeyed to Hartford, Connecticut, for a political caucus with Connecticut Republican leaders; at the Hartford Club he agreed to a routine press conference, and was questioned on his reaction to a recent statement of Dwight D. Eisenhower that America's six NATO divisions in Europe could be cut to one. Answering the question by Chalmers Roberts, chief of the national bureau of the *Washington Post,* Goldwater observed that the six divisions could "probably" be cut by "at least one third" if NATO "commanders" in Europe had the power to use tactical nuclear weapons on their own initiative in an emergency. Sensing danger, Goldwater's press adviser, the sage and genial Tony Smith, abruptly ended the conference. But the damage had been done, and the controversy arising from it, bungled to an unbelievable degree by Goldwater later, was to be central throughout the whole of the next year's campaign.

What exactly had Goldwater said? And what exactly did Goldwater mean?

In Roberts' notes on the conference the word is still scrawled: "commanders" (in the plural).[1] The next morning it was a front-page story in the *Washington Post*—interesting but not yet critical. The Kennedy assassination erased politics for a month. In January, however, when politics resumed, they resumed in New Hampshire—and by now

---

[1] It was months later (actually not until the start of the election campaign itself) before the Goldwater campaign managers came up with the gloss that might have defined him out of his later trouble. If, indeed, said his campaign managers, Goldwater had used the word "commanders," he had not meant divisional or field commanders—he had meant the succession of NATO "commanders in chief"; he had used the plural to describe the succession of Eisenhower, Ridgway, Gruenther, Norstad and Lemnitzer, since 1950.

the Rockefeller research staff had noted the story, clipped it, and decided to make the issue of the "bomb" central to its primary campaign in that state. Among all the blunders of the New Hampshire campaign (see Chapter Four), none, certainly, was more important than Goldwater's failure to scotch the issue then, in that forum, at that time. For the more Goldwater refused to clarify what he meant by "commanders," the greater concern could be roused in the public mind by his rivals. Did Goldwater mean that any tactical unit commander could use the weapons of holocaust? Under what circumstances? To destroy a bridge? Or an enemy concentration? Or a city? And if so—who then held the trigger on the implacable escalation of destruction?

If one is to cut through the semantics of the debate, one must not only try to define the facts about nuclear war as known at the Pentagon, but also stress how difficult these facts are to come by.

Briefly, in the summer of 1964 the United States enjoyed (if that is the proper word) a spectacular (but probably superfluous) margin of superiority over the Soviet Union in strategic nuclear missiles: 18 Polaris submarines with 288 Polaris missiles were already operational, with 23 more such submarines and 368 more missiles programed. On land, American intercontinental ballistic missiles outnumbered Soviet intercontinental ballistic missiles (but *not* intermediate-range missiles) by 4 to 1; and the bomber force of America, 1,100 planes (500 of them on fifteen-minute alert), outnumbered Soviet bombers (estimated at 250) by more than 4 to 1. These strategic forces, as all men here and abroad know, are clearly under Presidential control at all times. Goldwater never questioned the rightness of this control.

The facts about American *tactical* nuclear weapons are much more difficult to ascertain. So is Goldwater's position. The Army has at least seven types of nuclear tactical weapons plus an anti-aircraft nuclear missile in the works; the Navy has several nuclear weapons, chiefly concentrated on anti-submarine destruction; the Air Force has a panel of low-yield weapons in addition to the weird power of the strategic weapons outlined above.[2]

[2] Even the most cursory catalogue of known Army nuclear weapons gives some idea of the complexity of their control and orchestration. The smallest of our nuclear weapons is the Davy Crockett, a recoilless rifle that can fire a nuclear shell with a minimum blast of 40 tons of TNT for 2.5 miles. More important is that old workhorse of American battle, the 155 mm. howitzer, which can fire both conventional and nuclear weapons with a blast up to 100 tons and has a range of 11 miles. Both of these "little" weapons must be motor-drawn or motor-borne. The army's 8-inch howitzer fires a 2-kiloton nuclear warhead for distances up to 10-plus miles. From there, swiftly, the range and blast increase: the Little John rocket will go 12 miles with over 20 kilotons (just how much more is classified); the Honest John rocket is bigger and will go as far as 25 miles; the Sergeant, also

It was the Army's ground weapons that engaged Goldwater's earnest study. But his reflections could not in any way be translated into the idiom of public campaign discourse. *How* and *where* these weapons shall be used on the battlefield is a matter of the instrumentation and decision of a field commander; but *when* he shall use them is necessarily one of the most secret areas of American command and decision.

The history of these weapons is worth recalling. They were developed and designed in the early postwar period for use in Europe—at a time when both France and Germany were exhausted and the British and Americans, between them, mustered only three divisions for the defense of the Thuringian ridges. It was the fertile mind of Robert Oppenheimer that fostered the development of the tactical nuclear weapons as a response to the threat of Russian manpower—at that moment overwhelming. Only American blast power could protect from Russian manpower the ravaged and slowly healing Atlantic cradle.

In the fifteen years since then, all has changed. Today the Eastern European base of the Russians is far less secure than our Western European base. Europe has recovered. French and German manpower is now, or shortly will be, fully capable of engaging Russian manpower presently on the line; America has multiplied her divisions in Europe to 6, and highly mobile back-up has jumped in America by 4 times (8 divisions as against 2). The tactical nuclear weapons of 1949 and 1950, so essential at that time, can be considered now either a military luxury, an embarrassment—or a peril. Response with such tactical nuclear weapons in Europe must inevitably invite counter-response from the Russians—and how far is such a response to be taken? Does it go from a strategic bridge to a village concealing a regimental combat team that must be obliterated? To a rail junction at an important town? To a city like Breslau, turntable of the Russian southeastern command, which could be wiped out far in the enemy rear by a Pershing rocket with its

---

100 kilotons plus, will go 135 miles; and the Pershing, America's biggest "tactical" rocket, will heave a 200-kiloton warhead 400 miles. The last three rockets can deliver warheads many times more violent than the primitive Hiroshima bomb of 20-kiloton blast.

There are several technical points to be made about the use of the word "conventional" in referring to such weapons. The first is that, though it is true the smallest warheads can be tooled down to a blast power equal only to the blockbusters of World War II, it is uneconomical to do so if we can use much cheaper, old-fashioned explosives to do the old-fashioned conventional job. The second is that, though war-game theory can establish certain principles as conventional assumptions of deterrence or non-deterrence, the use of tactical nuclear weapons is far more complicated; no real doctrine exists to guide our soldiers in their behavior or response on the nuclear ground battlefield. Thankfully, war games have not yet been played with actual nuclear weapons to establish how they do fit into the command orchestration.

400-mile range? And if we take out Breslau and the Russians take out Düsseldorf—what next? London for Leningrad? Washington for Moscow? Just what is tactical use of a weapon—and at what range and kilotonnage does it escalate to that terrible moment when the underground silos in Siberia and the Great Plains silently roll back their concrete lids and loose havoc on a world to be incinerated?

Investigation at the Pentagon carries one only so far—and then, this side of the wall of secrecy, there is only surmise and speculation. Doctrine, public statement, reality and known fact all lead one to believe that the President of the United States *now, alone,* has the power to release any nuclear weapon, large or small. Yet emergency procedures do exist—"permissive action links"—in which Presidential permission has already been granted, *in advance* for emergency use of such weapons. The use of anti-aircraft nuclear missiles (or anti-missile nuclear devices if such exist) is already permitted to the commander in chief of the North American Air Defense Command (NORAD) if he cannot instantaneously reach the President and enemy missiles are already descending on American cities. It is believed that the commander of the Sixth Fleet in Asian waters has certain emergency permissions; and it is almost certain that the Supreme Allied Commander of NATO, U.S. General Lyman Lemnitzer, has similar emergency provisions. And if, by stealth or trickery, an atomic bomb is smuggled into Washington and beheads the entire American government—such a contingency is also foreseen.

Necessarily, such procedures are secret; and thus, necessarily, the issue, once raised, could only place Goldwater at a disadvantage. To this disadvantage Goldwater did his best to add. A devoted patriot, a dedicated soldier and airman, Goldwater as an Air Force Reserve Major General has access to high-security information; in his military opinion, more "leeway" must be given tactical commanders in the field to respond to an enemy with nuclear weapons in an emergency—blast is better than blood. But just how much leeway should be given, Goldwater in debate would never make precise. Military security imprisoned him in silence as to specifics, yet his pride insisted that he must make clear the difference between himself, a technical military expert, and the warmongering Barry Goldwater described by Democratic propaganda.

Thus trapped, he would have been wiser to leave the matter alone. Yet constantly, over the counsel of all his advisers, he continued, as we shall see, to belabor the matter, digging the trap deeper and deeper with each statement. Duty and patriotism urged him to debate and clarify in public a matter which he, as a soldier, felt had to be clarified, yet must have known, as a soldier, was necessarily secret. But if he merely wanted public acknowledgment of existing permissive procedures at the Pentagon—then, in effect, he was supporting the administration's practice and why had he brought the matter up? And if he wanted greater latitude of choice for subordinate commanders—just how far did he want to go?

Any such debate was bound to put Goldwater on the defensive —for he was, in effect, challenging not just a political antagonist but the whole United States government and its energetic spokesmen. The more he attempted to explain the matter, the worse his predicament became. It did little good for him to assail McNamara as he did. For example, Goldwater introduced the term "conventional nuclear weapons," a term new in history, for only two nuclear weapons (at Hiroshima and Nagasaki) have ever been detonated in combat conditions. The Pentagon responded by pointing out that the *average* blast of tactical nuclear weapons in the field was several times the frightening but primitive Hiroshima twenty-kiloton blast. Angrily, Goldwater pointed out that there were little infantry-size weapons that went down to a fraction-of-a-kiloton blast. But even a fractional-kiloton blast, the Pentagon pointed out, was devastating; to call any nuclear detonation a "conventional" weapon of killing, they insisted, was absurd. Squirm as he might, Goldwater found himself described as a man fascinated by killing.

Another candidate might have been able to sterilize even such an exchange. But in Goldwater's case every exchange rang with an echo of his own past—a past he could not shake. Three political books, hundreds of columns, a thousand speeches on patriotism and Communism had written the profile of a man whose views had not altered since, in his fundamental work, *The Conscience of a Conservative,* he wrote: "A craven fear of death is entering the American consciousness. . . . The Communists' aim is to conquer the world. . . . Unless you contemplate treason—your objective, like his, will be victory. Not 'peace,' but victory." At San Francisco, in accepting the nomination, he had insisted that Communism must surrender; his book *Why Not Victory?* had as its chief thesis the thought that only force will cause Communism to recoil. His press conferences and his quick rejoinders on a hundred occasions provided the Democrats, as they had provided Scranton and Rockefeller, with a score of choice quotes which proved that he did indeed sincerely believe that to avoid war one must first be willing to risk war.

The issue, in short, was not at what technical level nuclear weapons would be unleashed or which general might make command decisions.

The issue, as it rose steaming into politics and as deftly molded by the Democrats, was quite valid: Just what attitude *should* Americans take toward the use of nuclear power? How ready should government, as a whole, be to use nuclear weapons and risk the escalation to destruction?

The issue was never, as Democrats tried to define it to their advantage, a choice between war and peace. But it *was* a choice, nonetheless—between peace and *risk* of war. Later, as we attempt to explore the political reaction to the issue, we shall try to measure the response. But none of the stabbing slogans that blossomed in the wake of Goldwater's journeyings ever caught the essence of this prime issue as did the one suggested to the Democratic National Committee by an anonymous North Carolinian: IN YOUR HEART, YOU KNOW HE MIGHT, read the last of the variations on the enigmatic Goldwater catch phrase. And in the end, because they had been persuaded that indeed he *might,* millions of Americans voted against the Republican.

Goldwater's second great issue was an equally complicated one, equally overlaid with subtleties that had to be violated with a crude yes or no vote: this was his crusade against the central government—against an all-dominating, all-entangling Federal bureaucracy of Washington. Whenever a political reporter journeys with a Republican candidate—right, left or center—he knows the one sure-fire catch line is the attack on the government in Washington. All Republicans praise the flag—and denounce the government.

A shrewd candidate has, indeed, a very sound and forward-ringing issue in denunciation of this centralization—if he handles it skillfully. The American government is, to a large extent, stone-hardened in fossil structures and fossil theories that descend to it from the days of the New Deal emergency thirty years ago, too many of which have outlived their usefulness and hinder, rather than help, the bursting new society Americans have made.

Yet how to remove these fossil relics from American life is today one of the most technically complicated problems both of government and of politics. The farm subsidies, for example, have now developed into one of the most grotesquely unreasonable patterns ever stimulated by necessity and goodwill; yet to abolish them overnight would be as irresponsible as it would be cruel, and as heartless as cutting off a man innocently addicted to drugs from the habit with cold-shock treatment. Subsidies to American shipping, to American aviation, to American peanut growers; depletion allowances for gas and oil barons; labor legislation and many labor privileges—all have become obsolete.

All Americans, generally and as a matter of principle, will denounce distant Federal paternalism—but will scream if any tampering threatens the benefits which they personally get out of it. Thus, the roar that comes when any speaker denounces the distant Federal Government is a heartfelt response. Yet when the meeting is over and the hearers wander away to examine the general proposition, they very frequently have second thoughts; for what the other fellow calls a subsidy may seem to him a constitutional guarantee against need; what cramps one man is another's crutch; restrictions placed on one individual guarantee someone else's freedom; one man's open cook-out pit is someone else's smog; and the boss's annoyance with the paper that clutters his desk is balanced by his secretary's need for unemployment insurance.

The problems of governing a technological civilization like America are complicated enough in themselves. But they are fused by emotion into a general resentment of all forms of increasingly impersonal control over an increasingly accelerating complexity. The emotions of normal people resist the general condition of a Digital Society—digits for the boys who are drafted, digits for Social Security and income-tax people, digits on credit cards and union cards, digits replacing familiar telephone exchanges, the electronic recordings that answer the telephone at airports and railway stations. And the center of the digital web seems to lie in Washington, where more and more computers more and more rapidly chew up digits, to spew them out again in controls and directives that seem to promise man a more and more digital future. And to all of these confused emotions a final exasperation is added by the new Supreme Court—a Supreme Court which has abandoned the old concept of the judiciary as a balance wheel against excess, and replaced it by a concept of the judiciary as a propulsive wheel that speeds Washington faster and faster—from abolition of school prayers in village schoolhouses to legislative reapportionment.

For these confused emotions, no man over the years has been a more perfect voice than Barry Goldwater. Over the years he has denounced farm subsidies outright; advocated the abolition of Rural Electrification; urged the selling of TVA; indicted the National Labor Relations Board; excoriated the Supreme Court; riddled the bureaucracy with scorn and contempt. When over and over again during the campaign he would promise, "I will give you back your freedom," his sincerity would evoke the wildest response. His misfortune was that he could not make clear just how, and by what degrees, he would free the American people from paternalism and central government without exposing them at the same time to some personal loss.

It was here, in his confusion, that the Democrats lanced him. Out of the vast mass of his many statements and speeches, they chose to hook and hang him on one issue: Social Security. Goldwater was general in his denunciation of big government; the Democrats chose a specific for response, and they could not have chosen better.

Of all the enterprises of big government, none except the income tax touches more people than Social Security. But whereas the income tax takes, Social Security gives. In 1964 the Social Security system paid out more than $19 billion to recipients of its insurance; in February of 1964 over 19 million people received its old-age or disability benefits—and another 90 million people had "insured status" with the Social Security from which they expected, in their turn, someday to draw a benefit.

To attack Social Security is thus, in the highest political sense, dangerous. Nor did Goldwater ever attack it—just as, indeed, he never demanded war. What he did instead was to ruminate aloud about what might be done with it—just as he ruminated out loud about the use of nuclear weapons. His campaign statement officially declared that *"I favor a sound Social Security system and I want to see it strengthened. I want to see every participant receive all the benefits this system provides. And I want to see these benefits paid in dollars with real purchasing power."* But it was his earlier ruminations, rather than this final official statement, that frightened people. Just as his attitude toward war exaggerated and amplified every statement he made on nuclear weapons, so his general attitude toward government exaggerated every fear of 100 million Americans who get, or hope to get, Social Security payments from Washington.

It was headlines rather than statement that shaped the issue of Social Security against Goldwater. He had again and again, in columns, speeches and articles, toyed with various ideas of modifying Social Security. But it was in Concord, New Hampshire, in the fateful primary of that state, that on January 6th, in response to a question, he said one way Social Security might be improved would be to make contributions to it voluntary.[8] Now, a voluntary Social Security system is statistically and actuarially a silly idea; if young family workers at the height of their earning power can voluntarily withdraw, there will be nothing left in the till to pay the old people who have already contributed their mite and gone into retirement. The headline of the Concord *Monitor* the next day, GOLDWATER SETS GOALS: END SOCIAL SECURITY . . . , was a flat distortion of his remarks which rankles in his memory more than any other item except the "trigger-happy" charge. (Goldwater believes that Rockefeller leaders planted the headline, which is not true.) Yet, in essence, this

[8] The actual statement he made in his press conference of January 6th, 1964, as quoted in press dispatches and never denied, was: "I would like to suggest one change, that Social Security be made voluntary, that if a person can provide better for himself, let him do it." This was the statement chiefly used against him in the campaign. It was, however, no spur-of-the-moment rejoinder to a question. No less than seven times Goldwater was on record, in some form or another—in interviews, articles, and the *Congressional Record*—with one or another variation of this idea, which he never repudiated.

suggestion of Goldwater's, if taken seriously, could only lead to a crippling and ultimately a collapse of the Social Security system as Americans now know it.

Republicans (and Democrats too) might recognize that the vast and tangled jungle growth of Federal bureaucracy is a menace to American values and American options; but if, as his opponents—Rockefeller, Scranton, Johnson and Humphrey—were quick to point out, it meant forfeiting their Social Security, what then? It was all too clear, even early in the campaign, that millions of modest Republicans in retirement—in southern California, in Florida, in New Hampshire, in Pennsylvania— were unwilling to risk those monthly checks that could run from as little as $40 to a maximum of $254 for a family.

And, again, the subtleties of the great issue of central government versus local or private initiative could not be separated out. Again, on this issue as on the bomb issue, America had to answer yes or no.

These were the issues that translated best, however coarsely, into the campaign clash. But there were other issues, equally weighty, with which neither of the candidates could grapple because these issues were even more subtle, even more complicated.

The issue of civil rights was as central to American concern in the campaign of 1964 as the issue of war and peace. Yet both candidates jointly decided to exclude from the campaign dialogue as far as possible any implied appeal to racism—and, accepting this exclusion with high principle and great responsibility, Goldwater took the loss.

Goldwater had, of course, begun by painting himself into a corner on this issue too. He had, out of conviction, voted against the Civil Rights Bill in June—against the warnings, it must be noted, of his political advisers. Thus he had pinpointed himself as the outright anti-Negro candidate of the campaign, clearly on record in an area where the decisive morality of America was against him. Now the exclusion agreement cramped him in exploring the subtleties and shadings of this morality.

The subtleties and shadings of white morality on the problem of Negroes are most realistically examined by the coarse measurement of politicians at the three rungs in the ladder of fundamental Negro need: jobs, housing and education.

Generally, across the country, politicians know that their white voters will accept Negroes as equals on the job—or, at least, as entitled to equal opportunity on the job. There is little danger of serious backlash at this first level of demand. No Congressman who voted *for* the Civil Rights Bill in 1964 was defeated for re-election; eleven of the twenty-two Northern Congressmen who voted *against* it did suffer defeat.

The next level of Negro demand, housing, is more dangerous; in general, if politicians can keep the issue within their city councils or at a deputized level, they will vote for open-housing ordinances; yet they know that when the issue has been brought to the people by open popular referendum, the people invariably, without exception, have voted against open-housing laws.[4]

The final level, education, is the flash point of peril. Politicians everywhere flee involvement in the integration of schools as if it were instant contamination. School boards are customarily removed from voter control, and politicians are thankful for it; yet one must recognize that all across the country—from Max Rafferty in California, to Ben Willis in Chicago, to Louise Day Hicks in Boston—the most solidly entrenched school officials are those who defend the neighborhood school. New York is perhaps the outstanding example of political cowardice on integration of schools in big cities. From top through bottom—from Governor Rockefeller through State Commissioner of Education James Allen, through Mayor Robert Wagner, through the vacillating Board of Education of New York City—all are in favor of integration. Yet none dares either to defend or to denounce publicly the neighborhood school, or to let the people of New York City vote on the matter. Instead, as in a game of snap-the-whip, the hapless new superintendent of New York City schools, Calvin Gross, an administrator with no political sense or drive, was flung to the wolves and dismissed for failing to solve a social problem which political leaders themselves shrank from facing.

What is involved in the present development of the civil-rights problem in America is the nature of all the silent, unrecognized neighborhood and ethnic communities that make up American urban life. The morality of the Civil Rights Act of 1964—and the absolute political need for it—cannot be questioned. But the next question—which the campaign of 1964 might have illuminated, but did not—is how men once freed for civil equality in the general forum of employment, opportunity and politics shall go about living together, or apart, in communities of their own choosing. (See Chapter Eight.)

This, along with war and peace, was a central issue of the campaign of 1964. Discussion of this issue probably obsessed American conversation in the summer and fall of 1964 more than any other. From coast to

[4] We have already mentioned the defeat and repeal of the Rumford Fair Housing Act in 1964 by the astounding margin of 1.5 million votes in California, or almost two to one. No new statewide open-housing law has been passed in these past two years of cresting civil-rights activity in any state of the Union; and propositions to open housing to Negroes have been smothered or defeated by the legislatures of Rhode Island, Ohio, Michigan, Iowa, Indiana, Illinois, Washington and Wisconsin.

coast, and from drawing room to corner bar, from union beer hall to old ladies' home, probably no question agitated Americans more than the question, "What Do They Want?"—"they" being the Negroes. Lyndon Johnson and Barry Goldwater each made but one major speech touching on this issue, Johnson in New Orleans, Goldwater in Chicago—but, for the most part, they recognized it as too difficult and dangerous to debate in public.

But the race issue as such, a cardinal manipulant of emotions, was left buried—left to work its own results in the states of the South that Goldwater carried. And left to work its way silently in the big cities, where the white workingman took out his fears on local candidates who threatened to open and perhaps destroy his neighborhood—but in his national choice, forced to buy either the Johnson package or the Goldwater package, chose Johnson.

There remained, finally, an area of contention and difference that is so new it still lacks an appropriate name. One can call it, perhaps, the issue of quality. Quality was what John F. Kennedy was all about, in its classic, Greek sense—how to live with grace and intelligence, with bravery and mercy. Quality was what he sought in his brief three-year administration; but it had not been an issue in his campaign of 1960, except as his personality and eloquence gave an example of it.

Yet in 1964 two men with far less of the sense of quality than John F. Kennedy succeeded, together, in making quality the fourth and last of the major issues of the campaign. Goldwater called it the "morality issue"; Johnson called it "the Great Society." During the campaign neither could define what he meant—but they were bringing into engagement what in another decade, if peace persists, may well be the central issue of American life: What is the end of man? What is his purpose on earth? How shall he conduct himself with grace and mercy and dignity? No other society in history has ever been rich enough to face such a problem or to discuss it in the political forum; quality has been something left to the church and the philosophers; all other societies have been too deeply involved with making, getting, existing, subsisting and defending themselves. Americans in the 1960s are faced for the first time with the problems of abundance—the purpose and style of life in a society whose great majority has been relieved from real want and thus freed to express itself.

There was, of course, want enough in America in 1964—and to this the Democratic candidates, Johnson and Humphrey, constantly addressed themselves in their poverty speeches and program. But this want, though real and all the more galling because of its contrast with surrounding affluence ("the prison with glass walls" is how Eric Sevareid described the condition of the poor), was less than the giant fact of prosperity. There was no doubt that John F. Kennedy and his economists had brought about the first fundamental change in American economic policy since Franklin D. Roosevelt—and the nation glowed with a boom that was one of the world's wonders. The boom terrified Europeans, angered the underdeveloped in the world, baffled the Russians. In America, in mid-campaign, the newspapers gave only scanty paragraphs to a settlement won by the automobile workers in Detroit—guaranteeing retirement to auto workers at age sixty at $100 a week; abroad, the statement could not be believed and could be taken as propaganda in a country like France, where a full colonel on active duty receives the same base pay. Luxuries undreamed of five years before were now mail-order items. A record eight million new cars were purchased; pleasure-boat sales were estimated at 400,000 a year; new dwellings were started at the rate of 1.5 million a year. The *increase* alone of America's gross national product in the four years of the Kennedy-Johnson administration was greater than the *entire* gross national product of Germany in 1964—by $122 billion to $100 billion. Wages rose, profits rose, the stock market rose, vacations lengthened. The conscientious pointed out, quite rightly, that one fifth of the nation lived in poverty. But the other four fifths *were* the majority, freed apparently from the curse of Adam and invited to contemplate their souls and identities.

This contemplation of personality had led to a strange country: it was as if a radioactive dust, called money, was in the air, invisible but everywhere, addling or mutating old habits of life. PROFIT AND LOSS FACTORS IN RELIGIOUS PUBLISHING, read a headline in a trade journal of publishing. U.S. crime in the one year rose 15 percent. Assault and theft both rose sickeningly. When this reporter was young, fifteen years ago, the cliché was that New York had one murder a day; in 1964, New York, with a smaller population than it had in 1950, had 637 murders— or almost *two* a day. (In the first six months of 1964 the incidence of forcible rape in that city rose 28 *percent* over forcible rape incidents the year before; in Phoenix, Arizona, the hometown of the morality candidate, Goldwater, the crime rate was even worse.) Sociologists called their conferences week after week, and from their fascinating sessions emerged a profile of sickness and disturbance that was appalling: syphilis was rising, from 6,251 cases in 1957 to 22,733 in the year ending June 30th, 1964. (The Public Health Service pointed out that these were only *reported* cases, and estimated the more accurate figure would be 200,000, half among those under twenty-four.) Dryly, the service pointed out that syphilis had reached epidemic proportions in thirty cities—and, without editorial comment, ascribed it to a decline in morals among young people. Drinking was up, reported another group: 71 percent of all adults in America now drank hard liquor, and among doctors, lawyers, journalists and professional people the effective figure was approximately 100 percent. (Hard-liquor sales rose to a record of $6.5 billion in 1964—and the three largest companies had increased their annual gross by an average of 24 percent in five years.) Homosexuality had become a major problem in big cities, and local laws were being modified, not to eliminate it but to make it more permissible. Beach riots, resort hooliganism, raids on house parties were staple fare in the newspapers. In 1960 I remember reading for the first time of a morbid incident in New York where crowds, gathered to watch a suicide, had urged him to "Jump! Jump!" Now in 1964 one could hear or read about such cases in a number of large cities—as well as instances where men and women looked on or heard women scream and stood by and did nothing as murder and violence proceeded before their eyes.

America in 1964 was a perplexing place indeed. There was no doubt that American genius was surpassing itself in all the old measures of progress. The Japanese might be abreast of America in electronics and ahead in shipbuilding; the Germans were abreast of America in steel technology, the British in motor design, the French in biologicals. But no other nation possessed such capacity over the entire range—in steel, in electronics, in instrumentation, in plastics, in automation, in computer technique, in film direction, in chemicals, in medical innovation, in any exploration of man's mastery over matter.

But did this make Americans happy? Here the evidence was varied and could be read either way. A study of suicide and mental health conducted in 1964 brought up the estimate that one in forty Americans—no less than 5 million—had attempted at one time or another to commit suicide; and the suicide rate was gently rising—from 10.1 per 100,000 (in 1954) to 11.0 (in 1963). More Americans lived longer than ever before—the number of Americans over sixty-five was estimated at over 17 million, as compared to 12.3 million in 1950. But who would care for them? Where would they live? The average sixty-five-year-old now has a life expectancy of over fourteen years. How could the agony and loneliness of old age be comforted without destroying the vitality and resources of younger families whose energies and efforts were bent on their children? Should government involve itself in the care of these aging citizens? Should a compassionate society tax itself to ease their woes, or would such a step lock citizens in another system of digits and taxes? Goldwater's response to this unsettling experience of America was to mourn. "What's happening to us? What's happening to our America?" he would ask wherever he marched.

And here again the Democrats hanged him, not on fact but on attitude.

For what was happening was not all bad. Newspaper headlines warned of the alarming rate of dropouts from schools and colleges. But when one read the texts beneath the headlines, the statistics told another story; of 1930's fifth-graders, less than half finished high school; of 1950's, almost three fifths; of those who were in fifth grade in 1954-55, 636 per thousand had graduated from high school by 1962. Thus, the drop-out rate was falling—only the need for education was growing faster than the effort to catch up. There was, indeed, a school crisis, but whereas twenty-five years earlier only one in three high-school graduates went on to college, now more than half go to college; and their numbers had jumped from 1.5 million in 1940 to an expected 5.2 million in 1965. Americans were spending more on hard liquor—but per-capita consumption was not increasing. Americans were merely buying more expensive brands. Crime was up—yes. But not felonies. The shocking figures came in the shrinking big cities with their tensions. Overall, however, the new America was much safer than thirty years before—with murder down from 8.9 per 100,000 in 1930 to 5.1 in 1962.

The Democratic answer was most masterfully given, as we shall see in later chapters, by Lyndon Johnson himself. But what the two parties were debating in this issue was the very nature of American experience in this half century; and of this experience I remember best as a campaign sampling the first journey of Hubert Humphrey to his native place in South Dakota. For, had Johnson pondered for years on the proper choice to demonstrate the golden side of the American experience in the half century, he could not have chosen better than he did in his Vice-Presidential candidate.

I journeyed to Doland, South Dakota, with Hubert Humphrey at the beginning of his campaign on a plane named the *Happy Warrior*. (The journey began with a champagne breakfast paid for by the Democratic National Committee—a far cry indeed from the last journey on cold coffee in West Virginia in 1960.) In Doland, where his father had owned a drugstore, Hubert Humphrey had spent twelve years in the single schoolhouse, and they remembered him still as Pinky, the mischievous boy. Humphrey visited the neat red-brick schoolhouse and recognized the huge silver loving cup on the second-floor landing. He recalled how Julian Hart, the minister's son, and he had taken all the alarm clocks left for repair at his father's drugstore, set them to go off at two-minute intervals, then stuffed them into the enormous loving cup; how

they had clanged out to disrupt the entire school in mid-morning; and
how old Jeanette Higgins, the principal, had shrieked at him.

He stood on the main street of the town. There was a park now,
with Chinese elms. New fluorescent lights hung over the little street.
There was a library, left by old Doc Sherwood. There was a supermarket. Television antennae brought national television. It was not a prosperous town—and the young people, as Hubert himself had done years
before, were leaving it. But it was a good place to live in, far better than
when Hubert Humphrey's father had eked out a miserable living there.
And the opportunities it offered its young people were no longer circum-
scribed by the flat prairie—they were nationwide. In the old Humphrey
drugstore where he had once jerked sodas, the present management had
hung up a fly-blown mileage chart which announced the distance from
Doland to Ashley, North Dakota—142 miles; to Yankton, South Dakota—179 miles; to New York—1,217 miles; to Washington, D.C.—
1,342.

Humphrey stood on a little platform, greeting old friends ("Remember when we punctured the rain barrel?"), and then began to remi-
nisce. And as he reminisced, it was obvious he was not defending things
as they were in Doland or as they used to be. He loved Doland. He was
offering Doland not mourning but challenge, not yesterday but tomorrow. His twin themes were opportunity and government. And he bound
them both with the gladness, not the peril, of American life:

> I want to say that there are many improvements. First of
> all, the band is a lot better than I remember. They look better
> and they play better, and I can remember that I was in the
> local band here for a while. . . . When they got hard up for
> a bass drummer, the best you had to do at that time to qualify
> was to keep time, and since you were setting the pace, you
> could always blame others if things didn't come out right, you
> see. . . .
>
> I look down this main street and . . . I remember when
> they built that hotel over there. If my father were here, he
> could tell you he remembers too, because they lost some
> money. But it is there. That is part of the way we built Amer-
> ica. Some people had to take a chance, some people had to be
> willing to risk a little money, a little time, a little energy.
>
> I would like to say that the great treasure of this com-
> munity has been in its people. This isn't from an esthetic point
> of view in terms of beauty. . . . I suppose you wouldn't say
> that Doland would stack up with Rome, or Washington, D.C.,

or Paris. . . . I have bragged on this town, as I have said, all
over, but I wonder if we have ever stopped to think for a moment what is so different about us. Because people look so
much alike all over the world.

We have learned how to govern ourselves, making our
mistakes but learning from them. We have learned that if we
are going to have . . . [a government] that offers a future to
each generation, you have to keep open the doors of opportunity.

There isn't any opportunity for the illiterate, and there
isn't any opportunity for the uneducated. People who are uneducated nowadays . . . they are prisoners. They are like
slaves. So, education, in a free society, where you have freedom of choice, where you have opportunity to make choices
. . . is the secret of power, to freedom, to the good life.

Now, let me say that this system includes something else.
It includes respect for one another. The only aristocracy that
we have in America is the aristocracy of achievement and
merit, not because of your color, not because of your creed,
not because of your place of birth, but because of you. One of
the commitments of my life has been to help eradicate in
America these false barriers, these false standards of discrimination, of bigotry and intolerance that have denied so many
people a chance to give of themselves to their country. We
Americans need everybody helping out. . . . We need to set
an example of how we can bind ourselves together in common
purpose for great national and international objectives.

It was Humphrey, the next day, who out of his own past, suddenly
made the life of America that he was defending, and its direction, come
alive in a glorious passage of nostalgia and hope.

After a day of boyhood recapture in South Dakota, he left his native state the next morning with a prop stop at Rapid City. Rapid City is
the site of the Ellsworth Air Force Base, the most complete center of
nuclear destruction in the United States (command center for B52s,
KC1350s, 3 Titan launching sites, 150 Minutemen missiles); the base
personnel make up the fifth largest community in all South Dakota. After
leaving this terrifying and beautiful place, the plane passed over Mount
Rushmore with its stone carvings, and Humphrey remembered how as a
boy he would come here on camping trips in the summer; then he explained the needle-point mountains in terms of erosion; then he was sad
that this was such poor country, so difficult to make a living in. And that

either tourism or the government must be its salvation. From that to the importance of defense industries and how when the government established a defense complex these days it was like the Queen of England giving the South Seas Charter in Elizabethan times—a defense complex was a charter of development. He went on to the impact of defense complexes and scientific research complexes in the rebirth of the Boston area, and at that point, someone asked him why Boston, with such intellectual resources, had such bad politics. This led him to the ethnic tangle of American municipal politics, and thus to his father—the first man in Doland, South Dakota, who had voted Democratic. "I remember Al Smith," chuckled Humphrey. "My father was one of only five voters in Doland who voted for him."

"Some day I think I'll write a book about Dad," he said, peering out his plane window as we left South Dakota for Colorado, and with his audience huddling over him, he went on about Dad Humphrey.

Dad Humphrey had been, obviously, Doland's village radical. Mom Humphrey had been a good Lutheran, and Dad "used to drive her crazy" by quoting Bob Ingersoll to her. Then one day he saw God and joined the Methodist Church. But really joined it—he would preach on Sundays, and he had the biggest Sunday-school class in the entire county. Some days as many as 100 or 150 people would come to his Sunday-school class in Doland. When radio came in, Dad Humphrey's horizons expanded. All day Sunday they would tune in the radio and hear the divines: S. Parkes Cadmon or the Reverend Cannon or Harry Emerson Fosdick.

But politics was just as important to Dad Humphrey as religion. Sometimes after church on Sunday he would come home with as many as ten people and tell Mom they were eating there; and from religion they would go on to politics, until midnight sometimes. (Dad never went to bed. He told Hubert, "Never go to bed, stay out of bed as long as you can; ninety percent of all people die in bed.") Dad was a Democrat —he had become converted to the Democrats just as he had become converted to religion. William Jennings Bryan had converted him to the Democrats. And everybody else in town, and all the relatives, were Republicans. Even his own brothers. Uncle Harry was, of course, a scholar and not interested in politics. Uncle Harry went on and became the Chief Plant Pathologist of the Department of Agriculture in Washington. Uncle Harry was interested neither in politics nor money—only agriculture. Uncle John used to say, "You could put Harry sitting down on a pile of gold and he'd lose it all in a week."

Uncle John and Dad Humphrey were, however, small-town businessmen—full of schemes and business deals. Together they dreamed up the scheme of "green stamps," which later became, in the hands of more successful administrators, trading stamps. They went as far as opening an office on Wall Street—and they went broke; all they had left of the venture was cartons and cartons of unused stamps, which they stacked in Hubert Humphrey's home (he held out his hands showing how many) "as big as a grand piano."

When Dad Humphrey became a Democrat, the two brothers broke. They argued over everything. Rural Electrification, Missouri Valley Authority, price supports. "They got so mad at each other they didn't speak for years." But since the two brothers were close and both loved politics, they finally compromised. One year they would go to the Democratic state convention together, the next year to the Republican state convention together.

Business kept getting worse in Doland. They bought the hotel together, and they lost their money. And Dad's attitude toward business didn't help. He would have periods when he had a special interest in something. Late in the twenties he was in his music period. He got the Victrola concession for Doland, and he stacked the store with Victrolas. Then he'd buy cases and cases of Victrola Red Seal records and tell Mother that he was doing it for business. But, really, he bought the records for himself, and would take them home and play them and tell Mother he'd just made a bad business deal and overbought for inventory. Once, so carried away by music was he that he got up in the middle of the night, got in the old car and drove all the way to New York just to hear the Metropolitan Opera. Later on, when they were wiped out in Doland and moved on to the big city of Huron, South Dakota (population then 11,000) and business got better, Dad went on his poetry kick. He bought radio time with his own money on a Huron radio station to read poetry to the people—he felt they needed it.

Mom Humphrey must have had a hard time of it. But the two parents obviously loved each other. They disagreed on politics. Mom Humphrey felt that Dad was "mean" to people when he talked politics. But then, Dad did not take her politics seriously. Once he called the children in and said, "Now, look, everyone is a little nuts. The man down the street is a little nuts. I'm a little nuts. You're a little nuts. Even your mother is a little nuts. Your mother's a lovely woman, a fine faithful woman, she's my sweetheart. And any time you argue with her or you don't like it here—*get out!* But there's only one thing about her I want you boys to remember. Sometimes she's *politically unreliable!*" (At this point Humphrey added, as if defending his father, "You know, she voted for Harding, and Dad could never forget it.")

Humphrey had now been reminiscing for a full hour and his plane

was coming into Denver. It had been a hard life, all in all. Times had
gotten better after they all moved from Doland to Huron, but old Dad
Humphrey never trusted banks. Whenever they had any savings, old Dad
had put the savings either into government bonds or into inventory. He
loved to have his savings in inventory under the drugstore. ("I love to
have it here where you can look at it.")

Thus it was that Mom Humphrey (who at this writing, at eighty-
four, lives in a nursing home in Huron) never got her rocking chair.
There never was enough money to get her the rocking chair she wanted.
Which led Humphrey on to another thought—about money. On which
he disagreed with his father. "I believe in debt," he said. "I tell my kids
to borrow money. Look at me now, I've got two houses, both unoccu-
pied, the kids are away, and I'm on the road most of the time. The time
for people to have good things is when they're young and when they need
it, that's the time they can enjoy it. My son-in-law talks about buying a
house he can afford. I tell him you won't like a house you can afford. Get
something with a couple of big mortgages and then work like the dickens
to pay it off. When you're fifty they'll all be gone. Enjoy it now."

By then the plane was coming down into Denver and Humphrey
was scheduled to address Denver newsmen—where he spoke of the
Issues, with remarkable eloquence and staggering fact, answering questions on beef imports, the drought, the bomb, the income tax, rioting in
the streets, the John Birch Society, and Vietnam.

Johnson and Humphrey, Goldwater and Miller, all believed that the
purpose of America was to enrich the individual life. Something, perhaps, was wrong with the condition of that life in 1964. But Goldwater
and Miller saw what was wrong as the government; and Johnson and
Humphrey saw the government as the chief means of dealing with the
wrong.

Perhaps both Democrats and Republicans were wrong, as they tried
to explain America to Americans; perhaps the nature of life in the abun-
dant society requires deeper thinking than can be done in a political campaign. Yet if the approaches of both sides were unsettling to people and
unsatisfying to thinkers, Goldwater managed to bring to his approach a
particularly joyless quality. What the Democrats offered was offered with
glee, gusto and the colors of the rosy-fingered dawn.

Goldwater could offer—and this was his greatest contribution to
American politics—only a contagious concern which made people realize that indeed they must begin to think about such things. And this will
be his great credit in historical terms: that finally he introduced the con-

dition and quality of American morality and life as a subject of political debate.

Goldwater caused nerve ends to twinge with his passion and indignation. Yet he had no handle to the problems, no program, no solution—except backward to the Bible and the God of the desert. Fiercely proud of his sturdy grandfather who had struggled, fought, wandered and, by manliness, made civilization grow on the Old Frontier, he could not quite grasp the nature of the newer enemies on the new frontier of life. Proud of his handsome, clean-lined family, proud of his radiant children, proud of his family's war record and patriotism, he had in him neither the compassion nor the understanding to deal with the faceless newer enemies of the Digital Society.

These, then, were the issues—profound, moving, deeply dividing.

In 1960 either Richard M. Nixon or John F. Kennedy could well have campaigned under the other's chief slogan: Nixon could easily have bannered his campaign with "Let's Get America Moving Again," and Kennedy could easily have accepted "Keep the Peace Without Surrender." It was an inner music of the soul that separated them, an outer style of leadership which urged the Americans to the choice of 1960.

In 1964 it was otherwise.

The gap between Johnson and Goldwater was total. Though as masculine Southwest types they used the same language, the same profanities, shared the same drinking style, indulged in the same homespun metaphors, these similarities were meaningless when compared to the philosophies that separated them. How these philosophic differences were translated by organization, by orchestration, by campaign planning, by the combat leadership of the two chieftains is the story of the final election round of the campaign of 1964—to which we proceed next.

CHAPTER ELEVEN

BARRY GOLDWATER'S CAMPAIGN: "WHAT KIND OF COUNTRY DO YOU WANT TO HAVE?"

O F BARRY GOLDWATER's campaign it may be fairly said that no man ever began a Presidential effort more deeply wounded by his own nomination, suffering more insurmountable handicaps. And then it must be added that he made the worst of them.

There is an almost irresistible temptation to tell the story of the debacle as comedy—as does one of Goldwater's campaign aides who, asked when they knew the tide had turned, said it reminded him of the old football story: the tide turned when the other team first appeared on the field.

There is an equal temptation to tell the story clinically—as a political basket case. For by the time the Goldwater men were ready in September to begin their national campaign, they had already been so thoroughly sliced up as to remind one of men whose limbs are sheared off so quickly by machine blades that they do not sense the loss until it is over. If I had had a pint of brains, recalled Goldwater after the election, I should have known in San Francisco that I had won the nomination and lost the election right there. As it was, he went on recalling, it wasn't until August that I knew it was hopeless.

It is best, perhaps, to tell the story in the mood of the macabre, for Goldwater's campaign was hopeless to such an extent that it not only destroyed him as a political figure but profoundly undermined the worthy cause he had set out to champion, the cause of American conservatism, which was deprived of the voice and the clarity it merits. In 1964 that cause was exposed as formless in ideas and hollow of program.

One must begin with the political theory that accompanied the cause Goldwater championed. The theory held that for a generation the American people had been offered, in the two great parties, a choice between Tweedledum and Tweedledee; and that somewhere in the American electorate was hidden a great and frustrated conservative majority. Given a choice, not an echo, ran the theory, the homeless conservatives would come swarming to the polls to overwhelm the "collectivists," the liberals, the "socialists," and restore virtue to its rightful place in American leadership. The campaign of 1964 was to be the great testing of this theory.

Coupled with this theory was a concrete electoral strategy—pegged, as all strategies must be, to the electoral map of America: 270 electoral votes needed to elect, out of a total of 538 in 1964. On the walls of the various Goldwater headquarters could be seen the maps that plotted this strategy. Goldwater planners began with the South: they would need to have, and they believed they would have, all eleven states of the Old Confederacy, with a Southern base of 127 electoral votes. To this they added Oklahoma, Kentucky and Arizona for sure—for another 22 votes. Beyond that, then, there was an assault area: the Midwest and Rocky Mountain states. How much could be picked up in traditional Republican states like Nebraska, Kansas, Indiana, Wyoming, Colorado, the Dakotas, the smaller Mountain states? Fifty or 60 more? Which then left five major states for consideration: Ohio, Illinois, California, Pennsylvania and New York. Pennsylvania and New York were to be kissed off. (Indeed, one of the causes of the spring breach between Goldwater and Scranton was a remark by one of the Goldwater leaders to a Scranton lieutenant at the Cleveland conference in early June that Pennsylvania was *already* written off.) But in Ohio (26 electoral votes), Illinois (26) and California (40) were 92 electoral votes apparently up for grabs—enough to give Goldwater an electoral majority without a single vote from the Eastern Seaboard, which he had once advocated sawing off and letting float out to sea.

Thus the strategy.

And beyond that there was organization.

The organization of Goldwater's campaign in Washington and across the country was absolutely first-class, except that it reminded one of the clay mock-ups of the new models in Detroit's automobile industry. It was meticulously designed, hand-sanded, striking in appearance—but it had no motor. It recalled the famous quatrain of the medieval poet

Samuel Ha-Nagid about the crippled man who carved on a wall the en-
graving of a perfect leg and tried to stand on it, but fell.

Goldwater is himself a neat, efficient man, and once his victory in San
Francisco had been assured in July, he set out to remake the Republican
headquarters in Washington. Now, all political headquarters are sloppy,
confused and disconcerting gatherings. They deal with emotions, dignities, uncertainties, the stuff of dreams and fear—all of which are materials resistant to logical administration. Political headquarters need not,
indeed, be as slovenly as they normally are in America, but I know of no
efficient political headquarters except in countries where ironbound ide-
ologies direct committed or terrified partisans. One remembers still the
offices of the Democratic National Committee in the fall of 1960, where
a shirt-sleeved Bobby Kennedy charged back and forth, dodging desks,
clutter and crackpots, as he flogged his captains on in the campaign that
won his brother's victory. Compared even to today's Democratic National Committee, the Republican National Committee was in early
1964 an institution of outstanding decorum. Yet to the Goldwater leadership it was a mess—as much for its slovenliness as for the character of
the Establishment Republican functionaries it sheltered; it must be
streamlined, made efficient and converted into the instrument of the new
conservatism.

Thus, reorganization of this Party headquarters was the first order
of business for Goldwater after San Francisco. Permitting himself a
much-needed vacation after the exhaustion of the primary and Convention efforts, he dispatched to Washington two young men—Dean Burch
(thirty-six) as new National Chairman, and John Grenier (thirty-three)
as new Executive Director, both of them handsome, neat and efficient
as the most attractive junior executives of American Tel & Tel or
General Motors. Their duty: clean up the headquarters, purge and restaff
it; plan schedules; have ready for Goldwater's return a complete layout
of trips, speeches and ideas; organize the speech-writing staff; set up the
finance committee; prepare for expansion (the staff must go up from 100
to 700 in a few weeks); establish communications.

For five full precious weeks—while the national audience was
theirs, curious, cocked to listen—the two remarkable young men quietly
concentrated on streamlining the national headquarters as if, on September first, a whistle would blow and only then the campaign would start.
They could boast quite rightfully by September first that, on paper, it
seemed the best-organized headquarters ever put together for a national
campaign. But this was work that should have been done long before
August and by others. The Republican Party had known no less than
twelve national chairmen in twenty-four years; Goldwater's lieutenants

had much lost time to make up for; but all the while they labored, the country's opinion congealed.

First there was housekeeping—offices to be torn down, partitioned, repainted, soundproofed to keep spies from bugging them. A map-and-chart room was set up, so elaborate, said an observer, that it would have taken a nest of computers to keep it up to date. Communications were reorganized: the ancient switchboard of the committee was supplemented by a new one; teletype machines were installed for instant communication with each of the fifty state headquarters; nine high-speed teletypes were set up that could pour out up to 8,000 words an hour to each of nine regional headquarters. ("They were so efficient," said one old politician, "that each regional director had a full morning's reading on his desk every day when he came to work; but the point was for them to get out and find out what was happening, not for us to tell them.") Quotas were assigned for votes, state by state, county by county. Committees were set up. Charts and diagrams of communication were outlined. A "Go-out" board required everyone leaving the office to make his whereabouts known at all times.

On paper the system was magnificent. Biweekly, Opinion Research Corporation of Princeton polled a national sampling. Each Sunday the strategy board was to gather with the candidates at two P.M. to study the polling and review the situation. The strategy board was a shifting group of people: it included, of course, Dean Burch and John Grenier for the new organization; it included old-timers like L. Richard Guylay, director of public relations for Eisenhower's victorious 1956 campaign (as well as Nixon's in 1960), and Clifton White, architect of the nomination, now director of the amorphous Citizens for Goldwater-Miller; it included, among others, former National Chairman Len Hall, as well as Ohio's chief professional politician, Ray Bliss; Tony Smith of the Senator's personal staff; and Ralph Cordiner, once Chairman of the Board of General Electric, as finance chairman.

The strategy board talked, ruminated, met on some Sundays for as long as four hours. But what it discussed was mechanics: schedules, tours, advertising, television shows, organization. And here largely the power of veto lay with the finance chairman. There is no businessman who, when introduced as financial chairman to a campaign, does not become overnight an expert on television, images and voter impulses. But Cordiner was exceptional. He and Goldwater had agreed that this campaign, for one, was going to be run on a balanced budget, and that they would end it in the black. Now, the sad truth about any major candidacy is that it must run on credit. Every headquarters is always destitute when it needs money most—in September; then in late October and early November money comes pouring in as emotions heat the common people and the arms of big contributors are twisted to breaking. Thus, since the headquarters was broke at the beginning of the campaign, Cordiner insisted that the precious advance time spots on national television for the last ten days of the campaign—booked in the early weeks with such difficulty from grudging networks—be canceled. When finally, in late October, money came pouring into Republican national headquarters like confetti and a desperate effort was made to reinstate these time spots, it was impossible; they had been pre-empted. Cordiner ended the campaign with a historic record for a financial chairman: the largest surplus in dollars ever shown by any campaign—and the largest deficit in votes and offices lost too.

The strategy meetings were the X mark on the spot where the Goldwater campaign broke down. Between strategy board and candidate there yawned the critical gap. Goldwater had promised that he would personally attend these sessions with Republican strategists. But neither he nor Vice-Presidential candidate William Miller ever attended any of them. What the strategy board discussed or recommended was usually heard by the campaign director, Denison Kitchel, and Goldwater's chief brain-truster, William Baroody, a thoughtful man but a novice in politics—who would normally depart after a few hours to proceed to their chief's apartment in the Westchester apartments and there mull over the problems as they saw them. The courtly and intelligent Kitchel was probably the only man who could really influence Goldwater all through the campaign; yet he was no Bobby Kennedy, who had had full power of attorney for his brother and had been there, in 1960, translating candidate to headquarters at every minute. Kitchel was a dedicated man—but normally he was off with Goldwater on the cross-country treks of the campaign trail, and unavailable. And thus, except for Tony Smith, there was little personal contact between the candidate in the field and his streamlined organization; and it is the candidate, not the organization, who must make a national election campaign.

This is an iron rule of politics: the higher the office, the more important the candidate is—and the less important the organization. At a ward level in big cities like Chicago, a good organization can run a baboon on its ticket and carry him to victory by muscle alone; at state level a governor must score through on how he projects his proposals, on who he is and how he conducts himself; and at the supreme level of Presidential politics, the candidate and his behavior outweigh all other elements of his campaign. Organization and money are indispensable, but it is the candidate's words, his travels, his statements, his behavior, that the nation watches. His top team must be his *personal* team. He is the individual bayonet point of a mighty movement; the national press corps follows *him* from dawn to dusk; the television networks invest the greatest part of their news resources and energies in trailing *him*. One unguarded remark of a candidate, or one felicitous thrust, will reach more people than any accumulation of position papers or pamphlets. For it is not what gets written down in conditioned and qualified prose in platforms or pamphlets that counts; it is what gets into people's heads to move their emotions. And only one man can do this—the candidate. "The point of the matter is that the team doesn't run for the Presidency," proclaimed John F. Kennedy in 1960 from a platform in Los Angeles; "one man runs for President."

Nothing, therefore, in any Presidential campaign, is more important than the speech writers who accompany and serve the candidate personally. And no man, at this superior level of thought and direction, was more alone, more undermanned and understaffed, more deserted by friends, allies and associates, than Barry Goldwater. One man of unquestionably superior intellectual quality served on his speech-writing staff in Washington—Professor Warren Nutter of the University of Virginia, who by October was pale and haggard with exhaustion. One excellent political public-relations counsel, Tony Smith, in poor health, bore the main weight of the statements and reactions which a candidate must spout twice a day. And one accomplished writer—Karl Hess—accompanied the candidate on the road, editing and rewriting in weariness the erratic flow of research material that came to the campaign plane from Washington. A bitterness still lingers in several of Goldwater's personal entourage at all those eloquent conservative and right-wing intellectuals who had so urged the great cause on their champion. "Where were they?" said one to me. "Where were they when we made our charge up San Juan Hill? They blew the trumpets—but when we charged, nobody followed. Where was Buckley? Where was Kirk? Where was Davenport? Where were Bozell and Burnham? At least you got to say this for a liberal s.o.b. like Schlesinger—when his candidates go into action, he's there writing speeches for them." [1]

[1] In fairness it must be stated that the use of these men was several times contemplated by the senior level of the Goldwater team. They decided, however, that men like William F. Buckley, Jr. (editor of the *National Review*), James Burnham (one of its chief ideologues) and the others were "too arrogant, too cold, too intolerant, and they couldn't talk to people." However it was, whether unwilling or uninvited, the conservatives produced from the arid and negative pages of their writings and ruminations no concrete proposals or positive suggestions that could give their philosophy and Goldwater's any quality of promise. Liberal intellectuals do not have to be invited to contribute; Democratic headquarters in any campaign is usually overwhelmed by well-meaning liberals with ideas that run from the flossiest and most fanciful to the most practical and useful;

Several other vast problems faced the candidate which he and he alone could solve—and must solve if the tremendous odds against him were to be narrowed in the 110 days between San Francisco and the election. All were bungled. Not all the blame for this bungling is Goldwater's alone, but they demonstrate how little the most energetic and efficient organization can do if the leader and his flag cannot show the direction it must go.

The first of the problems was the deep division inside the Republican Party. Eisenhower, after his traumatic clash with Taft at Chicago in 1952, had taken the initiative within hours of his nomination and marched over to the loser's suite to earn Taft's goodwill; Kennedy, in 1960, had overridden the counsel of his closest devotants to choose Lyndon Johnson as his running mate and heal the wounds. In the politics of consent, no victory is ever permanent unless the victor makes it firm on a base of persuasion. But Goldwater was an ideologue; and he delayed for weeks—until too late—a gathering of Republican leaders at Hershey, Pennsylvania, on August 12th, which had been scheduled to bind the party together again. A 122-page transcript exists of that two-and-a-half-hour "harmony" session. Not even the fact that it is marked SECRET makes it interesting enough to reproduce or quote. No more dreary document has turned up in recent American political history than this transcript of a dozen powerful men—among them Eisenhower, Nixon, Goldwater, Rockefeller, Romney, Scranton—churning up talk without even a vagrant idea of how this election might be won, how they might combine to offer a common policy whereby the United States might be governed; nor even any deep indication of desire to do so. All the dissident governors made their points, left Hershey and then ran their own campaigns, avoiding appearances with Goldwater as if he were cursed with political halitosis. Bruised and humiliated in the primary fighting and at the Cleveland conference and San Francisco Convention, Goldwater had consented to the Hershey conference for appearance's sake. He brought no thrust of leadership to it.

The second problem was the devising of some respectable program, some orderly yet emotional vision which he could offer in terms of specifics, which would make the country understand how the old virtues might refresh the new society. There was, of course, as we shall see later, a generalized strategy of seven key issues; but (except for an income-tax proposal) they were all negative; the program of the Goldwater leadership consisted of indignation, not proposals. I mentioned to one of the Goldwater brain-trusters the emotional dedication of a Democrat like

the surliest and most limited Democratic politicians always find the liberal campuses a supermarket of ideas where they can shop freely and pick what they want.

Lawrence O'Brien to the single cause of Medicare; and the response came: "I can't think of a single thing we feel that strongly about." Again, no thrust.

And, finally, there was the hazard of the bomb, the "trigger issue," which Goldwater must somehow erase or neutralize, as Kennedy had neutralized the issue of religion in 1960. Yet here again the candidate seemed so stunned, so shocked by the attack on him as a killer, that he could not clear his mind to guide a counterattack.

The contrast between organization and thrust in a campaign can best be developed, perhaps, if at this point we leap ahead momentarily to Lyndon Johnson's campaign and see how the management of its television efforts differed from Goldwater's. For of all those matters in which organization is important, the direction of television in a political campaign in modern America is incomparably the most important. Here is where the audience is; here is where the greatest part of all money is spent; here is where creative artistry and practical commercialism must join to support the candidate's thrust.

The Goldwater television operation seemed in September, on paper, far superior to the Democratic operation. The Republicans had retained the Interpublic Group of New York—the largest marketing-communications group of companies in America or the world. The media specialists of Interpublic worked through Lou Guylay of the strategy board, who was, after the campaign, to rejoin Interpublic as an executive. But it was not until August 20th that its task force got down to work. Then, with the most sophisticated techniques, the specialists prepared their plans. Computers analyzed the last three national campaigns by party, state, ethnics; great charts were prepared showing "swing states"; overlays on the maps designated the two hundred major TV markets and how the largest audience could be reached at the smallest price. One billion seven hundred million bits of information were fed through the computers. But what were the media specialists to say? What was the message? This they did not know, for they could never reach their candidate, or even Kitchel, his alter ego; they had to work through Guylay, who required Cordiner's financial approval to move.

The Interpublic Group made a much more impressive executive presentation of its efforts than did the rival managers of the Democratic television campaign—Doyle Dane Bernbach of New York, the fastest-growing major advertising agency in the nation.[2]

----

[2] Doyle Dane Bernbach Inc. is responsible for a number of advertising adventures that will go down in the folklore of merchandising: the introduction of Volkswagen to America, the surge of Avis Rent a Car, the fragrance of El Al Israel Airlines advertising, as well as Levy's Jewish Rye. As imaginative a group as has

The happy-go-lucky attitude of the Doyle Dane Bernbach group was more earthy. It meant to go for the heart, and it knew the message to be delivered. From Washington, via Moyers, came one clear directive: Attack, jolt Goldwater, put him on the defensive from the beginning. At least three of the group's efforts deserve to go down in history as masterpieces of political television. First was a one-minute film shown only once: the famous Daisy Girl spot (run off on NBC's *Monday Night at the Movies* on September 7th), which began with a close-up of a towhaired moppet plucking petals from a daisy, babbling her count as she went, until the film faded through her eyes to a countdown of an atomic testing site and the entire scene dissolved in the mushroom cloud. The film mentioned neither Goldwater nor the Republicans specifically—but the shriek of Republican indignation fastened the bomb message on them more tightly than any calculation could have expected. The next, ten days later, was a portrait of a deliciously beautiful little girl innocently licking an ice-cream cone, with a gentle, motherly voice in the background explaining about Strontium-90 and pointing out that Barry Goldwater was against the Test Ban Treaty. This, as cruel a political film as has ever been shown, was also aired but once. And then, finally, there was the spot on Social Security, in which the fingers of two hands tore up a Social Security card. This, shown over and over again during the campaign, probably had greater penetration than any other paid political use of television except for Richard M. Nixon's Checkers broadcast in 1952.

The DDB executives had a clear message to transmit, and they needed no great organization charts to translate the message into emotion. They scored through on the opposition candidate personally—and jolted him. Goldwater was so lacerated by the cutting edge of the two "bomb" shows that they probably provoked him, as we shall see, to even greater exertion in defense of his frightening position. They scored through, with even greater effect, on the Republican Party generally, down to the county level. Over and over again, as Goldwater traveled, he would be brought up short by local Republican chieftains insisting that he must explain himself *again* to *their* people on the bomb issue and Social Security. Goldwater was running not only against Lyndon Johnson and the Democratic record, but against fear itself. He was responding to the thrust of the opposition, not attacking as a challenger must.

---

ever accepted difficult accounts, it has always specialized in the positive. The campaign of 1964, its first adventure in politics, gave its staff an opportunity to turn their fertile talents to a campaign of attack and made them as happy as a dog in a meat market.

It was much pleasanter traveling with Goldwater in the campaign than in the primaries.[8] Affairs moved on schedule, trains and planes arrived and departed when they were supposed to, the organization directed from Washington functioned well.

But the voice of the candidate was a blur.

There was still, perhaps, some lingering chance of a close race when at the end of September, trying valiantly, Goldwater embarked on his four-day whistle-stop tour through the Midwest.

This was to be the Sunday punch—Ohio, Indiana and Illinois were three states the Republican absolutely had to crack if he was to add anything in the Midwest to his expected Southern base. The polls, both public and private, were heavy against him (Gallup Poll: 65 to 29) and he had said several times to friends, who had relayed the thought to the reporters, that if there were no upturn by mid-October, then it was over. If there *was* to be a mid-October upturn, *now* was the time to move it.

But there was an air of timelessness over the entire Goldwater campaign from beginning to end, and the great whistle-stop tour seemed to proceed in a timeless void. There was no taste of fall in the air, no sense that the campaign was approaching climax. The days were hot, even sultry; the dust from the drought-dried farms hung over the lazy train stops; and the candidate's little talks might have been given in June or October, in September or November—or in 1960, or 1936. The twenty-two-car train, reminiscent of an older and more civilized pace in American politics, was the kind that Franklin D. Roosevelt (who had tracked more than 350,000 miles in 399 train trips during his long Presidency) or Harry Truman (who did over 21,000 miles by train to deliver his 300

---

[8] The improvement in Goldwater's press relations in the election campaign was in remarkable contrast to his press relations before and during the convention. Two new public-relations men shepherded the trailing press corps for him— Messrs. Paul Wagner, an urbane professional public-relations officer, and Vic Gold, a conservative intellectual who had gone down into the trenches for his hero. Gold, in particular, made every trip gay—clucking, fussing, mothering correspondents as if they were his children. Gold had joined the campaign to help Goldwater save Western civilization, but discovered his candidate's chief need was some way of getting through to the press. This Gold provided; he carried their bags, got them to the trains on time, out-shouted policemen on their behalf, bedded them down and woke them up, and before they knew it, the correspondents, about 95 percent anti-Goldwater by conviction, had been won to a friendship with the diminutive intellectual which spilled over onto his hero. The press buses following Goldwater might be booed by Goldwater partisans, and on one occasion stoned. But within the buses and on the plane a warmth developed that Goldwater himself came to share. The teasing of Gold and Wagner never stopped; correspondents preponderantly wore the little gold souvenir Goldwater had prepared for them which read EASTERN LIBERAL PRESS, but at the end the correspondents permanently assigned to Goldwater were almost protective of him and felt guilty each time they must report those things they could not avoid reporting.

"Give Them Hell, Harry" speeches of 1948) might have loved. Christened the *"Baa Hozhnilne"* (Navajo for "To Win Over") by the candidate, adorned with bunting, pulling eighteen dormitories behind the locomotive, with two cars for VIPs and a lounge car and staff car for Goldwater, the long train chugged through the parched farm belt, and the curious crowds gathered in healthy numbers, up to 5,000 or 10,000, at the rear platform wherever it stopped, to see the bronzed and handsome candidate step out and speak from behind the enormous white shield which bore the legend: IN YOUR HEART YOU KNOW HE'S RIGHT!

The crowds here in the old Midwest were supposedly people of the old faith; here, if anywhere, a conservative candidate should score. But the appearance of the crowds told much of the tale. The old folks were still Grant Wood in style: overalls and sometimes high shoes for the men; the women with hairnets pulled tightly over set curls, their housedresses neat, their shoes flat ground-grippers, their sweaters well mended. The middle generation was more modish—the latest shoes (sometimes unfashionably pointed); the men in business suits or leather jackets; hair neat on the men and teased on the women. But the youngest generation, in teens and twenties, might have assembled in Scarsdale, New York, or Marin County outside San Francisco: they wore the same sneakers, the same shorts (of the same Madras material), the same blouses and skirts; the girls' hair was cut in the silhouettes of young moderns everywhere. There was nothing country-style about the young people. National television, Sears-Roebuck and Montgomery Ward, and the new shopping centers had penetrated and homogenized them; they were already deeply contaminated by the collectivism that Barry Goldwater felt to be decay.

There was, of course, a theoretical as well as electoral-vote strategy to this trip: Goldwater this week was supposedly burying the "bomb issue," preparatory to going over in the next few weeks to "the offensive."

This was the high strategy, and the set speeches reflected it—but reflected it in an awful, half-blind manner. Six thousand citizens of Hammond, Indiana, gathered to hear him—largely Eastern European mill workers who bore such signs as TITO KILLED U.S. FLYERS; SERBIAN DEMOCRACY SUPPORTS GOLDWATER; TITO MURDERER. To them Goldwater promised liberation of Eastern Europe—and they roared approval. To them he insisted that only victory could end Communism—and they roared more. "Our government expects the Red Chinese Communists to explode a nuclear weapon sometime in the near future," he said. "We are at war in Vietnam. We are at war regardless of what the administration says. . . . This country must always maintain such superiority of strength, such devastating strike-back power, such a strong network of allies that the Communists would be [committing] suicide for themselves and their society if *they* push the button. . . . We will insure that the Red leaders always recognize that if they ever should push the button, we would destroy them." And so on. One of the correspondents, Charles Mohr of *The New York Times,* counted, and then gave up after counting partway through the speech that Goldwater had used almost thirty times such phrases as "holocaust," "push the button," "atomic weapons." [4] It was a thoroughly bellicose call to man the ramparts—or the *Nikes.*

The central point of the strategy of this speech had been to insist that Goldwater would never "push the button"—that Khrushchev was the "button pusher." But the tone, and the foreign policy of immediate, uncompromising and aggressive challenge to the Soviet Union, could not but be reported with all their appropriate warlike and martial quotes. And it was always the threat and the risk inherent in Goldwater's policy that underlay American alarm; the Democrats needed merely to paste the bomb against the background of such a speech and they were home free.

The impromptu speeches, off the rear of the train or at close-by fairgrounds, were, however, more exciting and more illuminating, as they always are—for it is always in the weariness of the stretch between the set speeches that the candidate shows his heart to the people and exposes himself as he is. And Goldwater was preaching Freedom.

Goldwater was a late starter; he spoke poorly, as most of us do, at breakfasts and early-morning meetings while still drowsy from the night before. But as the day wore on, his fine, dry voice would grow crisp and pick up the pounding rhythm, the yearning sincerity and indignation that characterize his best style. He would begin with the normal spatter of quips—that Lyndon Johnson had so much power and wanted so much more power that Democrats didn't know whether to vote for him or plug him in; that there was so much dirt swept under the carpet of the White House that it could qualify for the soil bank; that Lyndon refused to debate the issues and was going around the country dedicating dams, but he, Goldwater, had more questions than Lyndon had dams (it was sometimes "Light-Bulb Lyndon" or "Lyndon-Come-Lately" or "the interim President and his curious crew"); that the country needed less wishbone and more backbone—and then into the theme of freedom.

[4] A rare example of organizational foul-up marred the occasion. Hammond, with its twin city, Gary, is a center of Indiana backlash. Organization had, however, managed to find three Negro Republicans who consented to sit on the platform with Goldwater on this occasion. The band honored their presence by closing with the playing of "Bye, Bye Blackbird."

He meant it when he spoke of freedom—government was an enemy, government was taking away liberty. He commiserated with the farmers for living under this "mess of oppression." We had to end this "oppression" for the sake of the young people. ("If we can stop the concentration of power, they have the possibility of living all of their lives in freedom and living better lives than you and I have lived because of it.") The well-dressed young people would shift uneasily, and the farmers and their wives would peer blankly at him as he told them how they were being crushed. He would attack and denounce Lyndon Johnson's Great Society because it never mentioned the one essential ingredient that made America great—freedom. With a proper audience, he could swing out into genuine eloquence, as at the railway station in Effingham, Illinois, where he was introduced as the one candidate "who would not take us down the road to socialism." A wild yell rose from the crowd, saying, *"Give 'em hell, Barry!"* And then Barry proceeded to describe the dismal future that lay ahead if Lyndon Johnson were elected.

"What good is prosperity if you are a slave?" he asked as the little girls in their gold-and-blue uniforms jiggled and cheered for him.[5] "We have had rich slaves. What good is justice if you are a slave? You can find justice in some jails. And what good is peace if you are not free? . . . Without freedom you can't enjoy life. . . . If we have reached the point where we are willing to put our tails between our legs and say to the Communists . . . 'We are tired, we are tired of spending money, we are tired of worrying, we will go along with you, we don't mind being slaves.' " The thought trailed off as being too bleak to pursue, and he resumed, "I don't think I will ever live to see that day in this country. And I will promise you one thing as your President: this country will stay free, and it will stay free because your President will have respect for the Constitution. . . . We will place power back to the people . . . where it belongs."

One could not question the conscience of this conservative. Here was no ordinary politician pitching for votes; he was on a crusade to free America from enslavement. (In La Salle: "What would you do if I told you you had to line up at six in the morning tomorrow and get a num-

---

[5] The pretty little girls always used to cheer most loudly when Goldwater suggested abolishing the draft. He got attention from the young men who were going to be drafted, but the girls squealed with a mating sound for the man who was going to keep the boys at home. One of the oddities of the campaign was that after Goldwater had made his original proposal for eliminating the draft, he referred to it again so infrequently during the campaign. Although his draft proposal conflicted with his posture of soldier, it might have taken some of the curse of bellicosity off him—and it certainly touched the hearts of parents and girl friends.

ber?. . . Our freedom is being dribbled away, chiseled away. . . . Freedom is lost through inattention.") If only Americans would pay attention, they would understand this progressive enslavement. But apparently they did not. Somehow the prophetic message had no link, no connection with the life his listeners lived every day. Times were good and getting better, and they were working and taking longer vacations—the Prophet was not reaching them, Israel had grown fat.

The contrasts of the trip all told revealing stories.

There was the contrast between Goldwater and Charles Percy, running for Governor of Illinois. So many major Republican candidates avoided their standard-bearer whenever he entered their states that it was an act of grace when Charles Percy appeared in the late afternoon of Goldwater's second day in Illinois for three joint whistle-stop appearances. Young and handsome, Percy had done his homework well. At his whistle stops he would hook himself directly onto his local audience with a single paragraph of homely research—he would tick off the county's or town's population, its farm production, voting record and local problems, link them to the larger state problems, and only then hammer at his theme of boss domination by Mayor Richard Daley of Chicago over the Illinois Democratic Party.

Percy could clothe his attack in humor. Illinois politics at that moment was under the sentence of reapportionment in 1964; thus, unable to arrange local elections by traditional districts, Republicans and Democrats had agreed on a winner-take-all sweepstakes. All candidates for the lower house of their legislature were to run at large, on a statewide ballot. The ballot was an orange sheet more than a yard long, bearing two columns of 118 names each—or two thirds the 177 seats in the Illinois House. By scratching one cross, the voter gave all 118 to one or the other party. A sample ballot had been printed up for proofing purposes before the two parties chose the actual names on their slates; and the printer had set bogus names in each column. Except that in the Democratic column he had, by reflex, reflected reality in his choice of names. Percy would hold the enormous sheet out at arm's length, letting it flap, and read off the practice names: they started out with Alonzo Daley, went on to Amos Daley, through Mary Daley, through all the letters of the alphabet to Zacharias Daley—118 Daleys in all. The routine would bring the crowd to laughter, and then Percy would bore in on just who did they want to run the State of Illinois—a governor or Boss Daley of Chicago? Goldwater was apparently unaware of this local condition until late in the second day of his tour, and then his remarks, in contrast to Percy's, would be abstract, distant or furious. He could never make an audience chuckle.

Then there was the murmur of the crowds. Two issues seemed to bother them.

There was the Negro problem—and over and over again the perplexed remarks: "Part of that civil-rights law is nuts. It's okay to say you'll give a fellow civil rights, but once you tell me who can come into my restaurant and who can't—then, boy, there's trouble." Or "I'll tell you, mister, this Negro business bothers them, but people won't talk about it." Or when a civil-rights demonstration tried to outshout Goldwater, a bystander said, "This Negro situation got me all screwed up, they're overdoing it, just like they're overdoing it right here. Goldwater's got a right to be heard, that's what this country is all about." Yet Goldwater, with high responsibility, would not touch the issue.

The other issue was the bomb (see Chapter Ten). One did not have to buttonhole the crowds to hear about the bomb. The demonstrations spoke for themselves. Never in any campaign had I seen a candidate so heckled, so provoked by opposition demonstration within his own demonstrations, so cruelly billboarded and tagged.

The placards gave every variation on every theme, and at university towns particularly, they stabbed: VOTE FOR GOLDWATER AND GO TO WAR; STAMP OUT PEACE—VOTE GOLDWATER; GOLDWATER FOR PRESIDENT—OF HIS DEPARTMENT STORE; IN YOUR GUT YOU KNOW HE'S NUTS; YOU KNOW IN YOUR HEAD HE'S WRONG; YOU KNOW IN YOUR HEART HE'S RIGHT—FAR RIGHT; WAR ON POVERTY—NOT ON THE POOR; WELCOME DOCTOR STRANGEWATER; DULCE ET DECORUM EST PRO PATRIA MORI; KEEP YOUR ATOM BOMB IN ARIZONA; GOLDWATER FOR HALLOWEEN; BACK TO THE STORE IN '64; GOLDWATER IN 1864; IF YOU THINK, YOU KNOW HE'S WRONG.[6] Placards and signs, hoots and jeers accompanied him everywhere—part spontaneous and part developed by Lyndon Johnson's first creative contribution to Presidential campaign, the principle of "negative advance" (see pages 349-350).

The Goldwater placards could not match them: A CHOICE, NOT AN ECHO or HELLO BARRY or EXTREMELY FOR BARRY or the anti-Johnson signs: LBJ—LET'S BEAT JOHNSON or JOHNSON AND HUMPHREY SOFT ON COMMUNISM—IN LOVE WITH SOCIALISM or a dog picture and underneath L(ET'S) B(ITE) J(OHNSON); and, at the end of his trip, a huge hand-lettered sign: IT'S A DAMN LIE. BARRY GOLDWATER IS NOT ANTI-CATHOLIC, ANTI-JEWISH, ANTI-NEGRO, HE RESPECTS ALL HUMANITY AND IS ONE

[6] In other parts of the country, signs were cruder. In the South, students seemed to favor a big, black, simple mushroom cloud and the legend GO WITH GOLDWATER; a variation on such a sign would be: $AuH_2O + 1964 =$ and then a huge red-and-yellow crayon drawing of an atom-bomb burst.

OF FOUR GREAT LIVING AMERICANS. SO SAYS LAR DALY, ALWAYS AMERICA FIRST.

And there was a final undertone to the trip which I can only pinpoint by two placards, both of them at university towns. The first, at Athens, Ohio, seat of Ohio University, read: EVEN JOHNSON IS BETTER THAN GOLDWATER. And the second, in Illinois, was even more bitter. It said, simply, WE DON'T LIKE ANYONE VERY MUCH.

For the fact was that Goldwater was running not so much against Johnson as against himself—or the Barry Goldwater the image-makers had created. Rockefeller and Scranton had drawn up the indictment, Lyndon Johnson was the prosecutor. Goldwater was cast as defendant. He was like a dog with a can tied to his tail—the faster he ran, the more the can clattered.

This was the fundamental Goldwater problem—himself. How could he break the image that had been fastened on him first by Rockefeller, then by Scranton, then by Johnson, then by Doyle Dane Bernbach? The development of an image is a mysterious thing; once a public figure has been cast in a public role, it is almost impossible for him to change the character. It is as if once someone has assembled personality traits into a convenient pattern, no writer ever re-examines it; it is easier to use the accepted pattern. Occasionally one can trace the author: Westbrook Pegler, for example, created the figure of "Bubblehead" Wallace; Joseph Alsop cast Louis Johnson as "The Man Who Must Go." More often the pattern grows by itself and is then even more unbreakable: Adlai Stevenson as the Hamlet of politics and Eisenhower as the Patriot Above Politics were roles so quickly and completely cartooned and accepted that no one could trace their origin.

How could Goldwater erase the cartoon of himself? How could he break through the web of opinion-makers?

The best brains attempted to wrestle with the problem and came up with their varying solutions: speeches, position papers, gimmicks. Their most serious effort to erase the cartoon had come early in the campaign in a television show broadcast September 22nd, 1964.

The bomb issue, some had urged, could be put to rest by a television colloquy of Eisenhower and Goldwater, which they hoped would have as clean-cut an effect as Kennedy's address to the Houston ministers in 1960. The candidate had therefore journeyed to Gettysburg for a hastily arranged séance with a reluctant Eisenhower, and cameramen and TV experts worked with them all day. There emerged a dreary half hour, carved out of their conversation, which spoke only to those Republicans whose loyalty caused them to tune in. Eisenhower gave Goldwater his blessing and declared that talk of Goldwater's desire to use an atom bomb was "tommyrot." It was the kind of program that is meaningless —it editorialized rather than exposed. When Kennedy had discussed his religion with the Protestant ministers in Houston in 1960, he had answered hostile questions with hard response; the nation knew where he stood on Church and State when he was through. When Goldwater finished this broadcast, the nation still did not know how he felt about atomic bombs—except that Eisenhower thought well of him.[7]

Quits then with the nuclear or bomb issue. The Goldwater team had wrestled with it day after day; no one had a solution. Said Denison Kitchel, "When I went to bed, if ever I could have just a few hours sleep, I would lie awake asking myself at night, how do you get at the bomb issue? My candidate had been branded a bomb-dropper—and I couldn't figure out how to lick it. And the advertising people, people who could sell anything, toothpaste or soap or automobiles—when it came to a political question like this, they couldn't offer anything either."

A vast and desperate uncertainty settled over the Goldwater campaign in October. The leaders had decided after the Convention that they must go on the offensive, but the moment never came; they were defending. They had hoped originally, at the beginning of the campaign, that there would be a majestic dialogue on the two American philosophies, the conservative and the liberal. To structure this dialogue, seven speeches on major issues confronting the United States would be delivered in orderly sequence, exposing the gulf between the two sides. Goldwater would start in Prescott, Arizona, by promising abolition of the draft; he would go on to Los Angeles and suggest a five-year program of progressive and planned cutting of the tax burden; in Chicago he would discuss the Supreme Court and attack the arrogance of its pretensions; he would discuss national defense in martial Dallas; in Chicago, again, he would discuss civil rights and explore the meaning of freedom —and so on.

But they were baffled, and from the bafflement grew exasperation and frustration. For the Democrats simply would not join their debate. "Lyndon Johnson was so far away from it, we couldn't find an opponent. We were punching at a pillow," said Kitchel. And to their exasperation was added the echo the press gave of Goldwater's speeches. For the speeches, one by one, were mangled by the columnists and commenta-

[7] A few days later Lyndon Johnson revealed to a few newspapermen the Nielsen audience measure of those Americans who tuned in. Barry and Ike had scored a rating of 8.6 against the opposition programs, *Petticoat Junction*, at 27.4 and *Peyton Place* at 25.0.

tors, who needed only cite Goldwater's past record: there it stood, and inescapably they twined the record with the speeches.

Abolish the draft? A useful idea. But Adlai Stevenson had already suggested that in his 1956 campaign—and the Republican Party had denounced him for it. Cut the income tax? How could Goldwater advocate a five-year cut of the income tax when in February he had *opposed* the Kennedy tax cut? Goldwater had written three books, some 800 columns, given an estimated 1,000 speeches in the previous twelve years. He was on record against almost everything, and, as any daily writer, was trapped by written contradictions which, wrenched out of context, could be held against him. "Would that mine adversary had written a book," said Job many hundreds of years before, and Goldwater had written too much. The great programed delineation of issues petered off with a speech in Chicago on October 16th at a $100-a-plate fund-raising dinner—the best effort of his campaign.

How can we build a society of many races with liberty and justice for all? [Goldwater asked, and went on] . . . "We hold these truths to be self-evident," the Declaration of Independence says, "that *all* men are created equal."

Let us repeat, *"all* men"—not only Americans or Anglo-Americans, not Christians or Jews, not White Men or Colored . . . *all* men. . . .

The equality that is God's gift, however, is not the same as saying that all men's accomplishments must be equal, that their skills must be equal, that their ambitions are equal, or that their energies are equal. No, on those levels there is no equality. There is only opportunity. . . .

There are those who seem to denounce society as hopelessly evil because it is not perfect. On the other hand, there are those who tell us to be satisfied with what we already have. . . .

This all-or-nothing attitude is bound to end in disaster, and has already caused much harm to many innocent persons. . . .

Throughout this land of ours, we find people forming churches, clubs and neighborhoods with other families of similar beliefs, similar tastes, and similar ethnic backgrounds. No one would think of insisting that neighborhoods be "integrated" with fixed proportions of Anglo-Americans, German-Americans, Swedish-Americans—or of Catholics, Protestants and Jews. . . .

Our aim, as I understand it, is neither to establish a segregated society nor to establish an integrated society as such. It is to preserve a *free* society. . . . Freedom of association is a double freedom or it is nothing at all. It applies to both parties who want to associate with each other. . . . Barriers infringe the freedom of everybody in the society, not just the minorities.

Now the removal of such barriers enhances freedom. . . . But it is equally clear that freedom is diminished when barriers are raised against the freedom *not* to associate. We must never forget that the freedom to associate means the same thing as the freedom not to associate. It is wrong to erect legal barriers against either side of this freedom. . . .

One thing that will surely poison and embitter our relations with each other is the idea that some predetermined bureaucratic schedule of equality—and, worst of all, a schedule based on the concept of race—must be imposed. . . . That way lies destruction.[8]

By the time, however, that Goldwater made this speech, his campaign had already run its course; it was matched immediately by commentators with the fact that the man who made this speech—a genuine intellectual effort—had voted against the Civil Rights Bill only three and one half months before; and generally it was ignored.

If there was to be no dialogue on issues, what then was left? Confusion—and in the confusion the Goldwater people decided at the end to concentrate on the issue of American morality in an age of prosperity. They called this the "ethics" issue. They must take the offensive somewhere—and ethics would be the attack charge.[9]

---

[8] Perhaps the most necessary intellectual operation in American life is some redefinition of the word "freedom." I have attended as many civil-rights rallies as Goldwater rallies. The dominant word of these two groups, which loathe each other, is "freedom." Both demand either Freedom Now or Freedom for All. The word has such emotional power behind it that in argument, either with civil-rights extremists or Goldwater extremists, a reporter is instantly denounced for questioning what they mean by the word "freedom." It is quite possible that these two groups may kill each other in cold blood, both waving banners bearing the same word. The new technological society requires a commonly agreed-on concept of freedom just as much as did the Old Frontier.

[9] It was at this point that F. Clifton White, from the outer circle to which he had been dismissed, offered his analysis: Goldwater was running against fear—fear of the bomb, fear of the ending of Social Security, other fears. He must strike back with a counter-fear. American emotions must be galvanized by their fear of domestic violence. Out of his analysis came his film documentary *Choice*. His analysis was sound. But the film, as finally produced under the letterhead cover of "Mothers for a Moral America," was an absolute shocker. Naked-breasted women, beatniks at their revels, Negroes rioting and looting in the streets succeeded each

But "ethics" was a difficult proposition to define. For the media specialists of the Interpublic Group, it posed an impossible problem. Doyle Dane Bernbach might expose the hostile candidate in any degree of extremism it chose (there was always on Goldwater's record material to prove anything). But how does an advertising firm go about attacking a President? How does one besmirch the majesty and uniform of the office? How, for example, as one of the Republicans complained to me, do you ask how a Texas Congressman and Senator, never earning more than $22,500 a year, accumulates a fortune of $7 million?

There remained, then, only a nit-picking reconnaissance of the "ethics" issue—Bobby Baker, a gray, cheeky shadow on the President's past; Billie Sol Estes, whose intricate machinations spanned both the Eisenhower and the Kennedy years; and Walter Jenkins (who must be discussed in the context of Lyndon Johnson's campaign). It was decided that Goldwater must hammer here at the names. But Bobby Baker had never been found guilty of anything—he was, after all, a rabbitlike man with a poor boy's cupidity who had made a lot of deals and some money, and the only thing one might possibly pin on him was the use of "broads." But politicians know that denouncing a man for "broads" is a dangerous matter; at one time or another, too many Americans, Republican and Democratic alike, have sinned with "broads." Billie Sol Estes was too complicated for more than a reference in an occasional speech. And Lyndon Johnson had in no sense ever been guilty of the "hard take." The complicated higher finance of monopoly television stations and government franchises could not be made a clear issue for the girl behind the counter or the man in the mill.

The Goldwater strategy board hoped that it could provoke Lyndon Johnson to some vast intemperance, to some public outburst of temper and indiscretion. But Johnson was too shrewd for that. It was obvious by mid-October[10] that Lyndon Johnson was not about to be provoked into

---

other in a phantasmagoric film which, when shown to Goldwater, he flatly refused to authorize. He declared it to be an inflammation of racism, and he suppressed it completely.

[10] By mid-October the Goldwater headquarters had, in spirit, collapsed. Dean Burch decided then to cancel the public-opinion polls out of the Opinion Research Corporation—why should anyone pay for tidings of disaster? A final effort had been made to strain some optimism out of the polling results with a question worded, roughly, "Did the respondent himself know anyone else who might vote for Goldwater but was concealing the fact?" Respondents replied, to the degree of 45 percent, that there might be secret Goldwater sympathizers in their neighborhood, but they could not pinpoint them. Thereafter Goldwater headquarters flew blind; about this time a group of men gathered in the offices of Citizens-for-Goldwater and, at Clif White's direction, took the last hard look at the campaign. Their posting in mid-October was that they would carry only five states. They were wrong—Goldwater took Georgia unexpectedly, to make a total of six.

temper. Nor was he to be lured by any challenge. There remained, then, for Goldwater, only to grieve.

It was thus that one saw him again at the end of the next month—mourning for the American people, in whom he had lost faith.

Indeed, there was reason for his mourning. No upturn had followed his Midwestern jaunt. His faithful reported that Goldwater bumper stickers outnumbered Johnson stickers ten to one; but the Gallup and Harris polls showed little change in Johnson's vast lead (the Gallup Poll showed Goldwater down two points, the Harris Poll up only two points) and all about him his campaign was collapsing.

The Republican Party had split in two. In New York the Rockefeller troops ignored Goldwater as if he were spoiled refuse and concentrated on re-electing Senator Kenneth Keating, who openly disassociated himself from the national ticket. (In New York, Keating was to run a solid 860,500 votes ahead of his national leader—yet still lose to Robert F. Kennedy by 720,000.) In Michigan, Governor Romney cut loose from the national ticket entirely, refusing even to appear on a platform with the national candidate. In New England, except for Connecticut, it was fair to say that no Goldwater campaign existed at all as the New England Republicans took to their hills. In North Dakota, at a plowing contest, Senator Milton Young declined to share the dais with the party leader. The professionals knew weeks ahead that it was all over. At the end of September one of the best of them, Ray Bliss of Ohio, had already told a journalist friend, "As things stand right now, we face another 1936, and any goddamn fool that doesn't believe it had better. I don't know what there is now that can save this thing except maybe an act of God, and you know those things don't come along too often in a lifetime."

A proud man, Goldwater was galled by the public desertions and the bleak negativism of men who worked on doggedly but exhaled despair. Yet more depressing was the attitude of the nation's press and media.

The editorial support of the nation's press and national magazines is, generally, as solid a Republican resource as the labor unions are a Democratic resource. But, one by one, starting in August, the regional newspapers and then the national magazines—which, in Republican terms, have always been more Catholic than the Pope—cut their traditional candidate adrift or scorched him with editorial gunfire. Papers that had never before gone Democratic now lined up behind the Democratic candidate. The desertion of the Hearst papers, on September 18th, was the hardest blow to take—Democratic for the first time since

1932. By the end of September the best papers of the Republican heartland had checked in—sometimes at the rate of seven or eight a day—with their switches to the Democrats: the San Francisco *News*, for the first time since 1932; the Cleveland *Plain-Dealer* and *Press*, both for the first time since 1936; the *Rocky Mountain News* (Denver), for the first time since 1936; the Indianapolis *Times*, for the first time since 1936; the *Herald Weekly* of Camden, Maine, for the first time since its 1869 founding; the Syracuse *Post Standard*, for its first Democratic endorsement in its sixty-five-year history; the Binghamton (New York) *Sun-Bulletin*, for the first time since its founding in 1822. On and on the roll went, and there was more to come. In October the Detroit *Free Press*, although still strikebound, announced its support of Johnson; so did the Philadelphia *Bulletin;* so did the Baltimore *Sun*, Democratic for the first time since 1936; so did the New York *Herald Tribune*, which had been spiritual godmother of the Republican Party since Horace Greeley was midwife in chief to its birth. And the magazines: the *Saturday Evening Post*, then *Life*. In all the nation, only three major newspapers supported the abandoned Republican: the Los Angeles *Times* (which had opposed him in the California primary), the Chicago *Tribune* and the Cincinnati *Enquirer*. Nothing like this had ever happened before in American political history.

Crushed and punch-drunk, the Goldwater campaign staggered on. And as it went on, a vast incoherence wrapped itself about the entire proceedings. The campaign in September had begun in confusion: a speech on crime in peaceful St. Petersburg, Florida, whose retired elderly wanted to hear about Social Security (RIGHT CITY, WRONG SPEECH, commented the St. Petersburg *Times*); a speech scoring Johnson's anti-poverty program in stricken West Virginia; an anti-reapportionment stand in under-represented Atlanta; a swift and unsatisfactory reference to TVA (which Goldwater had once offered to sell for a dollar) in Tennessee. Now was the time to go over to the offensive—but no offensive came. And the great debate between conservatives and liberals, the great delineation of issues, the great description of a new America standing firm on old virtues? Where was it? No one could say.

So one visited him again to find the answers. Goldwater at the end of his campaign was entirely different from the tiger with black-rimmed eyeglasses of the public prints. He had relaxed, finally, into the charming person he can be, and whatever "kill" instinct he may have brought to his original candidacy for the Presidency had evaporated.

For the last two weeks of his campaign he had programed a swing through California and Texas, an appearance in Madison Square Garden in New York (his only major appearance in the enemy's lair during the entire campaign), another swing through the South and Midwest, a day in Pennsylvania, and then a wind-up in the West and in Phoenix.

When I rejoined his tour as he flew from New York (where he had gone to attend the funeral of Herbert Hoover) to Los Angeles, all tension was gone, and the traveling party had the intimacy of a jolly wake. Up front in the forward compartment of the plane the candidate sat in shirt sleeves, no longer scowling, but talking with easy courtesy to his companions and occasionally to members of the press. His seat faced forward against a bulkhead on which were three clocks (showing the three national time zones) and two signs: RE-ELECT GOLDWATER IN 1968, and the old one, BETTER BRINKSMANSHIP THAN CHICKENSHIP. Newsmen were friendly now. Among the press party, in his permanent seat, sat gentle Frank Cancellare, the UPI White House photographer; the wall beside him was hung with the keys to all the hotel rooms he had occupied in the long campaign—and Goldwater himself occasionally carried off a souvenir key to add to the collection.

Goldwater was pleased that they were making a record that night— it was the first time this plane, a Boeing 727, had attempted to fly coast to coast nonstop without refueling, and they were going to make it. There was another first too: as he flew over the Arizona desert the lights of Phoenix came up twinkling. In Phoenix his friends were holding a $100-a-plate dinner for the local Republican cause, and Barry was to address them from the air as he flew over—the first speech ever given by an airborne candidate in flight overhead. "I can see the lights of Phoenix very clearly," he said, speaking over the plane's communication system, "down to the left. I hate to say 'left,' but that's where it is," and went on wishing them well, and ended with a warm message to his brother Bob—he'd soon be back with them, he said.

He must have hurt inside, though, the next morning when he woke in Los Angeles and went to the railway station to begin a one-day whistle-stop tour through southern California. The regular Los Angeles Republicans had now written him off and were concentrating their efforts on the Get-Salinger, Elect-Murphy Senatorial campaign—and the organization had left the Goldwater rally in the hands of whatever amateurs wanted to organize it. No more than seven hundred people showed up at the station—at mid-morning in the heart of the city—an appallingly slim crowd at a moment when the final rhythm of the campaign should have drawn thousands. The pretty uniformed girls sang "Let's Carry Barry to the White House and Do Away with LBJ," and gravely Goldwater walked among them, shaking hands (with the policemen too), and then, from his rear platform, concluded his remarks by promising, "We'll give you back your government on November third," and the train shoved off.

Only four counties of California's fifty-eight were to give Goldwater a majority eleven days later, but that day he was journeying through two of the four—Orange and San Diego counties, rock-solid conservative. As he went on, he cheered up, for the crowds were of his own faith. He opened his first speech of the day at Pico Rivera with "ethics" —Bobby Baker and Walter Jenkins. But on his second stop, at Fullerton, he abandoned that and went back to "freedom":

> I just want to remind you, ladies and gentlemen, of one proven fact in history, because we conservatives are always blamed for wanting to go backwards. That is not true. We just want to take a look at what has happened before we try it again. Because everything that this administration is trying today has been tried not by just our own government, but in other governments of the history of the world, and I remind you, they have never succeeded. A government that is big enough to give everything that you need and want is also big enough to take it all away.

In Santa Ana he was warming to the crusade again: "What kind of a country do you want? Are you satisfied with a government—" NO! "Are you satisfied with a government that tends more and more and more to take the powers away from your city, your state and from yourselves?" NO! ! "Are you satisfied with a government that has not yet recognized that the one enemy to peace in this world is Communism?" NO! ! !

Through San Clemente, through Oceanside, through La Jolla, down to San Diego he journeyed. The palm trees waved their fronds by the broad ocean, the sun shone on him, and the handsome young married people, babies in arms, of this conservative centerland gathered to be aroused. But there was something lacking. They had come to march— and he invited them to mourn.

This impression of restraint, this sense of his own fear at what indeed he could arouse if he tried, grew as one followed the campaign in the last few days. He finished his California "thrust" at San Diego (where he delivered an oration he had originally delivered in Philadelphia three weeks before—by this time the limited resources of Goldwater's speech writers had been strained beyond their human energies) and was off to Texas.

In Texas the strategists had decided he should hammer on the theme of Hubert Humphrey—and how dangerous it was to have this "semi-socialist" one heartbeat from the Presidency. He flew the next day from Austin to Corpus Christi to Wichita Falls, and drove through nearly barren streets to address near-hysterical gatherings in the halls. And somehow, rather than firing his audiences, he banked the flame. The problem was evident as one watched: the storming crowds Goldwater could gather in an auditorium were more members of a sect than members of a party. If he stirred up such sectarians, he would lose the national audience; and by now he knew it was the national audience that counted.

I remember one meeting particularly.

In Corpus Christi he descended from his plane the *Yai-Bi-Kin* (Navajo for "House in the Sky") and traveled through nearly empty streets to an audience which, as he entered, rose in frenzy, hurtling back and forth the Goldwater chant of *"Viva Ole! Viva Ole! Viva Ole!"*

In the second row sat a lady-in-red of perhaps thirty-five years, crowned with a shock of black hair, a red-and-gold ribbon streamer saying GOLDWATER stretched across her bosom, two huge Goldwater buttons on top of that—and as Goldwater rose to speak she rose, too, sobbing, tears streaming down her face, her fists tightly clenched, pounding the air as she shouted, "We want Barry!" She was ready for political orgasm, but Barry knew the country was larger than she, and he was, by plan, supposed to pin Hubert Horatio's socialistic ADA on Lyndon Johnson. The lady leaned forward intently as Barry began, but somehow by the time he had reached the ADA's disapproval of censorship of pornography (". . . when I think of the pornographic literature that runs around this country today, when I think of some of the novels that I have read thinking that they might be good but find them nothing but the worst kind of filth . . .") her face had lost the intent and convulsive frown with which she had awaited him.

Goldwater went on next to the big government in Washington ("We have gotten where we are not because of government, but in spite of government"), and a round of applause burst, calling the lady-in-red to attention, and she clapped again, her face alert. Was real red meat coming?

But no. Goldwater went on to the national economy and the free-enterprise system, and the lady began to look around her, scanning the audience, not listening. A laugh aroused her (Goldwater had said: "In coming in, I saw those wonderful things that go up and down twenty-four hours a day—oil") and she paid attention while he defended the depletion allowance ("I know how people drill dry holes"). At this she smiled sadly, nodded her head, but did not clap. Occasionally, as Goldwater shifted his attack between Humphrey and the Federal Government, her attention would come back and she would clap her hands automatically, like a dog tugged by a string. He was on to a denunciation of the ADA-supported World Court before he finally lost her. She stooped down, pulled up the shopping bag labeled CARRY ON WITH BARRY and took out a camera. For the rest of the speech she was busy taking pictures of those around her, squinting at Barry for a long shot through her sightscope, joining the proceedings again only to participate in the final *"Viva Ole! Viva Ole!"* She had come ready for the call—but Goldwater was making no such call.

Why Texas should have been chosen for the great denunciation of Hubert Horatio Humphrey ("the mystery guest of the year," Goldwater called him) is unclear to this correspondent. Johnson was exceptionally sensitive about his Texas home base.[11] If there were any way of provoking Johnson, as was part of the original plan, Texas should have been the spot for it. But Goldwater chose to attack Humphrey instead. (In Austin, Johnson's home district, when Goldwater had asked the question that morning, "What kind of government do you want to have?" the galleries had yelled back, "Fairies, fairies, fairies!" and this had angered him. And when they had booed Johnson's name, he had scolded them: "No, he is your President. You shouldn't boo him. You can be quiet, but don't boo the office of the Presidency.") The previous day in California when someone had called, "Kill Johnson!" he had been obviously shocked. Goldwater's private opinion of Johnson was unprintable, but publicly he gave full, honorable respect to the Presidency.

Thus his trip to Texas. It closed with a speech at Wichita Falls (which concluded with a Mosaic outburst against the Federal Government: "Let my people go!") and he climbed back aboard the *Yai-Bi-Kin* for the return to Washington.

The swing had been a political nullity—but on the plane all was gay. The squawk box opened up and the familiar dry voice called, "Now hear this—I seem to have two sombreros here [at two Texas stops he had been given a gift of sombrero with blanket, or one more than enough]. The only way to get rid of it is to you fellows, so we'll just pull a name out of the hat."

A stewardess came round with yellow slips of paper on which the reporters wrote their names, and when that was done, the candidate pranced out of the forward cabin—a magnificent white sombrero on his head, a huge black-and-yellow-striped Mexican blanket over his shoul-

---

[11] The greatest single television exertion of the entire campaign was committed by the Democratic National Committee to make sure that Texas stayed with Johnson. Of the seventeen states in which the Democrats concentrated the Doyle Dane Bernbach spots, Texas was far and away first, with 1,345 time spots as against 823 spots in California and 410 in New York, each of which is more than half again as large in population.

der, his blue shirt open at the neck. *"Mi amigos,"* he began, "if I can call you that," and everyone laughed. The stewardess pulled a yellow slip out of the hat, and the winner was Cancellare, one of Goldwater's favorites, who immediately donned the huge white sombrero and draped himself in the Mexican blanket. Cancellare, a very dark, very small, very Mediterranean American, stood there draped in his regalia like a Mexican Kewpie doll, and Goldwater howled with laughter, then dashed back to get his camera to take a picture. Others took photographs of Goldwater taking photographs of the photographer, and the moment was one of high joviality until a correspondent, taking advantage of the relaxation, asked him to explain just what he had meant by his closing line in Wichita Falls—"Let my people go." At which Goldwater suddenly pulled himself into a withdrawal, realized he was still a candidate for President, and strode off, aloof, to his forward cabin, where he buried himself in a book as the plane sped the funeral route from Texas to the capital—while the reporters back of him sang the latest song of the trip:

> The man of our dreams has lost his hair,
> His glasses are blank and black,
> He hates to mess with the Eastern press,
> His knife is in Lyndon's back.
>
> The man of our dreams is free from care,
> He's certain he's going to win.
> The polls say he's not, but he's sure that's a plot—
> He's the sweetheart of *Yai-Bi-Kin*.

There was now, in the last week of the campaign, no apparent strategy—and if there had been, it would have been futile. Aloud to the press, the members of the Goldwater staff had to insist that the nation was about to witness the greatest political upset in history; but death was in their hearts and they were too smart to do more than make a bold noise and give the outer impression of victory. Their random travels now seemed to have no pattern at all; it was as though someone had dipped a brush into a paint can, then flipped it against the map—and where the drops spattered, Goldwater went.

His last week began in New York with a huge rally at Madison Square Garden. On Tuesday he was in Bristol, Tennessee; London, Kentucky; and Cleveland, Ohio (to attack the Civil Rights Bill). On Wednesday he was in Cedar Rapids, Iowa (denouncing American churchmen for messing in politics and not being concerned with the immorality of American life); Oshkosh, Wisconsin (where he pulled out a copy of the

Communist *Worker* to excoriate the Democrats for enjoying Communist support); and Belleville, Illinois. Thursday he whistle-stopped through Pennsylvania, a state which he had no prayer of winning, defending himself again against the Social Security charge. Friday he headed west—to Cheyenne, Las Vegas, Tucson and Los Angeles (this time his crowds in Los Angeles were good); he issued a mileage statement—more than 73,-000 campaign miles since September 4th—and then went home to sleep that night in Phoenix. But the next morning he was off to Texas and then Columbia, South Carolina, where he made a fourteen-state telecast to the states of the South. On Sunday he rested, taping a television show for Election Eve broadcast, a dull, contrived family scene. On Monday, the last day of campaigning, he flew into San Francisco to face a tumultuous crowd and there, at the scene of his great Convention victory, repeated word for word the same speech with which he had opened his campaign in Prescott, Arizona, two months before. ("Ladies and gentlemen, every word I have just spoken to you is from the very first speech of this Presidential campaign. The issues have not changed. I have not changed. The challenge and the choice has not changed. . . .")

From San Francisco, in early afternoon of the Monday before election, Goldwater took off for the village of Fredonia, on the Arizona–Utah border, where he had ended his two victorious Senate campaigns in 1952 and 1958. He shifted planes in Las Vegas, to descend over the lunar landscape of the American desert—the chalk bluffs, the sandstone spires, the arid tablelands—and arrived at sunset. Jimmy Breslin of the New York *Herald Tribune,* who was there, reported the last campaign rally in a superb dispatch:

Fredonia, Ariz. Nov. 3, 1964.
    The wind sprayed desert dust into the two bright portable lamps they had set up in front of the tiny stand. A lone bulb, over the doorway to the one-room brick office building, was obscured. High school kids in levis and windbreakers were sitting on the roof with their boots dangling in front of the bulb. Across the field beyond the lone runway, past the strands of cattle wire, the red sandstone hills were cloaked with cold dusk. The low cedar trees on them were dark knots now, and beyond that, straight out, where the hills rise into mountains, the sky was charcoal gray and pink and blue, with jet contrails mixed into it like a finger painting. An immense crowd for Fredonia, over 1,800 people, stood in front of the stand.
    They were silent, and the wind whipped at the autograph pads they held in their hands. These were Goldwater people,

and the country around them, empty and spectacular, was Goldwater country. And at sunset yesterday, here in this setting that he knows and that he belongs in, Barry Morris Goldwater made the last speech of his campaign to become the 36th President of the United States.

. . . he straightened and talked softly and with deep interest in his words.

Words to Live By

"I am proud and happy," he said, "to be back amongst the people who I am sure understand every word that I have been saying . . . because these are the words they have grown by, the words they lived by. I have been speaking about man, the whole man, about man's rights and man's freedom. I've been speaking about man's obligations to himself and to his family, to do things himself, to accomplish things himself, and only take help when everything else has failed."

. . . It wasn't a political talk. It was one man yearning for this America in front of him to become the America for everybody. And as he talked, with these tall men in cowboy hats, their hands hooked into the front of their belts, their tanned faces looking strong, standing and listening, and with the desert sunset at his back, there was, for a moment, meaning to Barry Goldwater.

But sunset and talk of the whole man at an airport in Fredonia, population 300, is not the nation. There are half men in this country and they see sunsets while walking through spilled garbage on 135th Street in New York. The right not to associate, which seems to have been the basis of Goldwater's last week of campaigning, is no answer for these people, and no matter how catching this scene yesterday became, and it was, at times, impossible not to be caught up in it, you kept looking at those mountains and remembering. . . .

The Simple Things

"I think from time to time about Pipe Springs down the road here a piece," he [Goldwater] said. "I think of the courage of those people who came here not knowing that the Federal government could help them, but doing it on their own, standing off all kinds of abusive action, standing off the weather, but finally triumphing in raising cattle where cattle probably shouldn't have been raised, and living their lives as they felt God wanted them to. These are the things, the simple things that I have talked about, and I will continue to talk about as long as I live regardless of what God has in mind for me to do."

. . . It may have been, standing in the light with the dust whipping by him, the only time in this campaign that Barry Goldwater really felt at home with what he was saying. He was not trying to work his way through the maze of problems which nations face. He was worried about Pipe Springs.

When he finished talking, he lingered for a while and told them about his friends in the area, and a fire he had once helped fight down at the lumber yard. Then he signed autographs and shook hands and finally they got him back to the plane.

It was dark now, the plane went down the runway, turned around and now came back. Came past the empty fields, past the rows of cars and pickup trucks, with their red taillights glowing, and up into the black sky and away from this little piece of America which he can understand.

He arrived late from Fredonia, home at last on Election Eve in Phoenix, and paused graciously to have a farewell drink with the press corps that had followed him for so long. The Challa Room at Camelback Inn served up enchiladas, tortillas, Mexican chili and tacos, but Barry ate nothing, only sipping his bourbon and water, moody. And he reverted now way back to before the beginning and the man who still gripped his imagination—to John F. Kennedy. He was discouraged by the failure of the campaign to spell out the issues. With Kennedy it would have been different. Jack Kennedy, said Goldwater, was honest— "I could tell him what I felt about his proposals. 'That's no good, Jack,' I would say, or 'That part is okay. That's fine.' And he'd listen. He would have debated me. It would have been good for the campaign. It would have been a good campaign."

Apart from this one pensive moment with a small group of reporters, he was for the most part gay and relaxed, and after about forty minutes left the party to go home to *Be-Nun-I-Kin* (Navajo: "House on the Hill").

All men who run for the Presidency of the United States are amateurs; there is no way of becoming a professional at it. And all of them, whether they win or lose, are forever altered in spirit and character by the ordeal. Yet Goldwater's campaign is more interesting than most, and one should reflect on it with a sense of compassion—a lack of which was one of his greatest flaws.

The first and most essential quality of a Presidential candidate, as Averell Harriman once pointed out, is that he should lust for the job—he should want it more than all things, with a passion surpassing all emotion and probably even all principle. This quality Goldwater lacked. He had wanted it in the fall of 1963, but then, after the assassination, repudiated this desire. His defeat in New Hampshire had led to another downturn of despair in which, momentarily, he discussed with his aides whether he should withdraw entirely—and he had been dissuaded by Kleindienst and White. Again, during the struggle over the Civil Rights Bill, he had discussed his withdrawal with Kitchel—he was certain to be tagged the anti-Negro candidate, and he did not want his candidacy to divide the country by racist clash. The violent battle of the California primary had finally fired his blood, and the vendetta with the Easterners through June and July had stimulated a temporary "kill" instinct that lasted through the Convention. But by September the appetite was gone and he was up against a man in whom the appetite for power was bred into the essential personality. From September on, Barry Goldwater was less a candidate for the Presidency of the United States than the spokesman, and the prisoner, of a cause.

Even so, he might have run a far better campaign—but the underlying theory of his cause misled him. Goldwater believed in the theory of the hidden conservative majority as the volcanic truth of American politics. All his own life, in his own political career, it had proven true; so he had extrapolated from that simple experience. He had been approached by a group of businessmen-citizens for the first time in 1948 and asked to lead the clean-up of Phoenix squalor and corruption; he had done so and won. He had been urged by dedicated conservatives to run for the Senate in 1952—and had won. Over the years from 1952 to 1960 he had gradually developed into the most popular speaker, after Eisenhower and Nixon, at Republican meetings and fund-raising dinners around the country. But always he had spoken to audiences of devotion, men who had come to hear one of his famous philippics; the sincerity of his message had always reached and moved audiences yearning to be moved.

The adventure of 1964 was something else again; it was a venture not in exhortation but in persuasion and leadership. Somehow, and above all, it was necessary for him to reach beyond the base of the faithful, to bring over and add to this base millions of uncertain voters whom he might educate or lead. But it was as if never once in the campaign had his men sat down and asked themselves, "What must we do to open their hearts and show he's right?" The theory of the conservative majority was a snare—for there is in America neither a conservative nor a liberal majority, but two dogmatic extremes which vie for a vast uncertain middle ground that can be tilted one way or another by events, by leadership, by campaign practice, by impact. The Democrats tilted the middle mass against Goldwater with prosperity and with a record of achievement, with the bomb and Social Security. Goldwater never came up with a counter-thrust of ideas and images that might tilt it, even partially, the other way.

The retrospect on any losing campaign always raises a whole sequence of "ifs." The greatest "if" in the Goldwater campaign is what would have happened if Goldwater had really limned out the causes of his indignation, made concrete and vivid the problems and dangers he saw. He could not have won, but he might have brought clarity to American affairs. If, for example, Goldwater had established a moral base for himself by voting *for* the Civil Rights Bill instead of against it—might he not then have been able to discuss the vast problem of racial clash and racial equality in America without the burden of implied racism? If he had tackled the problem of the bomb head on early in the campaign, choosing the most hostile give-and-take forum for a dramatic defense, and then been silent about it—might it not have spoiled the Democratic media offensive? If he had just once addressed the nation on Social Security over TV, explaining exactly what he did seek—could he not have eased the worries of hundreds of thousands of retired Republican pensioners? If he had attacked Social Security not for its structure or philosophy but for its bureaucratic administration—the red tape and the delays in payment—could he not have made a better offensive against centralization? If a good research staff had made concrete the grotesque evolution of some of the now obsolete New Deal inventions of the thirties—could he not have given his conservatism some smack of tomorrow rather than yesterday? If he had championed the fresher and more conservative approach to farm subsidies—that subsidies be based on the farmer, not the farm; that they be pointed to the farm family, not the farm factory—might he not have held the hearts of the Republican farm belt? If he had taken apart the foreign policy of America, admitting its triumphs and then pointing out its failures—would he not somehow have been more credible? If he had chosen to riddle the confused foreign-aid programs of the United States by pinpointing the mistakes and crediting the successes—could he not have explained his NATO policies better? If somehow he had described the state of American life as a tangle in which the good and the bad intertwine and snarl—might he not have had a campaign theme similar to John F. Kennedy's "This is a great country but I say it can do better"?

But it was all or nothing, black or white—and to him Washington and New York and the great cities were all, alike, Babylons of crime and decay and corruption.

He was a man of cause and of principle, but in his campaign he proved he was no leader, because he could make no bridge from his cause to the realities either of government or of politics. His campaign has been criticized as being run by amateurs—which is true, but is not a valid criticism, for all Presidential campaigns are run by novices. Professionals in a campaign are servants; they cannot tell a Presidential candidate *what* to do; they can tell him only how to do something once he tells them what it is he wants to do. And so Barry Goldwater indifferently used and received indifferent advice from old pros who were the best the Republican Party could provide for him. He was perhaps justified in dismissing Clifton White, who had brought him the nomination, for in White he sensed an appetite and purpose different from his own. But he could not guide by any sense of his own broader constructive purpose the professionals he later brought in.

And it was as a man of cause and principle that he was, indeed, portrayed by the press which he so bitterly criticized and so bitterly still remembers. For the press and the commentators and columnists whom he denounced had to deal with the whole record of the man. It was as a sharp, clearly limned figure of many writings and many passions that Goldwater entered the campaign; he would not repudiate them, nor would he explain them, and was innocently horrified when they were used to slash him to bits.

One must see him thus, lastly, as an innocent—not as a political figure. He was a wistful and pugnacious and aggressive innocent to whom the American people were overwhelmingly unwilling to trust the risk of war and peace in an age of complicated but instant destruction. Of his courage, honor and integrity there was never any question; but press and public alike found the innocent inadequate for the responsibility of war and peace in our time.

And yet—and yet, for all this, one cannot dismiss Goldwater as a man without meaning in American history. Again and again in American history it has happened that the losers of the Presidency contributed almost as much to the permanent tone and dialogue of politics as did the winners. Adlai Stevenson did, and so did Alfred Emanuel Smith. Such "creative" losers are rare in American life, and it is fruitless to search for a comparison to Goldwater either in these two great losers or in John W. Davis or James M. Cox or Alton B. Parker or Alfred M. Landon or the long succession of men who have faded into obscurity. One must reach all the way back to William Jennings Bryan to find a proper comparison to Goldwater.

Like Goldwater, Bryan arrived on the scene of politics at an indefinable moment of change as a raw and savage industrialism was altering all manners and morals. Many of Bryan's proposals were mad—silver at sixteen to one, a citizen army of musketeers—but the primitive emotions he expressed were real. Other men, both Republicans and Democrats, were to take these emotions in the first two decades of this century and give them voice, shape, form and, finally, institutional legislation which protected the new country and its old virtues from the savage greed and violence of its industrial overlords. The emotions Goldwater spoke for were just as real, but his proposals just as unreal, as those of William Jennings Bryan. He spoke, when he spoke most strongly, to the heart—against a more generalized greed, a more senseless mood of violence, a more subtle corruption than did Bryan.

Goldwater's foreign policy was one of peril, but at home he spoke of the quality of life; he brought matters hitherto non-political into the political discussion. America will be discussing such matters for decades.

History must record that Goldwater was the first to bring this quality of life into political discussion. And it must also record that Lyndon Johnson was not at all loath to take up the challenge.

CHAPTER TWELVE

LYNDON JOHNSON'S CAMPAIGN: "COME DOWN AN' HEAR THE SPEAKIN' "

T HE great theme of Barry Goldwater had developed gradually, so that it was not until October that one found the Prophet phrasing the question: "What's happening to this country of ours?"

But the great line of Lyndon Johnson happened, just like that, on a September morning as the street crowds boiled up around him in an engulfing demonstration at Providence, Rhode Island, halting his car and paralyzing his procession. Whereupon the President of the United States, transported, clambered atop his car, seized the bull horn and shouted: ". . . And I just want to tell you this—we're in favor of a lot of things and we're against mighty few."

For Johnson was Peace and Prosperity; he was the friend of the farmer and the worker, of the businessman and the teacher, of the black and the white; he was Mr. Responsible, Mr. Get Things Done, Mr. Justice-for-All, Mr. President.

But if his theme was simple, his organization was something else again. It was as bizarre as the organizational chart of the old China-Burma-India theater during the war, of which one befuddled colonel said, "To explain the C-B-I you need a three-dimensional organization chart with a wire framework and five shades of colored ribbon, which ought to indicate at least the simpler relationships." Or one might explain it as the Italian peasants described the Prince Torlonia hunting on his family estate: "First there is the Torlonia, and after that nothing, and after that nothing, and after that nothing, and then come all the rest." For this was Lyndon Johnson's personal campaign, personally directed in a manner not seen since the days of Franklin D. Roosevelt's later campaigns.

Nonetheless, a campaign so overwhelmingly successful—the most successful in all American electoral history—earns the most sober analysis and the reward of historical description.

The most graphic way to describe the campaign organization of Lyndon Johnson is to say that it was organized—like his White House—on a radial, not a pyramidal, model.

First, at the center of the wheel, were the three men with direct access to the President—Jenkins, Moyers and Valenti. The chief transmission belt for all presidential directives was—until his collapse, of course—Walter Jenkins. An accomplished shorthand expert as well as the closest associate of the President, Jenkins could take as many as forty-three messages from the President in a single telephone call, read them back word for word, and put them out direct to recipients, who, on hearing Jenkins, knew they were virtually hearing the President's own voice. The chief idea channel of the campaign was always Bill D. Moyers, frail, overworked and dedicated. And the ever cheerful, unwearying Jack Valenti was the President's shadow, companion, counselor and personal attendant. These three could be styled Team A, or the Household Guard.

From Team A, lines of control directly harnessed a number of critical activities. The most important lines gathered in Moyers' hands, for he was the in-funnel of speech writing. Speech material rose from every crevice and cranny of the governmental apparatus, from all departments —State, Defense, Labor, Agriculture, Commerce—down to the Bureau of the Census. The vast output of material flowing from these departments moved initially through the offices of Secretary of Labor Willard Wirtz (who, for a large part of the campaign, moved his desk from his chambers in the Labor Department to the Executive Office Building next to the White House). From Wirtz the material moved to the wordsmiths who operated under Moyers' general direction: Richard Goodwin of the old Kennedy speech-writing team first among them; then Horace Busby and Douglass Cater. From their editorial desks, finished speeches moved through Moyers and then to the President for final approval before delivery.

Team A also controlled the media campaign; it was to this team that Doyle Dane Bernbach (see pages 322-323) reported, and with Moyers (or his personal aide, Lloyd Wright) and Goodwin that the firm discussed television strategy. The enthusiasm of these three for the Doyle Dane Bernbach confections invited the New York advertising men to the vivid creations of imagination which so upset not only the Goldwater team but even the older Democratic politicians.

Team A was an all-purpose team. With Team A the President was at home, at ease and relaxed; with these men he could let off steam; with them he needed no good manners; they belonged to him.

But Team A was young; thus, Team A was subject to a senior court of review which may be called Team B. Team B was the trio of Johnson's oldest cronies—Clifford, Fortas and Rowe, three of the shrewdest lawyers in Washington politics. After the Convention these three met rarely as a group, but one or another was always cast as presiding appellate judge for any major speech, any major appearance, any major television show, any major change of schedule, any decision of policy. When a major crisis such as the Jenkins case occurred, these were the three first consulted.

Team C was the Kennedy operational team—or, rather, the Kennedy team minus Robert Kennedy. Any evening that all three were in Washington, Johnson would meet at 6:30 with Lawrence O'Brien and Kenneth O'Donnell to review the course of campaign. Officially, O'Brien and O'Donnell *were* the campaign directors (all state organizations were instructed to report to them) and they performed, as always, with superlative efficiency. In a six-week period on the road O'Brien summoned 23 meetings of 43 state Democratic organizations, from state chairman to county level, and at night Johnson waited for O'Brien's report dictated over long-distance telephone to the White House. Meanwhile, O'Donnell directed Johnson's travel schedule. And these two men, familiar with every nerve center of Northern and Eastern as well as Californian big-city politics, could and did make sure that every last resource of Kennedy loyalty was mobilized and in action, thrusting, pushing, striving for the last extra vote, needed or not.

Team D hardly ranks in importance with the other teams, but it was so fresh a departure in campaign practice that it must be mentioned. Team D called itself the "Five O'Clock Club," and each afternoon some twelve or fourteen of Washington's brightest young Democratic lawyers and junior officials met in the White House offices of Special Counsel Myer Feldman to discuss deviltry. The Five O'Clock Club was Johnson's original contribution to Presidential politics. Its franchise was to think about counteroffensive, or what may be called "negative advance." Its purpose was to precede, accompany and follow Barry Goldwater's campaign tour with instant, organized contradiction. Goldwater's advance schedules were always obtained by the club; so, too, somehow, were the texts of Goldwater's speeches hours before anyone else had seen them. Thus, well before Goldwater had arrived at his point of delivery, the material to refute or contradict him was generally in the hands of the local Democratic mayor, governor or Congressmen, so that, simultaneously with the appearance of Goldwater in any locality, the local paper carried in the same issue as Goldwater's speech a contradiction of it by a local dignitary or headline-maker or by a citizens' paid advertisement. So efficient were the men of the Five O'Clock Club that once or twice their contradictions appeared in local papers before Goldwater had delivered his text. Full of tricks, the club was also responsible for some of the best of the slogans, counter-demonstrations and hostile placards that greeted Goldwater wherever he went. This group reported directly to the President through Feldman.

Team E was the Democratic National Committee, directed by its Chairman, John Bailey of Connecticut. A hard-bitten and accomplished craftsman of big-city politics, Bailey looks upon all officials, elective or appointive, as "pols" like himself; in his life there has been only one political romance—with John F. Kennedy, to whom he gave his heart and to whom his heart still belongs. These were not the best of credentials to earn Lyndon Johnson's warmth, and Bailey described his own operation realistically as "the housekeeping job." Even in this realm, however, the "housekeeping" was of a new order of efficiency. Three years before, John F. Kennedy had insisted that registration be the main axis of new stress in the campaign of 1964; and this had been handed over to Matt Reese of West Virginia, who, in alliance with the COPE forces of the AFL/CIO, probably won some two million Democratic votes for the Democratic ticket of 1964 (see Chapter Nine). A new taping center provided speeches and programs that could be requested simply by a telephone call from any radio station in the country. Charles Roche of the old Kennedy team concentrated on marginal Congressional districts held by Republicans; and Louis Martin, one of the ablest though yet unrecognized political thinkers among American Negroes, mobilized minorities. Fred Dutton directed research. Margaret Price directed women's activities. And each morning Team E would meet at 9:30 in a group consisting of Bailey, O'Donnell, O'Brien, Martin, Roche, Price, Dutton, Feldman and Clif Carter. Feldman reported proceedings directly to the President; and so, also, did Carter.

One must not think, of course, that matters were as clear as they appear here on paper—for the various teams overlapped and interwove, sharing conferences, personnel and the attention of the President. Nor could the President be bothered with keeping all the channels clear while simultaneously he directed the affairs of the United States, domestic and foreign. If, for example, he was superlatively staffed in television—with production by Doyle Dane Bernbach, direction by Bill Moyers and review by Fortas and Clifford—it was not enough. Thus, he appointed to examine the entire operation, to give him the over-all view, yet another man, John Hayes, president of the *Post-Newsweek* stations, who would cast a final judgment on the effort of all the others and report his opinion to Johnson personally. If Johnson saw a billboard in Florida that pleased him, he would telephone personally and immediately ordering such billboards all around the country; and if ten days later he saw in California another billboard that pleased him more, he might telephone someone else to post that one nationwide; and if bumper stickers were short in Cheyenne or Chillicothe or Peoria, he would not waste time finding out who was in charge of bumper stickers—it would be a telephone call to Bailey to get on the ball, *now*.

And beyond all these teams was yet another force—Hubert Humphrey, the Vice-Presidential nominee, a team all by himself, plus his staff, his speech writers, swinging lustily around the country, spreading happiness and savaging Goldwater.

The thrust of political strategy could be examined best, however, not by reporting the individual groups or conferences or decisions but by plucking threads out of the tangle of activity to see what Lyndon Johnson sought.

One of the threads could be plucked by following the skilled and graceful operation of James Rowe. Rowe has been for thirty years an expert's expert in Presidential politics. To Rowe, in 1964, Johnson handed the bundle of activity called citizens' committees. In Presidential politics, citizens' committees are usually confused and frantic gatherings of goodhearted people who need political expression to let their hearts and emotions participate in the choice of a President. Normally, too, they are the greatest worry of professional year-round politicians, who value amateur energies and money highly—but who also know that citizens' groups can produce amateurs so excited by the adventure that they may stay in politics permanently, becoming professionals and challenging the entrenched leadership.

Rowe's citizens' committees were different—their target could be summed up in one word: Frontlash.

The Republican Party, obviously, was split in two; there must then be a guest chamber in the Johnson campaign for dissident Republicans —some place where their new loyalty to Johnson could be entertained, their fears of Goldwater given a chance to express themselves. This was Rowe's target.

Rowe's masterpiece was the National Independent Committee for Johnson and Humphrey, headed by Henry H. Fowler.[1] The National Independent Committee read like an honor roll of famous industrial and financial leaders. It included Robert B. Anderson, Eisenhower's second

[1] Henry H. Fowler has since been named Secretary of the Treasury.

Secretary of the Treasury, the man whom Eisenhower had listed as one who would himself make a great President; and John T. Connor of Merck & Company[2]; and Donald C. Cook of American Electric Power, who subsequently turned down Johnson's invitation to become Secretary of the Treasury. It included John L. Loeb, a partner of Wall Street's Carl M. Loeb, Rhoades & Company, a Republican, who in the spring had put up several thousand dollars for Henry Lodge's pre-Convention campaign; it included such awesome Republican names as Henry Ford II and two Boston Cabots; Thomas S. Lamont of the Morgan Guaranty Trust Company; Sidney J. Weinberg, a senior partner of Goldman, Sachs & Company. It included Maxwell M. Rabb, another Lodge campaigner, who had been secretary of Dwight D. Eisenhower's Cabinet; war hero James M. Gavin, former Ambassador to France; Raymond Rubicam, elder statesman of Young & Rubicam; Cass Canfield, of Harper & Row, publishers; and Goldwater's fellow Arizonan Lewis W. Douglas. It included technological industrialists of the new breed, like Sol M. Linowitz of Xerox, as well as powerful Southwestern oil figures such as Kenneth S. Adams of Phillips Petroleum. Under such an umbrella, any businessman in any small town anywhere could feel socially safe, financially secure, and clear in conscience in supporting Johnson and Humphrey too.

Rowe's committees covered the entire spectrum of American life—women's committees, labor committees, arts-and-letters committees, youth committees. The extremes of the span and the most colorful were, on the one hand, Rural Americans for Johnson-Humphrey and, on the other, Scientists and Engineers for Johnson-Humphrey.

Rural Americans for Johnson-Humphrey was an exceptionally able solution to a campaign problem. On the far side of the Alleghenies, where the government really twines deep in daily life, the Department of Agriculture, the REA and the Post Office swing far more weight than any big-city citizen recognizes. In precisely those towns of the West where the John Birch Society is strongest are also clusters of citizens most devoted to their Rural Free Delivery and Rural Electrification co-operatives. In every county, working through the Farmers Union and the Rural Electrification cooperatives, Rowe could mobilize thousands defending what Franklin D. Roosevelt had given them, what Johnson promised to protect and what Goldwater threatened to abolish.

Scientists and Engineers for Johnson-Humphrey was the other end of the spectrum. Beginning as an effervescence on the political scene, it operated so effectively as to lead many of the slide-rule thinkers to wonder whether they might not permanently enter politics and change them

[2] John T. Connor is now Secretary of Commerce.

—a problem to be considered in future campaigns. The Scientists and Engineers worked hard—they held meetings and rallies of their own, Nobel Prize winners licked stamps and sealed envelopes. But it was the dazzle of their names that was important. There were the great bomb-makers like Kistiakowsky of Harvard and Harold Brown of the Department of Defense; there were Jerome Wiesner, Dean of Science at MIT; Detlev Bronk, chief of the Rockefeller Institute; Emanuel R. Piore, Director of Research for IBM. Owen Chamberlain and John Rubel, Harold Urey and Herbert York were all members. Henry Ford II contributed $5,000 to Scientists and Engineers for Johnson-Humphrey. But the crusher was Dr. Benjamin Spock—baby-book Spock. (Inez Robb, the Scripps-Howard columnist, saw the recruiting of Spock as "the exact moment when all hope oozed away from the Republican candidate. . . . Millions of American mothers and grandmothers in the United States would as soon question Dr. Spock as they would Holy Writ.")

Another thread of strategy led from the office of Kenneth O'Donnell, for O'Donnell, even before the Convention in August, had been assigned the same task he had performed for Kennedy in 1960—that of plotting the campaign tour. He had begun with the reality of American electoral strategy, with the first questions of any campaign: Where do you plan to get elected? What states? What regions?—for that is where the candidate must go; and with the limits on his energies and time, he must resist the pressures of lesser contenders who need his help and the importunities of friends from sure states where time spent would be an indulgent waste. One gets elected—if one is a liberal Democrat—from the great quadrangle (Kansas City north to Minneapolis, east to Boston, south to Baltimore) plus California.

Here, in the original planning of the Johnson campaign, was where Johnson would fight for his votes.

It is the change in this original planning that reveals best the mood of the master. For it was obvious, immediately after Goldwater's nomination, that this President would not have to fight for votes and that he could thus design his victory to his own taste. There were the Senatorial and Congressional weaklings of his ticket whom he could help by flying visits; there was his own nostalgic connection with the South, for Lyndon Johnson wanted his Southland with him as much as Kennedy had wanted his New England; and then, as the polling results began to foretell the possibility of sweep, the horizons of Lyndon Johnson expanded still further until it was the entire country he wanted, all of it, as President of all the people. These pressures soon swelled the original schedule till it seemed like a storm assault on the whole nation.

One could trace, on a theoretical level, a strange parallel to the discussion of electoral strategy, an internal dialogue fascinating in terms of political thinking. Among the speech writers the dialogue was different from the dialogue among the strategists. What was the object of the Johnson campaign, asked some of the speech writers—to "broaden the base" or to "shape the mandate?" It was a luxurious internal argument, open only to men certain of victory. Should they stage a campaign whose purpose was to harvest the greatest majority ever in American history? Every poll and portent gave them the lock on Goldwater—should they go ahead and crush him entirely by expanding the safe middle ground of consensus to include the largest conceivable number of Republicans and Democrats? Or should they press the campaign in another way? To spend in advance some of this certain margin of victory by putting before the people such hard, cleaving issues as might lose a few million votes but would shape an explicit mandate to give the President clear authority for the new programs of his next administration?

This was a debate that interested chiefly the ideologues, for Lyndon Johnson decided both to broaden the base and to shape the mandate at the same time. He was for Peace and Progress—that was the mandate he shaped.

At the beginning, various strategists assured reporters there was a script: the President would be "Presidential" to start with—he would campaign from the White House, be on display, serenely coping with the business of the nation and the world, conferring honors or meeting great visitors in the rose garden; in this phase his field trips would be short, ceremonial, non-political. The second phase of the campaign would require the President to go over to the offensive—with an open attack on Goldwater and his recklessness. And, finally, at the end of the campaign, he must move to a positive stressing of the achievements of the Kennedy-Johnson administration—the prosperity of the country, the peace of the world, and the future that lay ahead in the Great Society.

It was a simple plan and, like most simple plans, sound—the trouble was that it gave only the vaguest frame of reference to one who tried to follow it. For the plan was always subordinate, at any given moment, to the instinct and impulse of Lyndon B. Johnson; and those closest to him recognized no plan at all but the master's wish at any given moment of campaigning. The master himself was the pivot of all action.

Lyndon Johnson was the Presidential Presence—and no challenger, at any time, can even approach the immense advantage that goes with being President. For, besides the majesty of the office, which cows the most hostile citizens to respect and attention, there are the facilities and the command that only a President can enjoy.

A President of the United States traveling in his own country has the touch of Merlin the magician. He wills—and he moves. At Andrews Air Force Base rests Air Force One, fueled for instant take-off, complete with all communications and office equipment; helicopters nest at Andrews to bring the President from his back lawn to take-off within ten minutes of call. Where the President moves, attention and service move with him. A White House transportation office of ten men makes sure that all who must serve or report Presidential doings glide with him as he goes. In New York the giant jets of the civilian airlines wait for the call of the White House charter office; overnight a Boeing 707 can arrive at Andrews from New York with two shifts of crew and six stewardesses, furnished with food and drink to comfort and feed one hundred men; and the scores who report the President's doings must be prepared, after an evening telephone call, to arrive at dawn in the barricaded alley between the White House and Executive Office Building and roll off on gray Army buses to Andrews, from where the plane carries them wherever the President may be going. Their advance reports will summon crowds by radio at any touch-down the President may make, banner his remarks in the afternoon papers, display him in the flesh on any evening television show; they conscript public attention and crowds for him.

To his pulse respond all the intricate energies and resources of the American military establishment. Signal Corps details will already be installing the telephonic communications at every stop across the country before he leaves his third-floor bedroom in the White House. The web of communications will be his on his journey to Andrews; will be functioning around the globe from his plane as he flies. When he descends, two, three, four telephones will be there as he speaks—one at the rostrum, one behind the grandstand, several more wherever needed. Trunk lines will be cleared in advance by both the Army and the civilian Bell Telephone System at all the towns and cities of his destination, reaching back into the Pentagon and the White House; from a large city as many as twelve such lines back to Washington may be cleared for President and staff.

For him the map, too, is different. Not only is every airport in the country open to him—private, public, industrial—but so also are the bases of the Air Force. At the gates of SAC, where sprout the huge signs saying PEACE IS OUR PROFESSION, unannounced intruders may be shot dead on approach. But when one travels with the President, the gates swing wide and one rolls casually down the field of droop-winged B-52s, their bomb bellies open on fifteen-minute alert, and one is a guest. And from any one of these countless Air Force bases the President may jump anywhere. No mountain village is too remote, no town too small to be reached. Almost anywhere in the country, at his desire, will wait the squadrons of olive-green helicopters that, like a little swarm of ugly beetles, lift his party from the larger fields which receive his jet plane and carry him directly to city park or construction shack or village square or high-school football field or wherever he may choose to dedicate a new dam or cut the ribbon on a new bridge.

For those who manage the formal campaign of a President, the map is a stage with scenery that can be mounted or dismantled at will for any speech the President cares to make, any impression he cares to give. For any pronouncement the proper place name can be arranged overnight as dateline. Texas is two and a half hours from the White House back lawn, Los Angeles four and a half hours. A new power project in the Columbia River Basin? The President is there within hours to cross into Canada, greet the Canadian Prime Minister and appear as international statesman at peace with our greatest neighbor. A new border treaty with Mexico? The President is at the El Chamizal strip at El Paso, hugging the Mexican President on the bridge and delighting both Mexicans and Mexican-American voters in Texas with his friendship. Power and bombs? The President is at the SAC base in Omaha, displaying America's bombers and missiles network to the Secretary-General of NATO, Manlio Brosio.

With facilities and staff like these, the formal campaign of 1964 proceeded almost flawlessly. The offensive against Goldwater rolled almost by itself—the nightly radio and television news spots; the organizational and creative deviltry of the Five O'Clock Club; the independent counter-offensive of Hubert Humphrey—all these held Goldwater in torment on the issues of bomb and Social Security, while the President was free to spread balm and peace and friendship and promises all around the nation. Never were Republicans denounced as such; the opposition was involved in its own civil war, and the President obeyed Napoleon's maxim: Never interfere with the enemy when he is in the process of destroying himself.

The formal strategists had little real need to book commercial television time either for spots or special programs. Presidents command time by their every action; they are news whenever they stir. And this President was not at his best in a half-hour face-the-nation address (his best stretch is the twelve-to-fifteen-minute discourse); his managers originally booked five half hours through the campaign, with a full hour reserved for Election Eve. They canceled at least one of these half hours (later reinstating it after the Jenkins case) and eventually cut their Election Eve hour to thirty minutes.

Only one of these formal television appearances need be remem-

bered or recorded; not simply because it was, technically, the end of the
first or "Presidential non-partisan" phase of the Johnson campaign, but
also because it was the best summation by the staff planners of what
their Democratic Party had to say about the candidate of the opposition:
that he was a radical. The idea had risen independently from several
sources in September as, on the one hand, Oliver Quayle's pollsters
brought back their countrywide soundings of popular reaction, and, on the
other hand, the speech makers tried to deliver the killing blow to Goldwater without elevating him to importance. From Quayle came the report that, generally, people were afraid of Goldwater, not only because
of the bomb but for all sorts of reasons—that he was "kind of radical."
From the speech makers came the thought that perhaps the easiest way
of getting at Goldwater was to accept the challenge on its longest-range
level. Was this quarter century of American experience really so bad?
Had America failed?

Sitting at his White House desk before the two flags, Johnson said
to the nation:

> We must decide whether we will move ahead by building
> on the solid structure created by forward-looking men of both
> parties over the past thirty years.
>
> Our prosperity is not just good luck . . . it rests on the
> basic belief that the work of free individuals makes a nation—
> and it is the job of government to help them do the best they
> can. . . .
>
> Today our whole approach to these problems is under at-
> tack.
>
> We are now told that we, the people, acting through gov-
> ernment, should withdraw from education, from public power,
> from agriculture, from urban renewal and from a host of other
> vital programs. We are now told that we should end Social
> Security as we know it, sell TVA, strip labor unions of many
> of their gains, and terminate all farm subsidies.
>
> . . . This is a radical departure from the historic and
> basic currents of American thought and action. It would shat-
> ter the foundation on which our hopes for the future rest. Too
> many have worked too hard and too long to let this happen
> now.
>
> . . . The choice is yours.

The formal campaign was efficient, no doubt of that. It was as if a
heavy mattress had been thrown over Goldwater at the beginning of the

summer and he lay buried under it, trying to wriggle his way out. All the Democrats needed to do was to rest heavy on the mattress, sprawl wide, bear down—and he was smothered.

But Lyndon Johnson's personal campaign was more than efficient; it was entrancing. To travel with him was to climb one of the rare heights of American political and dramatic art. It was like watching a great performer, at the height of his power, moving through a repertory and range that could not be topped—and yet seeing him top them again and again.

Not for years had a campaigner—not even Mr. Harry Truman in 1948—brought so finished a style of country oratory to a national audience. An old friend of Lyndon Johnson's once remarked that if Johnson had ever permitted himself, as a young man, the sort of fantasy dream which most of us pursue as we drift to sleep, his had probably been the fantasy that he was the true heir of Franklin D. Roosevelt; but that when Johnson woke in the morning, he woke to the knowledge that he was really the heir of Huey Long. For the Johnson style was shaped in the Old South, where, if the courthouse machines have not locked the race up in advance, one runs man against man, with victory going to the man who can out-shout, out-dramatize, out-campaign, out-smile and out-entertain the raw voters until they feel in their hearts that old Huey (or old Pappy, or old Ed, or old Hummon)—he understands them, he is one of them.

This style, at its best, is native American art, and it was probably in this idiom that America was governed for one hundred years. In an odd way, however, Johnson, like so many other Southwesterners, has been brainwashed by the Eastern press and the manners of the great leaders of the past thirty years until any reference to this homely style in print enrages him as mockery.[8]

---

[8] This curious Presidential sensitivity contrasts oddly with the attitude of Hubert Humphrey, who, like the President, is a small-town boy. Humphrey is probably, after Adlai Stevenson, the most popular living politician on Eastern campuses and relishes the company of Eastern intellectuals; but Humphrey takes from his own roots a sense of superiority—he can match the eggheads at *their* discourse, yet *knows* he can also reach the farmers in blow time with images that have never entered the head of an agrarian economist; Humphrey enjoys both worlds. Johnson, as a losing contender in 1960, would snarl to his friends about "those Harvard p——." But as Presidential candidate in 1964, what he sought above all was the speech of John F. Kennedy and Franklin D. Roosevelt; so that those who saw him in the formal addresses to the nation saw a heavy and somber man wrestling uncomfortably with phrases not quite his own, while those who heard him on the stump heard a genuine American artist. Not until after he had won the Presidency in his own right did he achieve the ease that made his message to Congress on Negro voting rights (March 16th, 1965) so noble and memorable a performance in the classic style of American Presidents.

One had been tempted, even before the campaign began, to make a catalogue of all the Lyndon Johnsons there were, for in the etymological sense of *persona* as mask, Johnson's *personae* were almost unlimited.

There was of course to start with, Lyndon Johnson as "Mr. President"—the solemn, grave man on television, talking of nuclear bombs, world peace, the public good, who spoke with ponderous gravity, licking his lips with pointed tongue between polished strophes written by speech writers, occasionally overstressing his "the's" and "and's."

There was occasionally the "Kindly Lyndon"—the man who at Kennedy's Inaugural had leaped from his seat to shade with his hat the sunstruck pages from which aging poet Robert Frost tried to read but could not for the glare. And the same man had found time, in the first two weeks of his turbulent take-over, to rush off to a Washington hospital to visit his ailing and aged first chief in the National Youth Administration thirty years before, Aubrey Williams.

And then there was the "Imperial Lyndon"—as at the Convention in Atlantic City, lounging in his box in the gallery, one long leg crossed over the other, leaning back in the reclining position of a Caesar Augustus at the Roman games, letting his eyes wander down over the floor and occasionally waving his imperial hand languidly in recognition of important delegates or dignitaries whom he would electrify when their eyes caught his.

But it was outdoors, on the road, away from the manicuring and restraint and monitoring of his blue-ribbon advisers, that he was most attractive. Raw and natural, casting away prepared text, reaching out for the hearts of the entire nation as he had reached out for hearts in his native land, translating the highest policy of the Western world into the simple speech of the old Texas Tenth Congressional District, the President of the United States could present a whole new series of *personae:* "Fair-Shares Johnson," "Preacher Johnson," "Old Doc Johnson," "Sheriff Johnson," "Uncle Lyndon," "Lonely-Acres Johnson," and several others.

"Fair-Shares Johnson," for example, arrived to open the Oklahoma State Fair in September, late in the afternoon after a hard day's campaigning in Texas and Oklahoma. When the helicopter deposited him on the fair grounds, a handsome roan horse was standing nearby. Nothing would do, then, but the President must mount the roan and, waving his cream-colored Stetson hat, urge the horse to a canter; arriving at the speakers' stand with a whoop, he announced that he had come "to talk to happy people. I came here to talk about what is right. I didn't come here to talk about what's wrong. . . . I'm proud of America and I'm proud of Oklahoma. . . . Seventy-five years ago, on an April morning, starting guns signaled the opening of this land. No one called it a give-away. . . . What came out of it? I'll tell you what came out of it—it gave us Oklahoma, and no one thought it would dull initiative."

Then after a few prepared passages, warming up as the audience warmed with him, he began to act out his "Fair-Shares" routine, identifying himself by gesture, drawl and voice with each of the partners in the American system he loved to describe.

"Our people," he began sternly, "believe that a capitalist ought to be able to take his dollar and invest it, that he ought to be allowed a reasonable return on that dollar without fear of having it confiscated." (To such a rural audience the President would not linger long over the capitalist on whom he would dwell with Eastern audiences.)

Then he went on—his mien changing to one of puckered lips and head-nodding sympathy—to the hard-working manager:

". . . [we believe] that a manager ought to come along and help invest that dollar in a producing enterprise; and that a manager that gets up at daylight and stays up until midnight and develops stomach ulcers while he is doing it—he is entitled to a two weeks' vacation, a bonus once in a while, a little profit sharing and Social Security when he is sixty-five." (When facing a heavily white-collar audience, "Fair-Shares Johnson" would add how that manager, "he takes his chances where he finds them and if he loses, he fusses with his wife about it, but he takes it in stride" . . . and go on a bit longer.)

"With that capitalist," he would continue, "and with that manager who manages, we have the worker who produces." Here his face would light up with a benevolent, radiant, protective smile. "He gets on the assembly line"—pantomime of laborer on the assembly line—"and every twenty-seven seconds he puts the rivets in the top of a car." Here the big hands would come up and the twist drill would drill frantically as Lyndon Johnson imaginarily drilled those holes and hammered in those rivets. "And every one that comes by, if he doesn't get them all in in twenty-seven seconds—if it takes twenty-eight to get them in—then that car goes out without a rivet. He does that all day long. But he is the greatest producer in the world, the American workingman. He doesn't ask much, either. He wants a little vacation. . . ." Here his voice would drop to a coaxing sound. "He would like to have a little sick leave, he hopes he has a little medical care. He has some things he wants, he wants a rug on the floor, and a picture on the wall and a little music in the house. . . . He doesn't ask for the world with a fence around it. All he wants is to make America a better land." (Sometimes he would include in this industrial pastoral a family picture of the workingman wanting "a place to take Molly and the babies when he retires. That is his great love. His boys go to war, and they fight to preserve this system, and he likes his boss and he respects him.")

"Preacher Johnson" was equally fetching. "Preacher Johnson" talked very slowly, even more slowly than "Mr. President Johnson" on TV—but with real quality, leaning one elbow on the rostrum, turning to face now the platform, now the audience, reasoning gently, urging people to be good, occasionally talking about the spirit of John F. Kennedy "up there in Heaven watching us." Sometimes "Preacher Johnson" could hush a huge auditorium and uplift his hearers, as in his oration at the Salt Lake City Mormon Tabernacle. But more often he offered just a plain, folksy Sunday sermon: "If you will just go back to the Good Book and practice some of the teachings of the Lord, if we will just follow the golden rule and do unto others what we would have them do unto us, if we just engage in a little introspection and look where we were and where we are, we won't be unhappy very long. We won't feel sorry for ourselves very long."

"Lonely-Acres Johnson" usually closed a speech by trying to make them understand the "awesome burden" of the Presidency. Sometimes it would be preceded by the "dawn alarm": ". . . not many of you get waked up in the night about Cyprus or Zanzibar or Vietnam, but I never send a reconnaissance mission out about eleven o'clock with our planes and our boys guiding them to take a look at what is developing and realize they have to be back at three-thirty in the morning, but what promptly at three twenty-five I wake up without an alarm clock, because I want to be sure my boys get back. And sometimes they don't get back. . . ." Generally, "Lonely-Acres Johnson" made his appearance at dusk or nightfall when the President's voice would send the folks off to home and bed to think about what they could do for their country in the voting. But as for him, he had to go back to those lonely acres in Washington, and those iron gates would close behind him and he would be left alone with that awesome responsibility in the corridors where Lincoln and Wilson and Roosevelt and Kennedy had paced before him.

There was Lyndon Johnson the practical politician, too, fighting for every last vote as if he were still running in a primary—as at Albuquerque, New Mexico:

> You know, down in Texas in 1941, President Roosevelt was very popular, and he asked me to run for the Senate because he said he needed me to help him . . . and I went down there and we worked awfully hard. The morning after the election, we led by five thousand votes, and three days later we lost by one thousand, three hundred and eleven out of a million and a

half. We lost by one thousand, three hundred and eleven. Then we waited seven years, to 1948, when we ran again.

The night before the election I met Lady Bird . . . and I said, "Come on, honey, we're going home and spend the night at the ranch."

She said, "Oh no we're not. I'm going back to Austin and I'm going to get our mother, your sisters, your aunts and your uncles, your friends and your cousins, and I'm going to take the telephone book and I'm going to assign one of them all the A's, and one of them all the B's, one of them all the D's, right through the Z's, and I'm going to call and say, 'Won't you please go to the polls and vote for my husband?' 'Won't you please go to the polls and vote for my son?' 'Won't you please go to the polls and vote for my brother or my cousin?' "

And she did that. And she went through that telephone book and we had another million and a half votes. We had the highest percentage of votes in Austin in that area we had ever had before, because so many people said, "Gee, I forgot to vote. I've been busy. I thought I'd vote but I went to the grocery store and I got busy washing the dishes and I haven't gone yet. Yes, I will go vote." We got out the highest percentage.

I won the nickname of Landslide Lyndon because I won by the magnificent total out of a million and a half of eighty-seven votes, and the Republicans have been talking about it ever since. I have been thinking about it ever since, because if it hadn't been for that extra work that she and my mother and my sisters and my cousins put in that day, Texas would have lost a good Senator.

The most impressive of all Lyndon Johnsons was, however, "Sheriff Johnson." This was a drawling, easy, country-style, no-nonsense candidate. He knew he was President and he wanted you to know it, but he was doing it easily and simply as he told you who was boss. The best performance of "Sheriff Johnson" came at Hartford, Connecticut, the insurance capital of America, the only New England state where there was the faintest trace of organized Goldwater enthusiasm. A group of sedate New England businessmen and financiers had offered Johnson their Medal of Freedom, which he accepted in Hartford's Constitution Plaza. He gave them first a bit of "Fair-Shares Johnson," then described the prosperity his administration had created for them, then ended his speech with an anecdote which I offer as a sample of the best Southwestern storytelling style:

> Now I am going to conclude by telling you a story. It is one of the real reasons I asked to meet you insurance people up here today.
>
> Remember back in the thirties when we had all of the problems on the farms, and all of the problems in the Stock Exchange, and we had men committing suicide and folks jumping out of windows, and banks popping like firecrackers, and all of the economic problems?
>
> Well, during that period Mr. Rayburn happened to be the author of the Holding Company Act. He came from a little red sandy-land farm down in Texas, and he was just a farm boy. But he had to author the Stock Exchange Act, and he set up the Securities and Exchange Commission, and he had all of those great financial reforms to work out.
>
> Right in the middle of it, Mr. Whitney came in and asked him if he couldn't come up to speak to the Bond Club in New York City. And he said, "Who is the Bond Club?"
>
> "Well," Mr. Whitney said, "it is a group of people who sell bonds and investments and who are financial experts. As a prerequisite to membership you have to have a good financial statement, and you have to have one million dollars or more. We would like to hear you and we would like to invite you."
>
> This fellow was testifying before Mr. Rayburn's committee and he kind of wanted to ingratiate himself anyway. But Mr. Rayburn said, "Thank you very much," and he said he would accept the invitation.
>
> The fellow who extended the invitation was almost sorry that he had ever asked him, because he did not think he would accept. But Mr. Rayburn went up there, and he was a little embarrassed in front of all those millionaires. So they introduced this sandy-land farmer from Texas who had put all of these reforms through. One was called the "death sentence," you remember, for holding companies, and this was going to ruin all of the power companies and destroy everything.
>
> So this fellow got up to introduce Mr. Rayburn and he said, "Fellows, I invited this man to come up here. I don't know why he came. But here he is, Sam Rayburn." And that was the only introduction they gave him. That is more than you gave me today. I just had to get up here and talk.

Mr. Rayburn got up and he said, "I came here for two reasons. If you don't know why I came, I want to tell you." He said, "All my life I have been poor. I have just had a modicum of wealth, and I have accumulated a little, and I haven't married, and I have worked for it, and I saved every penny, and I haven't spent it on my family, but I have never yet been able to acquire one million dollars. But I wanted to come up and associate with all of you fellows who have been successful and just hope that a little bit of it would rub off on me.

"So that is the first reason I came." And he said, "The next reason I came was to show you I ain't scared of you."

So I really have three reasons for coming up here today. I want to associate with you and learn things from you. . . .

Second, I want to commend you for the contribution you have made to your government during these ten months when we have had a real critical period on our hands. . . .

And, finally, I just wanted all of you to know I wasn't scared of you. . . .

The New Englanders laughed and applauded, and after he went on his way he was their President too.

Nor did he alter either politics or style, North or South. Journeying through the Southland ten days later, he arrived at New Orleans on October 9th and chose to speak in that city in the presence of other Southerners as a Southerner who had come to wisdom. Aching to win the votes of his home region, he nonetheless stormed, "Whatever your views are, we have a Constitution and we have a Bill of Rights, and we have the law of the land, and two thirds of the Democrats in the Senate voted for it [the Civil Rights Bill] and three fourths of the Republicans. I signed it, and I am going to enforce it, . . . and I think any man that is worthy of the high office of President is going to do the same thing. . . . I am not going to let them build up the hate and try to buy my people by appealing to prejudice. . . ." He rambled off at this point into a long anecdote about a Southern Senator of his acquaintance who at the end of his career had told the Speaker of the House, Mr. Sam Rayburn, "Sammy, I wish I felt a little better . . . I would like to go back . . . down there and make one more Democratic speech. I feel just like I have one in me. The poor old state, they haven't heard a Democratic speech in thirty years. All they hear at election time is Negro, Negro, Negro!" He was bellowing, his arms upraised and flailing, as he reached this point.

One correspondent in the audience swears that a towering Negro was so carried away by the revival attitude of the meeting that he yelled, "Knock yourself out, L-B Baby! Knock yourself out!"

To speak of such speeches as corny or phony is to misread entirely, I believe, the emotion of the man. Somehow it appeared that the man would not be content to win on technical points, but was trying to erase the flaw in the title papers of the Presidency which he had inherited by an assassin's bullet. And only the people could erase this flaw—only people shouting and cheering and seeking *him* as their President. So first he must seek the people. Thus, by September the poll results which had fascinated him and dominated his conversation in July and August had palled. It was crowds that fascinated him. And as the first few weeks of his "non-political journeyings" wore on, he was increasingly annoyed by the reporting that described the crowds, though healthy (greater than Goldwater's), as apathetic.

What Lyndon Johnson sought, perhaps, was the feel of the crowds that John F. Kennedy had assembled—that feel which was always half gaiety and half a sob of longing caught in the throat. If he could not have that, he would reach them anyway—he would make them respond. And he reached for them day after day in the best human language he had—the language of the tufted, dry hill country of Texas.

Then, overnight or over one weekend, the jump happened. In every campaign, as politicians know, there can come an unexplained quantum jump of attention when the crowds surge into the streets to cheer their candidate and give him love. It happened to Eisenhower in late September of 1952. It happened to Kennedy the first week in October of 1960. It happened to Lyndon Johnson on Monday, September 28th, 1964. His crowds, as I say, had been good and growing throughout September. But on the last weekend of September the Warren Commission issued its massive report. On Sunday afternoon the great television networks devoted hours to it; Monday-morning papers throughout the nation bannered the report, tearing open the scarcely healed wounds in the emotions of the American people, re-creating the black weekend of assassination over again. It was as if the nation hungered to see a President, real, live, healthy, in the flesh—as much as the President hungered to see them.

Lyndon Johnson had read the report himself on his ranch in Texas on Saturday and Sunday, September 26th and 27th; had flown back to Washington that night; and had risen early on Monday, September 28th, to give a day to campaigning in New England.

He arrived at the airfield in Providence, Rhode Island, at 9:30 A.M. on a cool fall day, and already some 3,000 people were at the airport, surging against the wire fence, girls squealing, children crying in the crush, babies held aloft, and boys chanting the particularly New England chant of "Two-four-six-eight, Who-do-we-appreciate?" The President's face suddenly illuminated. It was as if someone had turned the current on in the house. He paused only briefly to hug ninety-seven-year-old ex-Senator Theodore Green ("You gotta come to the Inauguration," he said to Green, "an' when you come, you gotta stay at the White House with Lady Bird an' me") and then strode directly to the wire fences. On the trip up he had complained to newsmen about their reporting of crowd reaction. Now he hailed them to the fence as he grabbed at hands, as girls burst into tears at his greeting, and said, "How's that for crowd reaction?"

There is no real way of measuring crowd numbers except by aerial photographs and analyses.[4] Some political observers use a modified version of Air Force jargon which describes cloud cover in a sky as two tenths, five tenths, until one comes to CAVU—Clear and Visibility Unlimited. From Providence on, Lyndon Johnson's crowds could be described as ten tenths plus—or, put another way, either in Providence, Rhode Island (population 208,000), or Hartford, Connecticut (population 162,000), one could see in the streets of his route more people than the census gave for the entire population of the city.

More interesting than the crowds, however, was the effect of the crowds on the man. The dark-blue open limousine of the President would plow through the choked streets, and several youngsters would slip from under the barricades and dart toward him; a breach would be opened; the police would frantically try to stay the breach, but the President would halt the procession, and then for hundreds of yards on either side of the breach would pour young and old in running streamers of people to clog around the car. And he loved it. Clambering up on the back seat of the limousine or sometimes helped to the top of a closed car, he would grab either microphone or bull horn and talk to them.

"I'm grateful to each one of you," he would say in absolute humility. "About ten months ago there was this terrible tragedy and we lost our beloved President John F. Kennedy. Give me your help, your hand, your prayers, and I'll do the best job I can as your President." Or he would invite them down to the Inauguration. Or, at the corner of Ninth and Broadway in Providence, when—by now long behind schedule—he tried to sum it all up in one sentence, came the grand description of his policy:

---

[4] Los Angeles Chief of Police William H. Parker declares that the easiest way of making a crowd estimate is to measure with the eye the square footage occupied by the crowd, then divide by two. Each human being, in his estimate, occupies a space one foot by two feet, or two square feet. This may be technically the best way of estimating crowds, but I have never been able to train my eye to judge acreage or footage in strange squares or plazas.

"We're in favor of a lot of things, and we're against mighty few." Or he would compliment them on their warm hearts. Or talk to them briefly, questioning them, cupping his ear for the reply: "I want to ask you just one question—are you going to vote Democratic in November?" And then, cupping his ear again, "I didn't hear you, did you say yes?" And they would bellow *"Yes!"*

What was fascinating was how completely the personality of the man dissolved the prudence and distance of the Presidency. The weekend before, he had read—as had every accompanying reporter and Secret Service man—the text of the Warren report with its criticism of the Secret Service for failure to protect adequately the life of the dead President. Secret Service men on this day swarmed about the President, desperate to protect him. But he would have none of them. He would reach down and clutch hands, first with his right hand, then with his left, then grab waving hands together in bunches and squeeze them. By evening his right hand was swollen and bleeding and a Band-Aid covered an angry bruise at the base of his thumb. But he wanted the people close; he wanted to press the flesh. He would direct the traffic policemen about him, ordering them to halt, ordering them to let crowds through, then stand above the mob and, when he was through, wave his hands like Moses parting the Red Sea, for he was late for his speaking—and the crowd would part.

He had made thirteen stops before perhaps half a million people in Providence before he arrived an hour and ten minutes late at a convocation of Brown University, where his euphoria subsided a bit. Donning academic black gown, he became "Old Doc Johnson" and told the students how important education was.

He arrived in Hartford from Providence two hours behind schedule —but the crowds, rather than fading, had grown during the long wait. And as he inched his happy procession toward the Hartford Times Square, he took a full forty-five minutes for the last five blocks. An ambulance howled to get through the wake of his procession to pick up a woman who had fainted. People hammered on the side of his limousine in frenzy and he grabbed back at them; the jumpers appeared—an empty pair of woman's shoes in his trail on the street indicated that one jumper had jumped right out of her boots. The Secret Service men might be in an agony of tension, but the President was enjoying himself.

Normally the *Queen Mary,* the President's open touring car, can hold five in the back seats and up to three in front. One by one, by the President's invitation, photographers clambered aboard with him until there were eight people in the back seats and seven people over the front seat and hood. I examined the battered *Queen Mary* when it pulled into the back court of the Hartford *Times,* and it was as scratched, scraped, dented, bumped and hammered as the most miserable heap ever turned over to a used-car dealer.

Every leader has his style—Goldwater and Johnson, Nixon and Eisenhower, Kennedy and Stevenson and Roosevelt; all were different, and in retrospect one remembers them by the quality of their wit or eloquence or wrath or dignity or straight talk. What Johnson was discovering was that he could be President and be himself too. Folks were folks, and the same country-style campaigning that ran down South would also work in the North. He was to go from Hartford to Burlington, Vermont; to Portland, Maine; to Manchester, New Hampshire; and to Boston, ending his day at two in the morning to the cheers of the same uproarious crowds. And from then on he felt free to talk to all other crowds, east and west, north and south, as plain old Lyndon Johnson. But somehow it was in John F. Kennedy's New England, with the pumpkins golden in the wayside mounds and the leaves turning, that Lyndon Johnson found the response he had sought. And because he had done it in New England, it probably pleased him more than if he had done it anywhere else. At one of his stops he urged one of New England's most distinguished Senators to join the press bus and tell the correspondents that these were crowds even bigger than John F. Kennedy had got (which was true). "If you tell them," said the President, "they'll believe you—they won't believe Jack [Valenti]." The Senator demurred.

Only one major jolt upset the Democratic campaign—dramatically sharp at the moment of impact and now only a detail: the Jenkins case.

On Wednesday, October 7th, Walter Jenkins, the President's closest personal associate, had attended a cocktail party celebrating the opening of *Newsweek* Magazine's new Washington offices. It was a distinguished gathering: no less than seven Cabinet members were present, as well as a constellation of Washington's journalistic luminaries. At about seven o'clock, after several drinks, Jenkins left the party and walked two blocks to the basement of the Washington YMCA, so notorious a gathering place of homosexuals that the District police had long since staked it out with peepholes for surveillance. Here, thirty-five minutes later, in the men's room, Walter Jenkins was arrested. Taken to a police station, he was booked with an old Army veteran on charges of disorderly conduct.

Within forty-eight hours of the arrest, someone in the FBI had unofficially leaked the story to friends on the Republican National Committee; simultaneously and circuitously, another channel informed the Republican National Committee of the happening. By Monday, five days later, Barry Goldwater, too, had been informed of what had happened, but he remained silent. Some time on Monday—it is impossible to say when—the story reached the Chicago *Tribune* and the Cincinnati *Enquirer* (both supporters of Goldwater), which chose not to publish it. The Washington *Star* learned of the event some time on Tuesday. On Wednesday, a week after the happening, the matter was known to all too many people—but not to the Democratic leadership—and the *Star*'s assistant managing editor, Charles Seib, with heavy heart, called the White House early in the morning, as he must, to check whether or not the story was true.

The telephone call, relayed to Jenkins by a press officer, must have come to him as a knife in the heart; for him it was the end. He left his office at the White House immediately to seek the help of Abe Fortas; Fortas, stunned by the incoherent revelation, summoned Clark Clifford, who also now learned the story for the first time. Together, that Wednesday morning, Clifford and Fortas began to make the rounds of the Washington newspaper offices, pleading with editors, on grounds of common humanity, that the story not be published. Those editors who had not known of the incident were alerted to it by the visits themselves. As the day wore on, the two eminent counsel of the President learned from newsmen that they were not dealing with an isolated episode: there had been a previous arrest for a similar episode in 1959. The charge at that time had read "pervert."

It was also apparent to Fortas and Clifford that Jenkins, flushed and incoherent, must be hospitalized. A doctor confirmed their opinion. By early afternoon Jenkins had already been hospitalized and the story was general throughout the city. By 8:25, when United Press International finally moved the story over the wires, it was public. And the nation, in mid-campaign, had to face the fact that the President's closest personal assistant, an attendant at National Security Council meetings, master of the inner chambers of the White House, was a sexual deviate.

For twenty-four full hours Republicans and Democrats alike held their breath to see how the nation would react. And perhaps the most amazing of all events of the campaign of 1964 is that the nation faced the fact fully—and shrugged its shoulders. Indeed, far more interesting than the episode itself were the human reactions of all those involved—the reaction of leadership under stress.

The first was that of Mrs. Lyndon Johnson—so tender, sincere and warm that immediately it set a tone of sadness and sorrow rather than one of scandal or hypocrisy. "My heart is aching today," she said from the White House, "for someone who has reached the end point of exhaustion in dedicated service to his country. Walter Jenkins has been carrying incredible hours and burdens since President Kennedy's assassination. He is now receiving the medical attention which he needs. I know our family and all of his friends—and I hope all others—pray for his recovery. I know that the love of his wife and six fine children and his profound religious faith will sustain him through this period of anguish." With the statement she set a climate of sympathy which all others had to accept.

The second was that of Barry Goldwater. The Republican candidate was in Denver when the news broke publicly. Here, if ever, was demonstration of his charge of "moral decay," of sickness of soul, of bestiality in Babylon. Yet the Goldwater painted as a killer by the Democrats could not bring himself to hurt an individual; urged by his young men to hammer the issue, to make the most of it as if a gift had been given him—he simply refused. He referred to the episode in the weeks following only rarely—and with conspicuous lack of relish.

The most obscure reaction was that of Lyndon Johnson. He had been traveling in New York State that day, and during the afternoon George Reedy had passed on to him questions about Walter Jenkins which the accompanying reporters, fed by the buzz-mill of rumor from Washington, were putting to Reedy. It was not until early evening, when the President finally arrived at the Waldorf-Astoria in New York City, that Clark Clifford, speaking from the White House, could give the President the full story—now confirmed in every respect. Yet the story was still publicly unannounced, and so, carrying the heavy burden, the President went first to visit Mrs. Jacqueline Kennedy, then on to a few quick and perfunctory remarks at Cardinal Spellman's annual memorial dinner for Alfred E. Smith. As he spoke, the full story began to spread on the wires, and George Reedy, weeping openly, confirmed it. The President had already, by then, commissioned Fortas in Washington to go to the hospital and seek Jenkins' resignation—which had been done.

What went on in Lyndon Johnson's mind is unknown, and can never be made clear; the man he had trusted most had broken.

The Executive decision at this moment of stress was calm and reportable; what private anguish he experienced he kept private. Exhausted by his campaigning, suffering from a cold and a fever, he went upstairs to the Presidential Suite of the Waldorf Astoria, and while his cold was ministered to, his mind functioned well. Should he invite a Congressional investigation of the Jenkins affair? Or should he call a special panel? Or order an FBI investigation? He discussed these alternatives with Fortas in Washington by telephone and decided to commission a full-dress FBI investigation. The only important matter was national security—had Jenkins' illness weakened or opened a latch on matters of security? The FBI would find out.

The next consideration was political. At midnight Valenti telephoned Oliver Quayle and asked him to appear at once at the Waldorf. Speeding seventy miles an hour from his Bronxville home to the President's suite, Quayle arrived in twenty-five minutes. A vaporizer was puffing steam at the President, who sat in bed in his pajamas, weary, his voice hoarse, his mood somber. Quayle was commissioned to do an instant poll of opinion—how many votes could such a sad affair shift?

All Thursday morning the polling personnel of the Quayle office manned the telephone, questioning cross sections of voters. All Thursday, too, Presidential advisers debated what the President should do. Should he cancel his Thursday-evening broadcast (to celebrate the anniversary of the Test Ban Treaty) and instead devote his time to clarifying the Jenkins affair? Or should he go through with the Test Ban broadcast as planned? By Thursday afternoon the telephone pollers had enough response for Quayle to reach the Presidential party in Brooklyn and report that they could detect no significant shift and that the wisest procedure was not to stop campaigning, but to carry on without elevating the affair to emergency.

Johnson's private ruminations are not known. Yet, had they been known, they would have been far more interesting than his outer actions. Johnson had few Washington friends, and his closest were those who did not serve him or take his money. But the record of those who had voluntarily enlisted in his service and to whom he had given his trust was a bizarre one. Bobby Baker, his protégé, had nonchalantly withstood the strain of service to Lyndon B. Johnson by slyly trying to parlay Johnson's authority into a little private fortune of his own. John Connally of Texas, who had been with him longest, had left to go back to Texas. George Reedy, another of the long-time servants, had preserved his honor impeccably but was visibly exhausted by the long adventure. Bill Don Moyers, his newest servant, a man of utter dedication, worked so hard he worried all who cherished him. But now gentle Walter Jenkins, closest of all to the President and his family, had collapsed in a psychiatric breakdown under the strain, a strain to which the President, his master, had long been blindly insensitive. One of Lyndon Johnson's closest present White House servants says of him that the President believes the essential margin he has over other men is his capacity to work harder—to work and work and work all through the day and night, beyond any other man's capacity. But the President makes all those who live with him work equally hard, without sense of the strain he imposes on others. The greatest casualty of this strain was, obviously, Walter

Jenkins. But whether any such reflections passed through the Presidential mind in the twenty-four hours of Wednesday-Thursday is total speculation. By Thursday the people had given their answer to Johnson, who was barnstorming New York in the company of Robert Kennedy. Mary McGrory of the Washington *Star,* who accompanied the President that day, reported:

> The crowd at Buffalo plainly revived the President, who had anguished much of the night over the sorry story of his aide. What the cup of coffee or the hair of the dog are for some men, the sight of a throng is for Lyndon Johnson.
>
> And by the day's end, he had received an unexpected assist from the Kremlin. By the time his public had decided who Walter Jenkins was and what his tragic fate had been, the news of Khrushchev's resignation had made the news of the Jenkins resignation something of a detail.
>
> And the multitudes in Brooklyn would have heartened a cigar-store Indian, which the President is not. For two incredible hours, they poured out of their homes and their bars and their launderettes. The women streamed out of the beauty parlors with curlers in their hair. The men stopped watching the World Series. Hassidic Jews with soft black hats and side whiskers waved wildly. Little boys in their temple skull caps ran after the cavalcade. Lyndon Johnson stood up at full height on the back of the car, giving his own special blessing, arms stretched out shoulder high, fingers rippling. Butchers with their gory aprons greeted him, and bakers with their puffy hats. Women held their children or their dogs high to see him pass. It was wall-to-wall people from Canarsie to Albee Square. In one great movement, as Representative John Rooney was clawing his way up the trunk of the car as the cavalcade entered his district, the President reached down and grabbed him by the scruff of the neck and hauled him aboard. It was hard to find anyone among the happy faces who was concerned about possible scandal in the White House. They only wanted to see the President and Bobby Kennedy.

The President returned to the White House late that night to cope with history. History would take little notice of the Jenkins case, and would not care at all that the Cards won the World Series that day by 4 to 3. It would be concerned only with the fact that during the fall of 1964—at some time after the Americans had already decided they would choose Lyndon Johnson as their next President, but before they had voted—there occurred in a forty-eight-hour period the following grand events: the deposition of Nikita Khrushchev as dictator of Russia; the detonation of the first nuclear bomb by the Red Chinese; and the overthrow by popular vote of the thirteen-year-old Tory government of England with the election, by the slimmest margin, of Harold Wilson as the Queen's First Minister.

Johnson returned to the White House late on Thursday evening and was up at daybreak, studying intelligence reports before gathering with State Department specialists, CIA executives and scientists to consider the meaning of Red China's bomb. He had an appointment at 11:30 A.M. with Ambassador Dobrynin of the Soviet Union, which would give him for the first time the sound of the new Russian government; this talk might be of critical importance.

In between, he slipped out of the office for a tree-planting ceremony on the south lawn a few feet from his office. Looking very tired and unusually somber, he made a little speech about the beauty of trees, and then took a shovel that was handed him. He hefted it in his hand as if it were a familiar, comfortable instrument and turned to the pile of dirt beside the open hole. With the grace of a good workingman, he put his foot on the shovel, leaned and heaved the shovelful of dirt into the hole. It felt good, and in a soft, gentle tone he said, "I did this for a living for two years from dawn-up to sunset," and he heaved another shovelful and another, as if he enjoyed the heaving of his big muscles, and then said, "My mother finally convinced me I'd enjoy life more if I used my head instead of my foot." He walked a few steps toward the Oval Office and then turned back to the little group of reporters and gently apologized: "I'm sorry I can't stay around and talk with you—Ambassador Dobrynin is coming over to see me and I got to go in."

He kept his speaking date in the Midwest that evening, but then he returned to settle down in Washington for four days, busy at his desk with matters involving the new Russian government, the new Chinese threat, the new British government, Vietnam. He explained these matters to the American people on television on October 18th, quite gravely and simply. The Jenkins case had evaporated as politics. And from then on the campaign was, in effect, over—it was all downhill, interesting only as the performance of a man.

Of the man it may be said that he could not stop; the appetite drove him on and on. "Come down an' hear the speakin'!" he would shout to the crowds who mobbed his way, or sometimes, "Bring your children and the family to hear the speakin'."

An unbelievable physical vitality fueled him. I remember, for example, a single day on which I tried to keep notes. The President awakened at the White House at nine (the previous night he had returned to the White House at 11:30 and done paper work until 2:30 in the morning) on Tuesday, October 27th. He read his newspapers in bed, looked at a preliminary draft of the Budget Message and some State Department dispatches. At 9:45 Jack Valenti arrived in the upstairs bedroom and they went over four or five drafts of upcoming speeches and put through seven or eight telephone calls (among them calls to the Secretaries of State, Defense and Labor and to McGeorge Bundy). Johnson gave Valenti about a dozen little errands to do; came downstairs to his office about 11:30 to confer with Britain's then (though briefly) Foreign Secretary, Patrick Gordon Walker; paused for a bit to promote John Glenn, astronaut, from Lieutenant Colonel to full Colonel; gave a full hour to a group of Latin American diplomats who wished to discuss Western Hemisphere problems; took soup at his desk for lunch; worked there until shortly after three; took a brief time out to tape-record several spots for television and radio; and then left by helicopter from the lawn to fly to Andrews Air Force Base and thence to Boston.

He arrived in Boston at 5:30, spoke for forty-five minutes; flew on to Pittsburgh and spoke another forty-five minutes; flew on to Evansville, Indiana, and spoke again; flew on to Albuquerque, New Mexico. (One had the impression that the President would prefer to campaign always in a westward direction if he could—the day grows by three full hours when one flies west, and is cut by three hours as one flies east.) He arrived at Albuquerque at 1:45 A.M. Albuquerque time (4:45 Washington time), and then from his hotel room telephoned McGeorge Bundy about the Russian and the Vietnam situations, talked to McNamara about the military budget, and to Rusk about Latin America. He was finally urged to bed at about 2:30 Albuquerque time, thus finishing a day of twenty-two hours' exertion. He slept four and a half hours and was off the next day to stump southern California (where, by this reporter's stop-watch count, he actually spoke for some five and a half hours of personalized campaign oratory).

Nor would he let up, even in these last ten days, on Barry Goldwater. Instinct told him to subordinate all campaign planning—even the theme of the Great Society—to the one issue of war and peace; and as he did so, he developed it in his own way with a style of his own that gradually, in the last few weeks, shaped it into a memorable passage of political prose. It varied from city to city—from Boston to Los Angeles, from Chicago to Pittsburgh—but the best rendition that I heard was in Los Angeles on October 28th, six days before the election.

After several rather short but fascinating autobiographical passages about his own life—delivered by "Old Doc Johnson," with an admonition that he didn't like "grouchy people"—he combined the styles of "Mr. President" and "Sheriff Johnson" and took off:

> Just because we are powerful, we can't just mash a button and tell an independent country to go to ——— because they don't want to go to ——— and we don't get very far rattling our rockets or lobbing them into the men's rooms or bluffing with our bombs.
>
> I saw President Kennedy in the Cuban crisis in thirty-eight different meetings, and we got up to the last hours. Khrushchev had his missiles trained on this country that would completely wipe out San Francisco and Los Angeles. There would be no life left. Those men stood there, one speaking for the United States and the free world, and the other speaking for the Communist world. They got eyeball to eyeball, and I saw the generals with their stars come into the room [the great hand reached up and stroked the shoulders where the stars had been] and the admirals with their braid [the great hand stroked the sleeve crusted with Navy braid] and the Secretary of State with all of his diplomatic experience. [Sometimes he would talk about the Secretary of State, who was a Rhodes scholar at Oxford, you know, and the Secretary of Defense, who was President of the Ford Motor Company, making half a million dollars a year.]
>
> I listened to every word. I never left home in the morning a single morning that I knew I would get back that night to see Lady Bird and those daughters. So, as a little boy in my country used to say, we were doing some pretty heavy thinking, because we were right up to the gun. But Mr. Kennedy put his knife right there in his ribs and held it, Khrushchev put his there and held it, and neither one of them shook, trembled or developed palsy; neither one of them wobbled. Our planes were in the air. They had their bombs in them. Our Navy was on the seas; they were ready. But Mr. Khrushchev finally decided rather than to see three hundred million people killed and the Soviet Union wiped out—and they could wipe out America too—that, humiliating as it was, it might be a little wiser to wrap up his missiles in those tarpaulions [here the hands bundled and wrapped the missiles] and put them on those ships and take them back home, and that is what he did.

. . . [the] first responsibility, the only real issue in this campaign, the only thing that you ought to be concerned about at all, is who can best keep the peace? In the nuclear age the President doesn't get a second chance to make a second guess. If he mashes that button [here the big thumb mashed and squirmed as it pressed an imaginary button]—that is it.

So we cannot make a foreign policy, we cannot keep the peace by bluff and bluster and by threats and ultimatums. We can keep the peace and we can only keep the peace by two methods: first, with a strong defense, and we are today, I tell you as your Commander in Chief, the mightiest nation in all the world; and second, we can keep the peace, in the words of the Prophet Isaiah, by reasoning together, by responsibility, by negotiation.

He was magnificent, though two hours behind schedule; and he would have continued to the rapt audience as long as it stood there, but Lady Bird passed him a note—and he realized that he must conclude.

Away from the public platform, he was a more reflective man—concerned, worried, perplexed. On the way into Los Angeles he had relaxed with a few correspondents and let his mind ramble around the furnishings of the campaign. A White House attendant was rubbing gobs of white salve into his bruised hands, and every now and then the President shoved great handfuls of white popcorn into his mouth. One reporter asked him where he would be on Election Night, and the President turned the question over to Lady Bird. Lady Bird thought they ought to go to Austin after the White House broadcast, and the President asked whether she'd really thought that through. The President's wife said, no, she just thought it was the best thing. The President turned to Jack Valenti and told him to call the Hotel Driskill in Austin and make a reservation—but, then again, he kind of thought that Washington was better, with all those TV facilities.

The President had the polls aboard and he ruminated aloud about them. On one sheet were several columns, state by state, with his private polls, the *Congressional Quarterly* poll, the *Newsweek* poll, the *Time* poll. The results were too good, he felt; state by state he would comment, "We just got no business carrying New Hampshire"; or on Texas, "They're far too high—I figure fifty-three—fifty-four percent is about what we should do." It was seven days from election. He gave Goldwater Arizona—and also Mississippi, Alabama, South Carolina, Louisiana, Georgia and Florida. (He was off by only one—Florida was to be his, not Goldwater's.) He felt he ought to carry Virginia—he'd like to get a helicopter and tell all those people in Norfolk and northern Virginia how he didn't let that pay raise die the first time Congress killed it. He felt good about Ohio, but Lady Bird didn't feel so good about Ohio, for some reason. He was going into Los Angeles and he warned the reporters it wasn't going to be so good—he'd been there over and over again and he'd never seen more than seven people in the streets (on this he was wrong, Los Angeles was great).

He returned to polls and predictions; he figured they'd spent half a million dollars (a gross overestimate, actually) on polls this season. But you know who'd taught him to use the polls? John F. Kennedy. Polls were valuable only to tell you what people were thinking about, like Social Security.

He talked about his Cabinet. It was a great Cabinet. He wanted to keep it. He admitted he'd had nothing to do with choosing it; President Kennedy had done that—but Kennedy had done him the courtesy of introducing every one of them to him over the telephone when he chose them.

A lot of things had to happen after the election. There were the Navy Yards—Brooklyn, Philadelphia, Boston and San Diego. Some of them had to fold. There were the task forces of the Great Society—they were going to tell him what was desirable and he was going to tell them what was feasible.

He talked about the Negroes. He could see those eighteen- and nineteen-year-old Negroes jumping right off the ground. The reporters couldn't see it; they were too far back in the procession. But some of them seemed to get right up into the air and walk on it, as if he were Jesus Christ.

He rambled with that astounding perception of Johnson's, a mixture of perspicacity and understanding and surgical nerve-sensitivity. There was Moyers—that poor boy had something wrong with his esophagus; the other night when Billy Graham was in, Moyers looked like he was dying, but just the same he wouldn't go home. Dick Goodwin: Dick wrote that Boston speech—good, wasn't it? And Wirtz did the Pittsburgh speech.

The eclectic mind of the President covered everything. Alabama—he was going to lose Alabama. But Alabama was in trouble. They had this Huntsville installation, the one that works on rockets. But you need scientists for the space program. Only you couldn't get the scientists, "the Einsteins," to want to go to Huntsville in Alabama. They think George Wallace eats people like them alive before breakfast.

Someone complimented Johnson on his Salt Lake City speech (and it was indeed very good). The President liked that. He knew in his heart he had more religion and compassion than that other fellow; and once we were elected we were going to investigate everything. "We're gonna *do* something about this morality."

Thus Johnson as the American people were about to choose him by the widest margin ever given a freely elected leader.

A television show on Monday, Election Eve, showed him leaving the White House lawn by helicopter for Texas.

Most candidates in recent years have wound up their campaigns in their home states, which is proper.

But Lyndon Johnson's staff and planners all stayed in Washington. They knew Johnson had won; they might as well watch television on Election Evening from Washington and rest with their families. Only Valenti of the campaign staff accompanied the President down to Texas.

So Lyndon Johnson was going home. Briefly some of his advisers had suggested that he end his campaign in the Dark City of Dallas, to take the curse of killing off it. But he chose Houston instead, and arrived home in Austin on a misty, rain-pregnant Monday night.

Here the crowd was good. He spoke from the steps of the State Capitol on a warm evening and he acknowledged that the campaign was over, the "rest was dedication." The University of Texas band blared for him, and the throng swirled around the live oaks on the Capitol grounds. "It was here," he said, "as a barefoot boy around my Daddy's desk in that great hall of the House of Representatives, where he served for six terms and where my grandfather served ahead of him, that I first learned that government is not an enemy of the people. It is the people. The only attacks that I have resented in this campaign are the charges which are based on the idea that the Presidency is something apart from the people, opposed to them, against them. I learned here, when I was the NYA Administrator, that poverty and ignorance are the only basic weaknesses of a free society, and that both of them are only bad habits and can be stopped."

He went from this speech to the Driskill Hotel's Jim Hogg Suite, where he had a little sip of Scotch and a few words about politics with old friends. When he came out, reporters clustered around him and he gave them a few of his last-minute worries. To his original list of doubtfuls he had now added several more, and he felt there was a real chance that Goldwater might carry Alabama, Mississippi, Louisiana, South Carolina, Virginia, North Carolina, Wyoming, the Dakotas, Ohio, Iowa, Maine, Vermont. But everyone knew that Johnson was poor-mouthing it, and that he was in.

Thus, at one o'clock in the morning he climbed aboard the Presidential helicopter that would chop its way to the ranch on the Pedernales River where he was born. When he was born, it was a good two-day buggy ride from the Pedernales to Austin, the capital of the state. As President, in 1964, in a helicopter of the United States Air Force, he could do the sixty miles in twenty minutes.

CHAPTER THIRTEEN

VOX POPULI—"FOR WHAT WE ARE ABOUT TO RECEIVE . . ."

I T was over before it began.

The issue had been decided long before—perhaps within minutes of the fatal shot at Dallas.

Or it had been decided even before that under the leadership of John F. Kennedy; or by the spring and summer boom; or by the spectacular performance of Lyndon B. Johnson during the transition; or by twenty years of growing awareness of what nuclear war could do. The decision had slowly deepened and strengthened as the nation watched the Republican proceedings in the snows of New Hampshire, along the marches of California in May, on the hills of San Francisco in July.

What remained to be settled was only how large, how broad, how deep would be Lyndon Johnson's sweep, and for this the networks and news services had finally provided America with its first true national reporting service.[1] For this purpose, too, some $15 to $20 million worth of electronic hardware had been assembled by computer technologists,

[1] Perhaps the happiest and most valuable result of the year-long cannibalistic, desperate and, at times, dangerous rivalry of the national television networks was the creation of the Network Election Service. The rivalry of the networks had, by the June 2nd primary of California, engaged them in a competitive game where some 50,000 poll watchers, at a combined expense of $1 million, were racing each other to the "call." The perspectives of competition were leading inevitably to a future in which all three networks would have to staff a majority of the 180,000 polling places in America at some incalculable cost to provide an artificial drama too disturbing even for the networks to contemplate. Five weeks after the California primary, in which CBS and NBC fought each other down to the wire almost as bitterly as Goldwater and Rockefeller, good sense prevailed and all three networks decided to pool their vote-counting resources in one cooperative group. These three nets later invited to join them the traditional vote counters of America, the Associated Press and the United Press International, and together they created the Network Election Service. The NES, under the direction of William Eames of

stuffed with some $2 million worth of data by precinct, county, city and state, sorting, subdividing, classifying all Americans into their income brackets, races, ethnic origins and religions, their rural, urban and sub-urban divisions. Scholars and analysts stood by in platoons to sift mean-ing from a gush of expected data, codified finally to strip all faces, identi-ties and names from the Digital Society in its moment of decision.

All, however, were superfluous on November 3rd, 1964. The tele-vision studios in New York slowly crowded with sweating men preparing for air time, when the nation would be told what was happening. But the youngest college-graduate office boy, early in the afternoon, could forecast the result as easily as the most senior commentator later in the evening. Shortly after three the first meaningless fragments of returns were already stuttering on the dials of the counting boards to tell the story. Normally, if there is to be any drama to the evening, these first totals must be Re-publican; but by four in the afternoon, wherever one scanned the first responses of the old Republican hamlets of New England, the tide was clear: Massachusetts—Johnson 28, Goldwater 18; New Hampshire—Johnson 35, Goldwater 29. By 5:30 Kansas had choked out the first meaningful large total. Kansas had been the most Republican state in the Union in 1960, voting for Richard M. Nixon by 60.4 percent. Kansas was still Republican at heart, said the dials, voting for its Republican nominee for governor, William H. Avery, by 70,000 to 56,000 for his Democratic challenger. But already these first Kansas returns were show-ing the Johnson sweep of the farm belt—Johnson leading Goldwater by 65,000 to 62,000. And if Kansas had given its heart to Big Daddy, who else could resist?

At six P.M., when the Network Election Service flashed its first pub-lic total with only one fifth of one percent of the vote counted, Johnson was leading by 77,572 to 74,139, or 51.3 percent to 48.7 percent. By seven P.M., when 2 percent of the national vote had been counted, the Democrats were checking in with their votes and the President led by 273,000 plurality, or 59.6 percent of the total. By ten o'clock the Presi-dential plurality had reached 3,906,000 and his percentage 60 percent.

CBS, Frank Jordan of NBC and Arnold Snyder of ABC, set up a tabulating center at the Hotel Edison, New York, whence a staff of more than 100,000 people was deployed to give Americans the quickest, most honest and most complete vote count in their history.

The NES is a departure in election service and politics of the most promising nature. Policed from within by the five news services, it also polices official count-ing everywhere across the country with its civic-minded student and housewife employees. It was successful in establishing independent poll watching and count-ing across almost all the country, the most notable resistance to the NES poll watchers coming in Cook County, Illinois, a bastion of old-fashioned politics and vote counting.

By then, of course, the impact of New York's predictable Democratic landslide was showing; but, more important, Indiana and Ohio had both fallen to Lyndon Johnson and Barry Goldwater's Midwestern strategy was dead. By midnight, with 65.6 percent of the national vote counted, Johnson's margin was 10,757,000 and his percentage of the total was 62. From there on, for the rest of the evening as for months to come, there was nothing to do except stand in awe at the results and pick what meaning one could from the local returns until historians should later describe the true meaning of the elections.

Lyndon Baines Johnson, 36th President of the United States, won re-election from the people of the United States by a margin of 16,951,- 220 votes out of 70,621,479, with 43,126,218 against Barry Goldwater's 27,174,898. This was the greatest vote, the greatest margin and the greatest percentage (61 percent) that any President had ever drawn from the American people; we shall live long before we see its like again.[2]

With him, Lyndon Baines Johnson brought back to Washington 28 Democratic Senators (for a total of 68 out of 100, a gain of 2) and 295 Democratic Congressmen (against 140 Republican Congressmen) for a gain of 37 seats. Only Franklin D. Roosevelt, in 1936, drew Congressional support greater than this. The Republicans gained in only one domain, adding one governor to their previous total of 16 to make 17; but they lost over 500 seats in the state legislatures around the country, and lost their control of both houses, in a year of redistricting, in 12 states of the Union.[3]

The results of the election were so overwhelming that they are almost as difficult to analyze as the election of 1960, equally perplexing in its fantastic closeness.

Certain gross facts protrude from under the tidal wave like mountain ranges.

The first of these, of course, was the voting of the South. Barry Goldwater won but six states—his home state of Arizona and five Deep South states, Mississippi, Alabama, Louisiana, South Carolina and Georgia. Here, unquestionably, race was the dominant issue. (In Alabama the name of the President of the United States was actually wiped

---

[2] See Appendix A for tabulation by state.

[3] The best state-by-state analysis of the elections that I know has been done by a group of young New England Republicans called the Ripon Society. Their 124-page report, *Election '64*, is an indispensable document for any further study of the 1964 election. The most technically sensitive analyses of the election by sociological and ethnic strata are the tables compiled, but as yet unpublished, by the joint efforts of Louis Harris & Associates, CBS and IBM.

off the ballot.) In addition to those five states, in no less than four other Southern states Lyndon Johnson could not have won except for the phenomenal increase and spectacular shape of Negro voting. In Virginia, North Carolina, Tennessee and Arkansas for certain, and probably also in Florida, Goldwater carried a majority of white voters and Lyndon Johnson's victory margin was brought to him by Negro voters. The Republican Party is now, finally, deeply rooted in the South, but these roots may nourish bitter fruit. It is impossible to conceive of any moderate Republican challenging the grip of the Grenier-O'Donnell-Yerger-Edens team on future Southern delegations to a Republican national convention; and to these, in 1968, will be added the conservative power in the Rocky Mountain states. One can foresee another bloody rending of the Republican Party ahead.

Another fact protrudes: though the total vote of American citizens in 1964 rose slightly over 1960,[4] the percentage of eligible voters who cast ballots dropped from 63.8 percent to 62.1 percent. The Kennedy-Nixon contest had resulted in the enormous numerical jump of 10 percent of American citizens who chose to cast a ballot; the Johnson-Goldwater contest failed to stir Americans by any similar profundity of emotion except in the South. In the South, where hundreds of thousands of newly enfranchised Negroes trooped to the polls, and were overmatched by the even greater numbers of white Southerners who hated them, the gross vote jumped by 20 percent!

But in no less than five major regions of America—New England, the Middle Atlantic states, the industrial Midwest, the Prairie states and the Border states—not only was the percentage down, but the actual totals were down, too. Remarkably enough, the landslide for Johnson was heaviest in New England, where the drop in total voting was sharpest. In Maine, where the actual vote dropped most (by 9 percent), Johnson won 68 percent. In Rhode Island, where the actual vote dropped 3 percent, Johnson won his biggest percentage—80.9. In Massachusetts, where the vote dropped 5 percent, Johnson won 76.2 percent. If the Goldwater theorists believed that there was a hidden conservative element in the Republican Party which would rally to the first clear conservative call, they overlooked the fact that there was an equally dedicated, and far greater, liberal element which would stay home—and it did.

Only by breaking the national vote up into its regional areas does one begin to arrive at a measure of how completely Lyndon Johnson, the

---

[4] The District of Columbia voted for the first time in 1964, adding 198,597 votes to the total. The actual gross national increase was approximately 1,788,000.

Kennedy administration and Barry Goldwater, all together, succeeded in shaking the traditional party loyalties of Americans.

Of the eight great regional groupings of American states, it will be remembered that Nixon in 1960 carried five, Kennedy three—but Kennedy had carried the Mid-Atlantic states to give him the critical margin (*The Making of the President—1960*, page 341). In 1964 Johnson carried seven of the eight regions; and the only one that Goldwater carried—the South—was one of the three that had gone Democratic for Kennedy.

One by one, each of the regions, except for the South, tells the same story.

In New England, where Johnson did best, the total vote was down most. But while Kennedy had pulled only 56 percent of the vote here (a margin of 603,587 out of 4,997,169), Johnson carried off 71.1 percent of the vote (a margin of 1,187,500 out of 4,783,582). This was to be expected. Thoroughly industrialized, thoroughly international-minded, New England was hopeless for Goldwater from the beginning.

The Middle Atlantic states—New York, Pennsylvania, New Jersey, Delaware, Maryland—gave equally predictable results, for Johnson's second highest percentage, 66.6 (Kennedy's was 51.5 percent.) Here Johnson's margin was huge—5,343,882 plurality out of a total voter turnout of 16,149,194 (down from 16,372,790 in 1960).

The Prairie states gave Johnson the most remarkable swing of all. The five farm states—Iowa, North Dakota, South Dakota, Nebraska, Kansas—had been in 1960 the most Republican of all the regions of the country, giving Nixon a percentage of 58.8. This time they swung to Johnson by 57.1 percent in a shift of loyalties which has generally been overlooked but which far outweighs the countershift of the old South. Here was where the choice of Hubert Humphrey probably paid off most. Johnson carried these farm states by 466,270 votes out of 3,178,101 (down from a voter turnout of 3,395,088 in 1960).

In the industrial Midwest the vote was also down—from 17,607,-696 in 1960 to 17,212,960 in 1964. Here, too, the drop in voting was accompanied by a vast shift in loyalties—where Nixon had carried this Republican heartland with 51.3 percent, Johnson carried it with 61.2 percent for a margin of 4,097,968 votes out of the 17,212,960 cast.

In the Border states (Kentucky, Missouri, Oklahoma, Tennessee, West Virginia), where anti-Catholicism had in 1960 led the voters away from their normal Democratic traditions to oppose the election of John F. Kennedy, Johnson reclaimed his own. He drew 61.4 percent of the Border-state votes, earning a margin of 1,224,876 votes out of a total of 5,732,596 (as against 5,837,945 in 1960).

Elsewhere in the country the vote was up—but this helped Goldwater little anywhere except in the South, where conservatives *did* turn out.

In the old South, ten states of the old Confederacy were the only regional group to give Goldwater a fractional percentage lead—49.0 to 48.9 percent of a total vote that had risen from 8,865,501 in 1960 to 11,168,021. Goldwater's winning margin here was, in actual votes, only 22,000—but, historically, the fact that any Republican could pull a margin in the old Confederacy closes a full century of change with paradox. History does not, however, stand still—and, with the certainty that Negoes, now largely deprived of the right to vote in Mississippi, Alabama, Louisiana, South Carolina and rural counties elsewhere in the South, will in 1968 be enfranchised by the new national voting laws, this new Republican ascendancy is anything but permanent.

Of the last two regions where the vote totals went up, Goldwater could find some comfort in one, the Rocky Mountain states—but scant comfort it was. In the eight Rocky Mountain states Johnson won only 56.5 percent of the vote, with a margin of 474,195 out of 2,830,743 (as against a turnout of 2,641,593 in 1960).

But the Pacific states gave Johnson a more commanding margin. Altogether, combining tiny Hawaii (Johnson's with 78.8 percent) and mighty California (Johnson's with 59.2 percent), Johnson won with 60.4 percent in the five Pacific states, or a margin of 1,960,966 out of a total of 9,367,685 (as against 8,733,361 in 1960).

One can, by using the Precinct Profile Data of the Harris-CBS-IBM group, refine other useful thoughts and insights out of this election.

The vote of the Negroes everywhere, North and South, was perhaps the most nearly unanimous expression of will in any community free of political surveillance anywhere in history. It makes no difference where one samples Negro precincts, rural or city, North or South, upper or lower class—they ran an average 94.9 percent for Johnson in the East, 95.3 percent in the South, 95.8 percent in the Midwest, 97.2 percent in the Far West. Some urban Negro precincts approached 99 percent for Johnson—which can only be interpreted as meaning that several people, by mistake, pulled the wrong handle in voting booths in several big-city ghettos. If the Republicans can do nothing to include the Negroes in their vision of America, they enter any future Presidential race with one ninth of the nation locked against them. Their alternatives now are clear—either to try again to divide the Negro vote with the Democrats, or accept the Negro vote as permanently hostile and make strategy accepting that hostility and appealing only to whites.

The ripples and bubbles of protest in the broad flow of the Johnson victory are so rare that one has to pore over charts for hours to find them. One finds them, indeed, in only one major cluster—in the rural South, where the drop-off of Democratic votes in Georgia, Mississippi, Louisiana, South Carolina and Virginia was, indeed, signficant. (An echo is found in the tiny—0.8 percent—fall-off of California farmers, so many of them Southern in origin.) In the Southern shift one sees a reversal of that pattern which first became evident in the days of Eisenhower and which Meade Alcorn, in Operation Dixie, hoped to make permanent— a Republican Party based on the new cities which would contend with a red-neck rural Democratic Party. The Republicans, this time, did moderately well among the whites in the Southern cities; but Johnson gained on them by adding Southern urban Negroes, who as late as 1960 had voted for Richard M. Nixon in large numbers.

Elsewhere in the nation one can discern strange faint ripples only with the utmost difficulty. In Polish working-class wards in the Midwest (Indiana, Illinois, North Dakota, Ohio) Goldwater managed to shave the Democratic percentages of 1960—but whether this was an echo of backlash or an ethnic identification with William Miller's handsome Polish American wife, one cannot say. Or one can note that in precincts sampling Idaho Catholics, Johnson ran 18 percent behind John F. Kennedy. But so overwhelmingly, elsewhere, did Catholics vote for Johnson, and by so much heavier margins than they had voted for Kennedy, that the Idaho situation must be considered an aberration.

One can find interesting patterns, too, by comparing result to program.

The traditionally Republican farmers of the nation offered, perhaps, the most interesting group reaction to sharply pointed issues that can be identified. Goldwater was against the whole structure of government control of the American farm economy as built over the past thirty years. Johnson was for it. Forced to choose, the American farmer went along with what government had given him. A survey of the farm vote by the national Rural Electrification Administration's cooperative associations provides some fascinating insights into farmer loyalties. In no less than *twenty-three* states the rural vote went more heavily for Johnson than did the state as a whole. In some states like Kentucky, North Dakota, Oregon, Washington, Texas and Maine the farmers voted by three to one for Johnson. In several areas the shift of traditional Republican farmers was astounding—in Maine, from 42 percent Democratic in 1960 to 74 percent in 1964, in New York from 20 percent Democratic in 1960 to 57 percent in 1964. The farmers, evidently, liked what they had—rural electricity, farm supports, soil conservation, the entire twined system of control and regulation which Barry Goldwater had styled a "mess of oppression."

One could go on. But every identifiable group, except for the Southern white, seemed swept along in the tide. The elderly of St. Petersburg, Florida, Republican for the past twenty years, shifted to Johnson to protect their Social Security. The suburbs, cockpit of future American politics where a decade ago so many seers had seen the Republican future, went overwhelmingly against Goldwater. And the workers in their unions, the poor in their strivings, simply fattened their traditional Democratic majorities. The Harris-IBM breakdown by economic class gives the voters in the lowest economic bracket these percentages for Johnson —in the East, 90; in the South (with the new Negro votes added in), 92; in the Midwest, 90.8; in the Far West, 89.3.

There would be much looking back in years to come. Such an election as this would intrigue future storytellers. Looking back, they might find in the election of 1964 the seed-names of some entirely new era— as do we who now remember that it was the squalid meat-rationing election of 1946 which sent on to Washington for the first time two young veterans of the Pacific war as freshmen Congressmen—John F. Kennedy and Richard M. Nixon. Or they might see it like the vintage year of 1948 that delivered into the Senate for the first time, or to national eminence, such men as Lyndon B. Johnson and Hubert Humphrey, Adlai Stevenson and Chester Bowles, Estes Kefauver and Paul Douglas. Some who lost in 1964 might stand again to fight another day—as Charles H. Percy of Illinois or Robert Taft, Jr., of Ohio; and some who stood against the tide, as John Lindsay or George Romney, might make American history by their future decisions. But more interesting than any of the known names, speaking for known qualities and purposes, were those young men who in the election of 1964, unknown to any national fame, unrecognized beyond their own districts, first entered the process of government. Among them might be some future President of the United States, sixteen or twenty years hence. But what that future President might do, and how he would make his mark, would be shaped by the answers to two great present questions: whether and how quickly the Republican Party could pull itself together as a cohesive force in American politics; and whether and how swiftly Lyndon Johnson could guide America to the uplands he styled the "Great Society."

For the elections of 1964 had left the Republican Party in desperate condition. Someone has said that an American political party is so low-grade a zoological organism that, like a worm, if it is chopped in two, one or the other half can wriggle away and thereafter regenerate itself. In

1964 the Republican Party had indeed chopped itself in two—yet in all the months since then, to the day of this writing, no one can tell whether the two halves can sew themselves together or whether enough vitality remains, in one or the other half, to find a direction in which they can invite the American people to move.

The Republicans suffer, first, from a general condition—a continuing failure to capture the imagination of the American people. The failure has been traced over twenty-five years by that senior analyst of American public opinion, Dr. George Gallup, whose figures indicate that the number of Americans who think of themselves as Republicans has declined from 38 percent in 1940 (as against 42 percent of voters who thought of themselves as Democrats) to 25 percent (as against 53 percent for the opposition) in 1964.

Republicans suffer, next, from a specific political ailment—the lack of any agreed purpose for their Party. The split of their conflicting purposes can be temporarily bridged, as it has been at the moment, by the appointment as National Chairman of a competent professional like Ray Bliss, a master of techniques and media. But the direction of an American party is never given it by its National Chairman nor by its paid staff. It is made by those who wish to use a party as an instrument of their vision. The Republicans, beneath their temporary truce, remain split in vision as in politics—split between the new Southern Republican party and its Far Western allies on the one hand, and the Eastern party on the other hand, with the balance held by the old Republicans of the Midwestern heartland. The Goldwater party could and did increase the Republican vote in the South by some 30 percent; but it paid by losing Republican strongholds everywhere else from Beacon Hill in Boston and Westchester County in New York to the farmers of Kansas, Nebraska and the Dakotas. Whether or not the two halves of the party can be sutured together will remain a continuing drama, to be reviewed again in 1968.

Good candidates can, of course, help the Republican Party. Good candidates in 1964 ran so far ahead of the Republican ticket as to tempt leaders to an almost undue stress on personalities. Eisenhower, in 1952 and 1956, had swept the nation by personality alone. Outstanding candidates of broad appeal proved again, in 1964, that they could defeat both zealots within and Democrats outside—as did Liberal Republican Daniel J. Evans in Washington State, who trounced the powerful Goldwater-Williams state organization (see Chapter Five) in the open Republican primary of 1964, then went on to win the governorship against an incumbent Democrat in the November election itself. But the triumph of outstanding Republican personalities—Lindsays or Javitses, Romneys or Chafees, Evanses or Murphys—are fleeting triumphs; they stand in isolation and can be defeated in isolation. The Republican Party requires a strategy; and no one yet has persuaded the Party of a strategy that can unify and move it.

For strategy must begin with a vision—a perception of the nature of America as it has changed, with all its promises as well as its menaces; and the greatest split in the Republican Party is not geographical, or even governmental, but between those whose dreams lead forward and those whose dreams lead back. There is still room in America for a new creation of the conservative imagination to re-order the growing complexity of American life. Barry Goldwater signally failed to deliver this conservative vision in 1964, offering little more than a Populism of the right; nor have the Republican governors, men aware of what is happening, been able in the months since the election to offer an alternative Republican vision which can lead their Party back to the mainstream where it must challenge Lyndon Johnson.

For it is Johnson who most threatens the Republicans. Lyndon Johnson is, to be sure, no visionary spectator, no philosophic analyst; he is an activist politician, an enormous pressure vessel in which all ingredients of experience are distilled. Yet out of this experience, finally, after thirty years of knowing the best and the worst in American life, Lyndon Johnson has finally distilled a vision—a vision which embraces the entire nation. And though it was probably true that Americans in 1964 had voted above all for Peace over Risk of War, the voting of 1968 and beyond will be moved by the visions that Lyndon Johnson has summed up in the phrase "The Great Society."

What does he mean when he speaks of a "Great Society?" Where have the ideas come from? Where do they lead?

To appreciate both the phrase "Great Society" and its meaning, one must appreciate that exquisite process whereby ideas, maturing over decades of thought and stimulated by decades of change, first find form in the words of thinkers and scholars and then, slowly, as they ripen, come to The King's Ear—to the man who must then decide to act on them or not. This marriage of ideas and decision is what makes history work. The decisions can be dated—but no one can say precisely where in the long flow of thinking any idea is born.

Generations before it became the pennant of the Johnson administration, the phrase "The Great Society" was in circulation among men of ideas trying to define a future where men would be free to strive, yet safe from harm. But it surfaces first in library catalogues as the name of a book written by an English thinker, Graham Wallas, who in 1914 wrote *The Great Society,* an attempt to greet the promise with which man's new mastery over material so tantalized Utopians. Wallas had taught in Harvard Yard; unaware of future high symbolism, he inscribed his book to one of his particularly brilliant former students; and Wallas tells young Walter Lippmann in the preface that this book is ". . . an analysis of the general social organization of a large modern state. . . ." (Years later Lippmann, by then a far greater influence on political thinking than his former teacher, returned the compliment with a far more influential book, *The Good Society*.)

Whatever the deep springs of the flow of American ideas, the relentless conquest of nature's materials continued to channel it. For, as reality changed, ideas raced forward to point out where reality led. One can find almost all the visions of the Johnson administration somewhere in the speeches of Adlai Stevenson in the American campaigns of 1952 and 1956. As early as 1956, Arthur Schlesinger, Jr., chief of staff of the brain-trust of Adlai Stevenson, wrote an essay entitled "The Future of Liberalism—The Challenge of Abundance," in which he tried to define the difference between the "quantitative liberalism" of the pre-war America and the "qualitative liberalism" that must be sought with America's increasing mastery over production and abundance—a distinction still central to the ideas formulated in 1964 as "The Great Society."

To the incubation of these ideas, one must add the immense healing influence of the Eisenhower years, the temporary stabilization of the postwar turmoil abroad and at home, the calm and the process of recovery from the turbulence and clash that accompanied and preceded the wars. Ideas and examination continued; Congress held formal hearings on The National Purpose; Eisenhower, as his parting legacy to the new President, was able to bequeath in 1961 an official paper that attempted to define the national goals. The great foundations began to underwrite and subsidize new studies and examinations of the condition of America changing. But the scholars and explorers talked apart, cloistered from politics, for, with the final acceptance by Eisenhower of the dominant liberal thinking of the 1930s and 1940s, this old thinking, accepted by all as orthodoxy, froze into dogma.

Thus, the great revolutionary in the flow of ideas proved to be John Fitzgerald Kennedy. He put ideas to work. He was the watershed.

It would be a tragedy to remember John F. Kennedy only for the romance he brought to the American capital; of that, other men will write. He deserves harder tribute. All things to come for the next twenty years will be built on the concrete base of achievement he left behind in his three-year administration (see Chapter One). Yet of all those things he did to provide the base for the forward movement which he promised Americans in 1960, two are primordial.

The first was the reorganization of the American economy. Alone among Presidents, Kennedy might have claimed that *his* was the first administration in American history never to have known depression or downturn; this burst continues as I write in the 51st month of unbroken American prosperity and growth, the longest such upward movement in the history of the country. Abundance is the most obvious of the legacies of John F. Kennedy to the future—and abundance is the first of the underlying assumptions of the Great Society. All things rest on it.

The second of the legacies was peace. Urged by thought and informed by fact, Kennedy led the American people to share a recognition which at any other time would have been considered heresy in an American President: the recognition that American power was, in fact, limited; and that peace must be made with the Russians. No peace could have been made without the magnificent reorganization of the Armed Forces which created the armor for lack of which peace might have been interpreted as surrender. The Kennedy peace was, as all know now, only a partial peace—a peace made as a first step toward sealing the breach with the Soviet Union and the establishment of a coarse but orderly truce inside the white world.

It grows more apparent each day, of course, that Europe, America and the Soviet Union together have not yet made their peace with the unfurling and frightening new societies of Asia and Africa which are still learning the art of government among men. But even partial peace meant reversal of the frightening and exhausting rivalry of the great powers and the distortion of life and science by preparation for war.

These two legacies together—Abundance and Peace—were what John F. Kennedy delivered to the Americans and to Lyndon B. Johnson. It was as if Kennedy, a younger Moses, had led an elder Joshua to the height of Mount Nebo and there shown him the promised land which he himself would never enter but which Joshua must make his own. For Johnson's must be the design that will organize the place in civilization that John F. Kennedy cleared.

Of this historic purpose and opportunity, it can flatly be stated, no one has been more conscious than Lyndon B. Johnson himself.

Normally, a President enters the White House from the heat of a campaign in which his scholars, speech writers and advisers have hammered out over a full year the ideas and phrases that have shaped, gradually, into his own program. Johnson had entered the Presidency otherwise. Through December of 1963 and January and February of 1964 one can follow him through his shock and his seizure of activity, slowly groping toward some way of stamping his own identity on a future that glowed with unprecedented promise; and, in his groping, seeking ideas.

One can date with some precision the first passage across the President's desk of the phrase that gave a frame to the ideas he was seeking, the phrase "great society." It came on March 2nd in the draft of a speech written for him by Richard N. Goodwin of the Peace Corps. Goodwin, only thirty-two, was part of the flow of ideas that comes from the fountain of Harvard Square; a summa cum laude of Harvard Law School (1958), a law clerk to Supreme Court Justice Frankfurter, Goodwin had been broadened and enriched by three years in service to John F. Kennedy, through campaign and administration. The draft needed by the new President, Johnson, was for remarks to be delivered at an Eleanor Roosevelt Memorial meeting. It was not to be another speech on economics, nor on civil rights, nor on foreign affairs, nor on specifics. It must have something to do with Eleanor Roosevelt's memory—but what should it say? The President wanted to talk about higher purpose; it was an occasion to discuss the quality of American life, paying tribute to the New Deal (in which Lyndon Johnson had grown up), yet to talk relevantly of new kinds of problems, too. "Today," wrote Goodwin for the President, "the problems of our society lie—like some giant iceberg—largely out of sight beneath the surface of abundance and might. One of the principal, primary tasks of leadership, in our day, is not only to solve problems, but to alert the nation to the need to solve them. . . . We now have the opportunity to move not only toward the rich society and the powerful society, but toward the great society. . . . [The] challenge is to ensure for all Americans a life as rich and productive as their talents and capacities will allow."

The draft speech was never delivered. But its thinking had pleased the President, and shortly thereafter Goodwin joined the Presidential staff. One story in the development of the Presidential thinking tells of the President splashing around in the White House pool a few weeks later with his young men—Moyers, Busby and Goodwin—still musing aloud about the need for ideas. He was well along now in carrying out the Kennedy program. Yet he needed to add something of his own; there had to be a body of ideas that bore an LBJ brand. He had already toyed with various phrases to explain himself; one had been "The Better Deal," echoing the long progression of the Square Deal, the New Deal, the Fair Deal. But it was an unsatisfying phrase. He wanted more.

On April 23rd, for the first time, the President tried out in public the phrase "a great society"—in a speech written by Bill Moyers, the twenty-nine-year-old chief-of-staff of his thinkers. Moyers comes of another tributary in the flow of American ideas: Oklahoma born, Texas raised and educated, trained as a Baptist minister, he is closer to being a man of religion than any other person in the White House today; when he speaks of man's spirit, it is because of his belief in it. Moyers' interpretation of the phrase in a speech before a dinner of the Cook County Democrats, the toughest band of hard-rock politicians left in the country enlarged its dimensions further—and the politicians cheered. Johnson tried out the phrase and its connotations no less than sixteen times through April and May until the press began to pick it up; and then the young men urged the President to convert phrase into doctrine, in a full-dress formal statement. Thus, on May 22nd, 1964, at the University of Michigan in Ann Arbor, in a speech written by Goodwin, the Great Society was elevated to capital letters in the text and hoisted by Lyndon Johnson as the banner over his purpose.

Universities are frequently the scenes of great departures in American policy. George Marshall had chosen a Harvard Commencement in June, 1947, to propose the Marshall Plan. John F. Kennedy had chosen the Commencement of the American University in Washington, on June 10th, 1963, to give the greatest of his orations on foreign policy, the famous address on the nature of peace and American power. Now, in 1964, Lyndon Johnson chose the Commencement of the University of Michigan to state his policy for Americans. He arrived by helicopter from Detroit on a warm and sunny day to find the stadium of the university crowded with 80,000 people. He took his place on a platform ornamented with evergreens and purple peonies, the space behind him draped in the blue and maize colors of the school. Dressed in black robes and cap, bending as an honorary degree was conferred on him and the blue and maize and purple sash of the university was hung about him, he joked for a moment like "Old Doc Johnson." But then, and for the first time in the campaign of 1964, he became "Mr. President," reaching for his own:

> The challenge of the next half century is whether we have the wisdom to use [our] wealth to enrich and elevate our national life—and to advance the quality of American civilization —for in your time we have the opportunity to move not only toward the rich society and the powerful society but upward to the Great Society.
>
> . . . The Great Society rests on abundance and liberty for all. It demands an end to poverty and racial injustice—to which we are totally committed in our time, but that is just the beginning. The Great Society is a place where every child can find knowledge to enrich his mind and enlarge his talents. It is a place where leisure is a welcome chance to build and reflect,

not a feared cause of boredom and restlessness. It is a place where the city of man serves not only the needs of the body and the demands of commerce, but the desire for beauty and the hunger for community. It is a place where man can renew contact with nature. It is a place which honors creation for its own sake, and for what it adds to the understanding of the race. It is a place where men are more concerned with the quality of their goals than the quality of their goods. But most of all, the Great Society is not a safe harbor, a resting place, a final objective, a finished work. It is a challenge constantly renewed, beckoning us toward a destiny where the meaning of our lives matches the marvelous products of our labor. . . .

We are going to assemble the best thought and broadest knowledge from all over the world to find those answers. I intend to establish working groups to prepare a series of conferences and meetings—on the cities, on natural beauty, on the quality of education, and on other emerging challenges. From these studies, we will begin to set our course toward the Great Society.

It was a speech of politics, of course. Yet it was much more than that—it was a perception of the nature of America changing. And we must see the Great Society, thus, as much more than a phrase of mood or a political label, but as Lyndon Johnson's Act of Recognition in the flow of history.

From such Acts of Recognition all great political departures begin. For Americans live today on the threshold of the greatest hope in the whole story of the human race, in what may be the opening chapter of the post-industrial era. No capital in the world is more exciting than Washington in our time, more full of fancies and dreams and perplexities. For the first time in civilization, man's mastery over things is sufficient to provide food for all, comfort for all, housing for all, even leisure for all. The question thus arises: What, then, is the purpose of man? How shall he conduct himself at a moment when he is being freed from want, yet freed to ask the tormenting questions of who he is and what he seeks and what his soul needs?

About these questions, if peace prevails, American domestic politics will turn for decades; and out of the answers and programs rising from these questions American life will be shaped. So large and hopeful is this future shape that it is essential, before examining what Lyndon Johnson's vision brings to it, that we digress at length to stress the two great menaces

—one foreign, one domestic—where the Act of Recognition has not yet taken place.

There is, first, a foreign menace, which rises no longer from the obsolete civil wars of Western civilization, but from the clash of Western civilization with the seeking peoples of Asia and Africa.

Technically, of course, the American people see it as the Johnson administration must cope with it—a dirty jungle guerilla war in Vietnam and Southeast Asia. Yet it is much more than that—and it would be futile to discuss it without discussing the most hateful aspect of it which diplomats and officials cannot realistically discuss in public; that it is, deep down, all across Asia and Africa, in the nature of a race war between colored peoples and white peoples. If this characteristic could be taken from it, it would be soluble.

One must understand the ideas of Africans and Asians—for they live in a world all of whose mirrors were fashioned by the people of the North Atlantic Basin. Their thinking is cramped as much by the way history is lit for them as by their common condition of poverty and striving. For no Egyptian can read an Egyptian work of scholarship on Japanese history; nor can a Japanese read an original Japanese scholarly history of the Arabs; nor can the Arabs read an original Arab history of the African blacks; nor can an African black read a history written by an African which gives an African appreciation and analysis of the history of China. All of these people must, when they try to understand each other and their common problems, read histories translated from Western tongues and Western scholarship—written by Englishmen and Americans, Frenchmen and Germans, Russians and Scandinavians. From this mirror of the general past offered by Atlantic scholars, they can learn only how Atlantic scholars told the development of the distant peoples. And all these histories are nailed together by the fact of conquest, the tale of the white man's guns, the cruel story of that brief take-over in the nineteenth century when white men reorganized other societies for profit and greed, in the name of God and duty. Africans and Asians are bound, all of them together, by their quivering indignation at what the white man did to their dignities and their nationhood. What divides them from each other, what gives them their own individual identities and creativity they cannot know—for the history they get of one another is of Western making and tells their story only as strife with the white man. They measure themselves not against the future or their own past, but against the white man.

This hate is there; how it can be quenched is unknown.

The perspectives of this reality are morbid and terrifying. Yet there is another, equally valid reality which offers some room for reason. It is to perceive the development of these other peoples not only in their clash with the marauding white man but in their struggles within. For, fundamentally, the quest of Asian and African villagers has not yet, even in imagination, approached the American quest for abundance; they are seeking the first preliminary to abundance—an orderly form of government. The nature of men is such that they crave government—the ignorant submissively require it to protect them against casual violence; the intelligent need government in order to plan their future in a society where a master logic prevails to which they can attach their individual logic and planning; and the virile and ambitious desire it because government is the form of art-in-authority which most satisfies men who want to lead because they were born to lead.

This striving for a form of government which will replace the primitive and worn-out inheritance of their own past is as much an ingredient —though unrecognized—of Asians' and Africans' thinking as their hate of white men. And this, then, offers the most difficult and most promising problem of American diplomacy in all the new states of the world: Who governs? Who leads? With whom can we deal? Who are the men who can mobilize the resources of their peoples, win the assent of the passive, the energies of the active, the imagination of the young and ambitious? For it is not in any sense a Western representative democracy that America must seek in these new states, but leaders who, in orderly fashion, can lead their peoples in those roads of change out of which, after generations or centuries, a Western liberty of the individual and democracy may develop.

This, for twenty years, has been the problem of American statecraft with the new states—bungled in China twenty years ago, brilliantly managed in Japan, handled with varying results in the new states of Africa. It is the central problem of Vietnam and the war that, at this writing, threatens to engage and once again warp American energies from consideration of the Great Society to a consideration of combat. With whom does one deal and negotiate in Vietnam? Can we create a leadership rooted in South Vietnam which can then negotiate its own differences and its own way to orderly change with the northern leaders of Vietnam? Can the shield of our armor and the strike response of our airplanes gain the time for us to let develop a Vietnam leadership with whom we can then settle?

This is a question for which, at the moment, there is no answer.

Neither is there, at the moment, any clear Act of Recognition which lights up the second threat to the Great Society—a threat as yet so formless that it can only vaguely be styled by some such phrase as the condition of Domestic Tranquillity.

And this, like the foreign threat, is envenomed by the poisons of racism.

For the surface reality would make it appear that the most pointed threat to the Domestic Tranquillity is the condition of Negro life in the great cities. The great cities can indeed no longer sustain or support, with their limited resources, the kind of family life which is spreading in so cancerous a fashion among the nether half of the Negroes in the cities. Of the Negro population explosion much has been written (see Chapter Eight); yet the *rate* of decomposition of decent family life among Negroes, rather than the numbers created by it, remains the dynamic.[5] For nothing in the present American system of law copes with these dynamics. And since no law, even one as important as the Civil Rights Act of 1964, can touch on, or relieve, the phrase-less inner torment of those who emerge from this decomposition, there are hundreds of thousands of young Negroes who believe that no law, no civilized rules can help them at all—or should govern their actions.

Many, thus, view the future of the American city as chiefly a problem posed by Negroes and the condition of despair among so many of them; they hope that enough Federal funds, combined with vigorous *internal* Negro leadership, may reverse the trend. Yet the problem of the peace of the city is *not* entirely a Negro problem. There is another general complicating factor which, though linked to the problem of the Negro community, has a reality of its own. This is a thought-mood new to America, among many leaders both white and black, that laws can be made or repealed in the streets. This is a very old idea, however, and the last time

[5] A study in rate and trend of decomposition is best suggested by examining the family welfare programs begun in the time of Franklin D. Roosevelt. In 1940, shortly after the Aid-to-Dependent Children program had begun, the record showed that one third of all children so helped by government (or 253,000) came from homes where the father had disappeared. To care for such children had been the noble aim of the original program. But by 1963 two thirds of all children nourished by such welfare—or almost two million (1,889,000)—came from homes where fathers were absent. And the charts of the early 1960s traced a new story. Always until 1959 the number of new applications for aid to children had followed, with roughly one year's delay, the graph of unemployment. When unemployment rose, the number of abandoned children rose, too. ("You had the picture," said a graph analyst, "of the guy sitting around the kitchen without earning a nickel week after week and then, after a year of it, she gets fed up and tells him to get the hell out and she goes on relief with the kids.") Up to 1959, when unemployment dropped, so would the number of abandoned children. But in 1959, for the first time, then more sharply in 1962, 1963 and 1964 this understandable correlation was uncoupled. Unemployment dropped sharply—but the number of abandoned or fatherless children rose on a curve of its own, and went on rising, and continues to rise. And where it rose most steeply, preponderantly, overwhelmingly, was among Negroes. It was as if, in the big cities, a new thinking had ruptured an old pattern —and the male became a bearer of seed, sown at will, for whose culture and nourishment the community at large was responsible.

it was seriously tried in a Western country was in the streets of the Weimar Republic, when Nazis, Communists, Socialists and Nationalists all together destroyed the peace of their cities in the streets—and brought Hitler.

The American acceptance of this old idea is fashionably called by some "direct democracy" or "the democracy of pressures"; but it has roots entirely different from those of the disturbances of the Weimar Republic. It is fostered chiefly by the remorseless depersonalization of life in the Digital Society, which strips personality and identity alike from adolescents on relief rolls and adolescents in mechanized educational factories. It erupts just as violently in a campus community, like that of the University of California in Berkeley, as in Harlem; and it expresses, in Berkeley as in Harlem, that same seeking for contact and response which modern life increasingly denies. This thought-mood has, so far, affected chiefly those who have made civil rights their banner, who find in this emotional issue a "commitment" of conscience by which they can express their own resentments too. But it expresses itself just as authentically in the revival of the Klan in the South and the new flourishing of the right-wing groups across the country. To which must be added the stimulant of television with its search for dramatics. Television creates news as much as reports it—it seeks viewer attention by sucking to its cameras demonstrators of all kinds, who there find the vivid, immediate attention of which they have felt deprived.

The conjunction of this thought climate and of the despair in so many urban ghettos poses potentially violent problems for the mayors of American cities. The intertwined complexity, administrative and technological, of American urban life is vulnerable to disruption as no other way of life ever was before; and the constitutional right of assembly has already been stretched in the modern American city to the point where "non-violent" sit-ins or demonstrations are physical disruptions of the nerve centers of orderly government. It grieves a reporter to notice that, for the past two years, one of the most beautiful public places in America, the Georgian City Hall of New York, has been provided with a permanent stand-by stack of gray police barricades so that emergency procedures can convert (and have converted) this peaceful place into a fortress within an hour. It is even more disturbing to hear the thinking of advanced advocates of "democracy by pressure," such as, during the New York water shortage: "Why don't we turn on all the faucets in Harlem and Bedford-Stuyvesant and let the water run out?" Or the ingenious scheme I heard in Chicago: "Let's go into Marshall Field's and all order C.O.D. packages sent to addresses on the South Side—I'll bet we can close down their entire C.O.D. operation." And it is most disturbing of all to notice the beginning of counter-demonstrations at the other end of the political spectrum; for if demonstration meets counter-demonstration in the streets, as in the Weimar Republic, then the hallowed right of peaceful assembly can be warped into the most dangerous threat to orderly government, and the greatest civil right of all, domestic peace, can be destroyed.

Of these two threats—the one foreign, and the one domestic—the Great Society of Lyndon Johnson applies itself only to the domestic threat. And so we must examine where it leads.

Nowhere in any file of American government is there a blueprint of the Great Society, or any program even as detailed as the Monnet Plan for post-war France. "The Great Society" as a phrase belongs on the same political shelf with such phrases as "The New Deal," "The Fair Deal," "The New Frontier." And it is best that at this point we dismiss the phrase "The Great Society" as a design for Utopia; for to oversell it is to demean and reduce Lyndon Johnson's own great contribution to current history: his Act of Recognition and the experience he brings to the visions he has taken from the flow of American thinking.

Somewhere in the course of 1964 Lyndon Johnson's thinking was subtly yet surely elevated, as happens to all Presidents, from the politics of means to the politics of ends. The politics of means was one that he had, as a Senatorial baron, learned as well as any other man—in all its squalor, all its sordidness and all the final immorality in which high principles are generally compromised in the Senate to a consensus in order to make law. Out of this sordid background of means and mechanics had emerged by 1960 a man considered useful as Vice-President; but, in 1964, emerged a President who, knowing the politics of means, could still remember the original dreams that had moved him as a hungering young man, could remember that beyond the goal of "making it" is what happens after you've "made it." What then?

Lyndon Johnson entered on his Presidency after Kennedy had "made it" for the country in terms of new prosperity. Always previously, the politics of ends had been frustrated for Presidents and for the nation by foreign wars, by depression, by limited resources. Now, in 1964, it seemed possible to close the gap between aspiration and reality. With wealth and power growing, it was possible to think of what one does with further wealth and power. And thus the supreme politician of means, seeing the opportunity and knowing the mechanics, became the communicator to the American people of a new search for ends. Politics and ideas had found a new meeting hour.

Some day, some historian of ideas in politics should find the time to tell the full story, with all its anecdotes, of this meeting: of the gathering

of task forces of American scholars and experts through the summer of
1964; of their work on their various specialties (education, cities, beauty,
taxes, foreign aid, transportation and other areas, to a total of fourteen);
of the binding of their reports by November 15th; of their re-examination
by governmental task forces in November and December of 1964; of the
boiling down of their ideas into three fat black books described as "an
encyclopedia of American problems, of what has to be done"; of the final
rendering down of these encyclopedias by December 25th into one volume of legislative proposals, supervised by Moyers, the thirty-year-old
thinker who is the chief companion of the conscience of the President;
and their final presentation to Congress as bills for enactment in the
spring of 1965.

Here, in this book of men in politics, there is no space for that long
story; and it is best to judge what came out of this meeting time of ideas
and politics by going back to the speech of Lyndon B. Johnson at Ann
Arbor in May, and seeing the development out of the original three "C's"
of that speech—classroom, city, countryside—the three main thrusts of
Lyndon Johnson as person and President.

Of these three great thrusts in the Johnson program, which must
become the departure in 1968 of political debate, education is, far and
away, the clearest in Presidential mind and in public approval. "Johnson,"
said one of the men who worked with him on the educational program,
"has a passion for education of the same order of intensity as Kennedy's
passion for stopping atomic testing." The passion can be explained in
many ways; it can be explained by the emotions of youth and background
and his never forgotten memory that, had he not gone back to school, he
would have missed all he has personally won from life. Southwest Texas
State Teachers College was his launching pad for all that came after;
and, had it not been for his mother's urgings, he might have missed the
moment. Or the passion can be explained as a politician's perception—
for nothing worries American families more, in this period of abundance,
than the education of their children; families move and resettle on the
basis of the quality of neighborhood schools, their nearness, the bus
routes; they worry as much, in this new time, about children's grades as
they do about earning a living. Personal emotion and political perception
alike urged Johnson to make this the chief of his thrusts. And both were
proved right when the massive Educational Act of 1965 became law with
almost no Congressional revision or amendments.

The Educational Act of 1965, the first flesh on the bones of his
proposals, projects the Federal government into American education at
levels never dreamed of before—down to the high-school and primary-
grade levels, with the prospect of Federal involvement in pre-primary

education, too. It reflects the recognition of change. Without more learning, Americans simply cannot live in the new world they are creating. In the last decade, jobs requiring a high-school education in the United States grew by 40 percent; but those open to people with no high-school education dropped by 10 percent. Four million more new children will have to find seats in elementary and high schools in the next five years; 400,000 new classrooms must be built for them. (The college crisis is worse—the post-World War II baby boom will add 50 percent more students to the present enrollment of American colleges in the next five years.) The first Johnson program, as he has fairly warned the nation, is only a beginning, a down payment of $1,200 million chiefly to schools in poverty-stricken (largely Negro) areas, but with a significant appropriation to examine the nature and style of education itself in the changing society: What new subjects should be studied? How should old curricula be revised? How can a child's mind most swiftly be lit by learning?

The first of the Johnson programs is already moving. Yet the perspectives it stems from and points to are near-revolutionary—and, as during the campaign of 1964, no one describes them better than Johnson in his own words. Offhandedly, one day in early March of 1965, he met a group of educators and told them about his bill in his own way,

> I don't know what will be written about my administration. Nothing really seems to go right from early in the morning till late at night. And if they do . . . someone approves . . . almost accidentally. . . . But I would hope that it would be said of this decade, if not of this administration, so far as the ancient enemies of mankind are concerned . . . [we coped with] those ancient enemies . . . ignorance, illiteracy, ill-health and disease. . . .
>
> The Education Bill we picked out can be improved. The T-Model Ford could be improved and has been. The first train that ran from Fredericksburg, a little town I lived close to, went to San Antonio, and has been improved a great deal, but I remember the story that they told about it the day it took off—from one of the founding fathers. Some said, Well, they will never get her started, and if they do they will never get her stopped.
>
> Now, we have people that could improve this or that. When I was a boy growing up we never had these issues of our relations with other nations so much. We didn't wake up with Vietnam and have Cyprus for lunch and the Congo for dinner. All we knew because the folks that kind of molded the

opinion and contributed to the political atmosphere, they kept us debating whether we were wet or dry, whether we were prohibitionists or anti-prohibitionists, and the fact one teacher had seven grades to teach in a school that was falling down, and a lady that was underpaid . . . my teacher—I sent her money to come here from California for my inauguration, because she held me in her lap when I was four years old, and she taught seven grades. We have baby-sitters to do that now, but we debated whether wet or dry. Later when I was in high school and college we talked about whether Klan or anti-Klan, and we chased them all around the country back and forth.

I am glad in our time we are talking about how to improve the soul and improve the mind and improve the body and to live and learn and expand and wipe away all of these ancient curses. Why, if we could find the answer to heart disease and to strokes and to cancer we'd save thirty-two billion a year and this whole Education Bill hasn't got but a billion two hundred million. Now, some of them are going to say, If it is a billion two hundred million this year, it will be more next year. Well, it will be.

Because I am not going to be proud to be President of the richest nation in the world when there are hundreds of thousands and millions of children that can't read and write. I am not going to be proud of the nation where disease is still rampant and many children live out a crippled life because they don't have adequate medical care in time.

We are going to improve our education. We are going to improve our medicine and our medical care. We are going to improve the economic condition of our people. We are going to live under the Golden Rule, do unto others as you would have them do unto you. And if you would take a trip with me to the slums of this city, or if you would go to my hometown and walk six blocks from the capitol of the State of Texas, you will see what I am talking about.

I went there one Christmas when my wife and children were away and I couldn't be with them, and I saw 119 people using a single outdoor toilet in the center of the metropolitan area. I saw 119 people drinking water from a hydrant that was located centrally. I walked into their little hovels and I saw a father dying with TB and three or four little children around him that were out of school because they didn't have clothes to wear and food to eat, and taking the TB very lightly that their daddy had.

. . . We got education, doing something about it and doing something about it in this new bill. We got disease and we are doing something about it. We are taking some of that money we have been putting in tanks and bombs and putting it in minds, stomachs and hearts, and of course they are going to find something wrong with [that] formula. They are going to try to bring up the old wet-and-dry fight, or the old church-and-state fight, or some other old fight that will prejudice you until your children grow up in ignorance, but the time has come when we no longer listen to those who oppose for opposition's sake. We are living in the twentieth century and don't ask a fellow about his own plan. They all have these excuses, they will produce another plan and they will offer another amendment and will try to defeat it this way or that way, and you get to be like the old man who said, "Don't get it started or you won't get it stopped."

The second of the new programs reaches further into the future—and is less clear because of its further reach. This is the program for re-examination of the life of American cities, acknowledged as tentative ("We are still only groping toward solution," said Johnson in his Message on the Cities. "The next decade should be a time of experimentation"). Already the single metropolitan belt from southern New Hampshire to northern Virginia holds 21 percent of American people on 1.8 percent of the country's land. Between now and the year 2000, 80 percent of all American growth is expected to take place in such metropolitan clusters; within fifty years 320 million Americans will live in such clusters—which, unless action is planned now, will be virtually uninhabitable then.

This second of the new programs, as it admits, gropes. It proposes a new Federal Department of Housing and Urban Development, which, if accepted, becomes the thin edge of another wedge. It proposes that such a department begin to think about, to urge, perhaps even to subsidize the search of the metropolitan areas and their suburbs for some reasonable, manageable form of modern government. It offers the opportunity to re-examine some of the dogmas of urban renewal; it offers the first faint possibility that low-cost housing may be provided to Americans in urban centers without tearing apart the homely neighborhoods in which most Americans prefer to dwell.

It rests, however, on one new and central recognition: that abundance, as it comes to Americans now from the market economy, congests and clogs the cities built by a market society. It recognizes an idea which, I think, can first be traced in the writings of John Kenneth Galbraith—that abundance, though it liberates, can also suffocate; and that unless a larger and larger share of abundance is used by the public authority to permit people to enjoy their private gains, these private gains may choke each other off. In the cities these problems have become acute: if all men can buy as many automobiles for themselves and families as the automobile industry can produce, they will clog every street and highway—unless some of the abundance is used by public authority to provide such streets, channels and highways. Thus the new experiments in a national transportation program. Or if all real-estate developers can sell as many new homes to all the people who become able to afford them, they will gobble up all the green space and open space about the swelling cities until there are no lungs through which the city can breathe, and no grassy places where people can lie down and look at the sky. Thus the suggestion that public authority now, with the help of Federal funds, buy open spaces in the metropolitan areas to preserve vistas and roaming areas for tomorrow's children.

The third of the new programs is the most difficult of all to define, for it reaches furthest of all: it is called the program of Natural Beauty. It is awkward, clumsy, primitive—but it is a first major attempt, by means of legislation, to fumble to whatever it is that lies beyond wealth, beyond power, beyond the raising of the Gross National Product, in figures, from $500 billion in 1960 to $660 billion in 1965, to $700 billion eighteen months hence. When one talks with the thinkers who have framed the program, it is something beyond the chapter headings under which the program is encased: Highways, Rivers, Trails, Countryside, Clean Water. It is, somehow, to make America, with all its congestion, a livable society—or, as Bill Moyers has put it, "to find out what happens after you release Americans from economic bondage—what do we do then? Can we get a release of the spirit?"

And here Lyndon Johnson, too, reaches the end of his frontier. Beyond his immediate proposals lie questions that we can only dimly see except from the reports that our scouts and scholars bring back to us.

Politics is a process which should slowly bring to public concern all the private worries and hidden hopes of individuals. Generation after generation, formal politics in America has been disrupted by the sudden delivery, center stage, of a concern that had once been entirely private and apart from government. Hunger and sickness were once matters of private charity; jobs and employment were once the responsibility of private industry. The beauty of the wilderness was once nature's endowment and nature's care; a man's wealth was once his own to spend; and the poor's misery, their own to bear. Government's first concern long ago in Western civilization, as it still must be in Asia and Africa today, was to provide discipline and order so that men could be safe. Each new concept of government offered by the extreme fringe of those who looked ahead was hailed with shrieks and mourning.

So, too, it will probably be as Lyndon Johnson moves through his Presidency to the far edge of the present plateau. For on the far edge of the plateau lie problems which we in this decade cannot conceive of as political.

Politics—in short-range American technical terms—will be made by Republicans and Democrats as they argue over the energy and the specifics with which Johnson invests his proposals to meet the new American reality. Here Johnson will prove a formidable opponent for the Republicans. Horace Busby, the Texan of longest present association with Johnson, defined once in conversation, without meaning to, the forbidding problem of the opposition. "Johnson," he said, "is a man without ideology. The New Deal needed an ideology—it needed enemies. They were fighting against the economic royalists. Their liberalism was an effort to gain the attention of the rich people to the problems of the poor. But Johnson sucks his feelings up from the gut. He was one of the poor. He can't see the need for any enemies. This man's passed beyond looking just for majorities; he's looking for consensus. What he wants most is that his programs should go down in history not as Democratic Party programs, but as American programs."

But politics in the longer-range sense—of how men and women combine to live well together—will be made by thinkers.

There, at the distant edge of current politics, are all the other problems that Americans may be approaching by 1972:

§ What may happen as mastery over material increases so fast that the production of goods is no longer as important as the providing of functions for human beings? If, for example, the logic of technology continues to wipe out useful work for people of less intelligence than the technologists—what then? There was an era of human life, for example, that lasted for five centuries and is still memorialized in the term "spinster," when an unmarried woman who could not plow or labor still had a useful function. Whatever female had life and could use her fingers could twirl fibers into thread and thus was useful; she became a "spinster" who could find shelter by friendly fireside or hearth, where she spun, earning her keep without loss of dignity. Then, in a five-year period in the eighteenth century, two Englishmen, James Hargreaves and Richard Arkwright, devised spinning machines that wiped out the function of spinsters entirely. Males enjoyed a longer span of usefulness; all through history, as long as animals provided motor power, males of low intelligence could take care of horses and cattle in the stables, where they were happy and useful—until the internal combustion engine arrived and wiped out *their* function. Today scores of thousands of those whom Henry Ford put to work on the assembly line of the internal combustion engine are similarly being wiped out of function. (In 1964 the automobile industry finally, with 7.6 million cars, broke its production record of 7.4 million in 1955—but it had reduced its number of workers by 14 percent, from 891,200 in 1955 to 771,100 in 1964; the trend in railroad work is even more striking—the number of rail workers has fallen from 1,359,000 in 1946 to 665,200 today.) And the larger number of people who work at desks, in shirtwaist or white collar, are also doomed in function as data-processing machines promise more accurate and swifter data to fewer and fewer executives who make more and more exciting decisions. How then does a society geared to production of goods make a place for people who refuse to let their identities be submerged in data-processing digits?

§ Or how does one conduct a public discussion of issues before the people of America in an age of television, in which, increasingly, the market and art insist that discussion be dramatic—and serious matters are permitted display or discussion only if they can hold the attention of Americans by visual or dramatic tension? If the forum of public discussion and public dreams is to be, as it increasingly tends to be, the electronic arena of television, is it wise to leave it concentrated in New York? Or is it wiser to disperse it, by law, into the less competent, less artistic hands of regional talents, hoping that dispersion will give diversity and that diversity will be more valuable than artistry?

§ Or how does one approach the problem of population and procreation, of love and morality? If, indeed, science offers a pill whereby sex becomes a sport rather than a sacrament honored by love, what happens to older standards of love? And if, indeed, the breeding of human beings becomes controllable, how many people are needed for a good society? And of what kind? And who decides?

§ Beyond that—the infinitely complex problem of control of climate and environment. If, as begins to be conceivable, the rains and the clouds and the winds and the streams may be controlled, who will decide what weather we need where? And by what standards?

§ And beyond that—the moon and space. How large a part of American energy should be invested in this exploration with no definable certainty except the certainty that it will change the lives of all our children?

All these problems seem to Americans at this moment beyond politics. Yet that most crushingly defeated of all candidates, Barry Goldwater, already sensed their approach in 1964 and moved people as deeply by his unease as did the victors Lyndon Johnson and Hubert Humphrey by their hopes.

It is unlikely that Lyndon Johnson will, in his administration, bring the American people to the solution of problems such as these that lie on the far frontier. Not all his volcanic energies—and he continues, today, to work from six in the morning until midnight day after day, as if each coming dawn would bring another election day—can make tomorrow come faster than it will.

What he has brought to the movement is in part his mastery over means combined with a growing ability to subordinate the means to greater ends. But more than that: from all the evil he has seen in his career, from all his animal sense of weakness in other men, he has apparently distilled a desire and a decision to let the good in the American spirit play freely over its future.

The ultimate paradox of 1964 was that the Americans had chosen one of the most passionately political of all Presidents, who proposed, if he could, to make the Presidency a non-political office; and, if he succeeded, he might set the stage for yet further Acts of Recognition by all those unknown younger leaders of the future, also chosen in 1964.

APPENDIX A

THE VOTING OF 1964

The table of votes by states, given below, is that of the Associated Press, and throughout this book, as in *The Making of the President—1960*, the final Associated Press figures are used as standard reference. No *official* vote count in America is yet prepared by any Federal agency. The Network Election Service of 1964, of which the Associated Press was a member, succeeded in bringing the unofficial totals closer together than ever before. The reader who is interested in digital detail should note that the highly reliable vote totals of the Congressional Quarterly Service differ from the AP totals: the Congressional Quarterly Service gives the total vote as 70,642,496; it gives Johnson's total as 43,128,873; it gives Goldwater's total as 27,176,873. But as these totals differ from the Associated Press figures in such infinitesimal detail, I shall stand by the traditional source for continuity's sake, as below:

	TOTAL VOTE	JOHNSON	%	GOLDWATER	%
ALABAMA	689,817	*		479,085	69.5
ALASKA	67,259	44,329	65.9	22,930	34.1
ARIZONA	480,783	237,765	49.5	242,536	50.4
ARKANSAS	560,426	314,197	56.1	243,264	43.4
CALIFORNIA	7,050,985	4,171,877	59.2	2,879,108	40.8
COLORADO	772,749	476,024	61.6	296,725	38.4
CONNECTICUT	1,218,578	826,269	67.8	390,996	32.1
DELAWARE	201,334	122,704	60.9	78,093	38.8
D.C.	198,597	169,796	85.5	28,801	14.5
FLORIDA	1,854,481	948,540	51.1	905,941	48.9
GEORGIA	1,139,157	522,557	45.9	616,600	54.1
HAWAII	207,271	163,249	78.8	44,022	21.2
IDAHO	292,477	148,920	50.9	143,557	49.1
ILLINOIS	4,702,779	2,796,833	59.5	1,905,946	40.5
INDIANA	2,091,606	1,170,848	56.0	911,118	43.6
IOWA	1,184,539	733,030	61.9	449,148	37.9
KANSAS	857,901	464,028	54.1	386,579	45.1
KENTUCKY	1,046,132	669,659	64.0	372,977	35.7
LOUISIANA	896,293	387,068	43.2	509,225	56.8
MAINE	380,965	262,264	68.8	118,701	31.2
MARYLAND	1,116,407	730,912	65.5	385,495	34.5
MASSACHUSETTS	2,344,798	1,786,422	76.2	549,727	23.4
MICHIGAN	3,203,102	2,136,615	66.7	1,060,152	33.1
MINNESOTA	1,554,462	991,117	63.8	559,624	36.0
MISSISSIPPI	409,038	52,591	12.9	356,447	87.1
MISSOURI	1,817,879	1,164,344	64.0	653,535	36.0

* The Alabama ballot listed no Democratic electors pledged to Johnson. Alabamians cast 210,732 votes for unpledged Democratic electors; had those votes been for Johnson instead, only 109,631 out of the total 70.6 million votes cast for President would have gone to minor-party candidates, marking a low for third parties in this century. As it turned out, third-party candidates polled only .5% of the Nov. 3 total.

	TOTAL VOTE	JOHNSON	%	GOLDWATER	%
Montana	278,628	164,246	58.9	113,032	40.6
Nebraska	584,154	307,307	52.6	276,847	47.4
Nevada	135,433	79,339	58.6	56,094	41.4
New Hampshire	286,094	182,065	63.6	104,029	36.4
New Jersey	2,846,770	1,867,671	65.6	963,843	33.9
New Mexico	327,647	194,017	59.2	131,838	40.2
New York	7,166,015	4,913,156	68.6	2,243,559	31.3
North Carolina	1,424,983	800,139	56.2	624,844	43.8
North Dakota	258,389	149,784	58.0	108,207	41.9
Ohio	3,969,196	2,498,331	62.9	1,470,865	37.1
Oklahoma	932,499	519,834	55.7	412,665	44.3
Oregon	783,796	501,017	63.9	282,779	36.1
Pennsylvania	4,818,668	3,130,228	65.0	1,672,892	34.7
Rhode Island	390,078	315,463	80.9	74,615	19.1
South Carolina	524,748	215,700	41.1	309,048	58.9
South Dakota	293,118	163,010	55.6	130,108	44.4
Tennessee	1,144,046	635,047	55.5	508,965	44.5
Texas	2,626,811	1,663,185	63.3	958,566	36.5
Utah	400,310	219,628	54.9	180,682	45.1
Vermont	163,069	108,127	66.3	54,942	33.7
Virginia	1,042,267	558,038	53.5	481,334	46.2
Washington	1,258,374	779,699	62.0	470,366	37.4
West Virginia	792,040	538,087	67.9	253,953	32.1
Wisconsin	1,691,815	1,050,424	62.1	638,495	37.7
Wyoming	142,716	80,718	56.5	61,998	43.5
TOTAL	70,621,479	43,126,218	61.0	27,174,898	38.5

APPENDIX B

THE CHOOSING OF LYNDON B. JOHNSON
(THE GRAHAM MEMORANDUM)

One of the saddest losses that America suffered in the tragic year 1963 was the death of the late Philip Graham, publisher of the Washington *Post* and *Newsweek* Magazine. Graham was a sparkling person—wise in strategy, quick in action, sharp in his opinions, witty in response. These qualities made him one of the great publishers of modern America. But, above all, he had a taste for men of talent—not only as a director of great journalistic enterprises, but as one of the more devoted observers and movers in the politics of Washington.

He cherished alike both John F. Kennedy and Lyndon B. Johnson.

Philip Graham bridged the gap between these two rivals as they arrived to contest each other for the Democratic nomination in Los Angeles in 1960. His account of the choosing of Lyndon B. Johnson is probably as close as we shall ever get to the truth of how the present President of the United States was introduced to the succession. Graham disagreed with various interpretations I had made in my account of that transaction in *The Making of the President—1960*. Shortly before his death he therefore offered me his own personal story of that introduction to power which in the twenty-four hours of July 13th-14th, 1960, made Lyndon Baines Johnson Vice-President. If ever, said Graham, you want to write about it again, this is the way it was.

Graham had written his account as a memorandum to himself within a few days of the events; he dates his passages. The marginal comments on his memo I have inserted into the text in brackets; the "Teddy" to whom these comments are made is me, but Graham did not live long enough to explain the two cryptic comments. The "Kay" referred to here is Katharine Graham, his wife.

NOTES ON THE 1960 DEMOCRATIC CONVENTION

San Francisco
July 19, 1960

We arrived in Los Angeles for the Democratic Convention on Wednesday, July 6th. On Friday, the 8th, the first of the candidates, Senator Johnson, arrived, followed on Saturday afternoon by Senator Kennedy, Senator Symington and Governor Stevenson (a non-candidate open to draft) in that order. There were large welcoming crowds for all of them.

We saw Stevenson, some two hours late, at a party given for him by Mrs. Meyer in a home she had rented in Pasadena. His plane had landed in Las Vegas for refueling, causing the tardiness. At the party I asked him, at Jim Rowe's request, if he would breakfast Sunday morning with Lyndon Johnson. This was arranged for 8:30 A.M. at a private home where his sons were, near the Beverly Hills Hotel where Stevenson was then staying. Walter Reuther was coming to see him at 9:30 and, since the labor leaders had agreed to support Kennedy at a Thursday meeting (a scoop by Joseph Loftus of *The New York Times*), Stevenson wanted to avoid a Reuther-Johnson confrontation.

Sunday, July 10

On the next day, Sunday, Kay and I went to see Stevenson at his Beverly Hills Hotel cottage. We learned that Johnson had canceled the breakfast at midnight (Rowe had acted on his own, and Rayburn and others later persuaded Johnson it was unwise). We also discussed with Stevenson the possibility of his nominating Kennedy. He was quite apathetic to this insofar as the Kennedy people had urged it as a means to avoid his being "humiliated" by an unsuccessful draft effort. Simultaneously, he seemed willing to be tempted if only this course were urged as a way to unify the Party. But he expressed the feeling that he could not nominate Kennedy unless Johnson and Symington concurred, because he had assured all three avowed candidates that he was not a candidate and would not help any candidate in the race.

Monday, July 11

By Monday it seemed overwhelmingly probable that Kennedy would win the nomination. In the course of the day, Joe Alsop and I began discussing the merits of Johnson as a running mate. At Joe's urging, I accompanied him to Kennedy's suite on the 9th floor of the Biltmore where, after considerable delay (and after much observation of the hubbub in what Joe termed "the antechambers of history"), Kennedy appeared and we went with him into a living room for the five minutes we had asked for.

At Joe's request, I did the greatest portion of our talking and urged Kennedy to offer the Vice-Presidency to Johnson. He immediately agreed, so immediately as to leave me doubting the easy triumph, and I therefore restated the matter, urging him not to count on Johnson's turning it down but to offer the VPship so persuasively as to win Johnson over. Kennedy was decisive in saying that was his intention, pointing out that Johnson would help the ticket not only in the South but in important segments of the Party all over the country.

Joe and I were a bit shaken by his positiveness (his brother Bobby had told me earlier that Johnson would not be considered). We were more startled by his aplomb when he began asking about the party Joe was giving that night at Perino's, saying it was the one event of the Convention his wife would really miss.

We then turned to Kennedy's hopes that Stevenson would nominate him. I told him of Stevenson's feeling of commitment and suggested I would inquire of Johnson's feeling on this—assuming that if Johnson saw the inevitable, then Clark Clifford could prevail on Symington to recognize the same. Kennedy asked me to pursue this and said he would hold up on selecting his nominator until he heard from me the next day.

On the basis of this I agreed with Al Friendly and Chal Roberts that the *Post* could write for Tuesday that "the word in Los Angeles is that Kennedy will offer the VPship to Lyndon Johnson" but forbade them writing more strongly in order not to embarrass the Kennedy confidence.

Tuesday, July 12

On Tuesday morning Kennedy phoned me and I told him that I had arranged luncheon with Johnson that day and would inform him of the results of the discussion.

I lunched alone with Johnson at about 1:30 and learned once again the wisdom of Robert Burns on the best-laid plans of men.

By clerical error, the Kennedy staff had sent to the Texas delegation a form telegram they were sending to all delegations. It was signed by Kennedy and asked for a chance to meet with the delegation. The Johnson people had seized on this as a last-ditch opportunity to push Kennedy into error and had telegraphed back requesting a "debate" at 3:00 before a joint meeting of the Texas and Massachusetts delegations. Johnson's suite was on the 7th floor, in the same corner of the Biltmore as Kennedy's 9th-floor suite. But in the tradition of modern publicity, no one from either camp had seen anyone in the other. The telegraphic exchange was equivocal, and when Johnson and I lunched (and, indeed, until after 3:00) it was uncertain whether Kennedy would agree to a joint appearance.

The possibility of "debate" had given Johnson a tremendous exhilaration. Once again he was a candidate for the Presidency with a chance, even an unlikely one. But he was also bone tired from the long Senate session, from the days in Los Angeles, and especially from a grueling morning during which he had spoken before three or four delegations.

In that atmosphere I could not possibly have asked him to "release" Stevenson so that he could nominate Kennedy. Instead, I listened to his ideas about the forthcoming debate, which seemed a bit harsh and personal, and I finally persuaded him at 1:50 (after receiving several pleading notes from his staff) to take a nap until 2:20. A Negro couple from his ranch were in the room throughout our lunch, and the three of us converged upon him, disrobed him, pajamaed him and got him in bed. As I was tucking him in he began talking in *ad hominem* terms about Kennedy, whereupon I told him "No, we're not going to say that sort of thing. We're going to talk about the world. Walter Lippmann came out for Kennedy this morning, saying you were an ignoramus about the world, and we're going to show him he's wrong." He was wonderfully surprised that "his friend" Lippmann would say that—really unbelieving. Then I assured him I'd jot down some thoughts for him and have him awakened at 2:20. He went off to sleep with a quickness I shall always envy.

In his living room I scribbled out some thoughts (and clichés) about the world situation and got them typed by a young man in the adjoining room. There was a lively chaos in the living room, with no one aware whether Kennedy would or would not appear.

In a thoroughly direct, if surreptitious, manner I finally phoned Kennedy's suite at 2:45 and could get only his press aide, Pierre Salinger, who said he thought Kennedy would appear.

At 2:50 I found Johnson still in pajamas and still sleepy in his bathroom, handed him the typed sheets, said I was still not sure Kennedy could appear, but suggested he scan the sheets in case the "debate" occurred.

At 3:00 in the Biltmore, Johnson appeared as though he had had a quiet morning in preparation for this occasion only. And soon after Kennedy appeared and made a fine speech. Johnson then rose and used my "high road" for his opening, continuing on *ad lib* for some thrusts at Kennedy which he kept within the bounds of propriety.

Meanwhile, I learned that Rowe and John Connally, with whom I had discussed the issue on Monday, had on Monday night told Johnson of the question of whether Stevenson should nominate Kennedy. Johnson's reaction was predictable—that Stevenson was over twenty-one and should be capable of making up his own mind.

On Tuesday evening Steve Smith, Kennedy's brother-in-law, reached me at about 9:30 or 10:00 at Joe Alsop's party. I told him why the telegram to the Texas delegation had made it impossible for me to present the Stevenson matter to Johnson. I also told him that I knew others had put the matter to Johnson. Since the man to nominate Kennedy had to be chosen that night, I urged Smith to have Kennedy put the matter up to Stevenson and to say that Kennedy had indications that Johnson felt Stevenson was free to make up his own mind. Smith said they were urging Stevenson but were having little luck.

Wednesday, July 13

At 5:00 A.M. Wednesday, I got a wild idea and wrote it out. It consisted of a message to the Convention from Kennedy, to be read by Stevenson on Thursday, asking the delegates to draft Johnson for VP. After writing it, I left Kay a note to call Smith in early morning and ask for a Kennedy appointment. She awakened me at 9:30 saying I had an appointment for 10:40, so I dashed off as quickly as I could shave and dress.

After a ten-minute wait, Kennedy came in amidst a tremendous bustle and suggested I talk to him as he drove across town to his next meeting. This was scheduled at the Virginia delegation at the Beverly Hilton (40 minutes away), but on the elevator he realized this would consume his whole morning and canceled this, heading instead for a farm caucus at the Statler-Hilton (5 minutes away). We rode alone in the back of a Convention car and I explained the idea, saying I had some longhand scribble I could leave with Sorensen or Bobby. "Leave it with me only," he said. And I agreed I'd get it typed and leave it with Mrs. Lincoln, his secretary. He calmly said he might be twenty votes short (the nominating session was to begin in four hours) and asked if I thought he could get any Johnson votes out of the Vice-Presidency offer. I said I could think of none unless Smathers would try to swing some of Florida's. He said he was going to see Smathers later but the trouble was that Smathers wanted to be Vice-President. At this point I said he'd never miss by any twenty votes, that his nomination was assured by *res ipsa loquitur*. Whereupon, in the midst of traffic jam and Convention hubbub, his face became a student's face and he asked, "What does that mean?"

About 1:00 that day I showed Al Friendly my typed "message from Kennedy" to check on my common sense. He liked the idea, so I left the pages in an envelope with Mrs. Lincoln. (In stubborn hindsight I think it was a good idea and that it might have been adopted had Stevenson not grown bemused by his chances of being drafted and in so doing infuriated the Kennedy group.)

[On Wednesday night John F. Kennedy was nominated; early Thursday morning, word spread that he had offered the Vice-Presidency to Lyndon Johnson. T.H.W.]

Continued in Flight, San Francisco to Chicago—July 24, 1960

Thursday, July 14

On Thursday, about 1:45 P.M., I went to Johnson's suite in the Biltmore. The Los Angeles papers and the press corps in general were prophesying Symington. On entering Johnson's suite, he at once seized my arm and took me into his bedroom, alone with Lady Bird. He said that Bobby Kennedy was with Sam Rayburn in another part of the suite offering him the VPship and he had to make a decision.

We sat on a bed, the three of us, about as composed as three Mexican jumping beans. Lady Bird tried to leave. Johnson and I lunged after her, saying she was needed on this one. I tried to duck LBJ's inquiry, but finally said I felt he had to take it. Lady Bird was somewhere between negative and neutral. At this point Sam Rayburn entered and said Bobby wanted to talk directly to LBJ. Lady Bird intervened, apologizing by saying she had never yet argued with Mr. Sam, but saying she now felt LBJ should not see Bobby.

LBJ asked my advice—the while all of us were pacing around the bedroom, in and out of the bathroom, etc. "No," I said, "you shouldn't see him," and then, repeating LBJ's expressions on the bed previously, "You don't want it, you won't negotiate for it, you'll only take it if Jack drafts you, and you won't discuss it with anyone else."

Mr. Sam seemed to think LBJ should see Bobby, but he also seemed to think he should turn down the VP. Finally, in that sudden way decisions leap out of a melee, it was decided. Mr. Sam was to tell Bobby LBJ's position and why he wouldn't see Bobby. I was to go phone Jack LBJ's position.

In the bedroom hubbub, Jim Rowe and (I think) John Connally had joined us.

I dragged Rowe along as witness and we went out through thirty or so press people in the hall, down into a vacant bedroom.

About 2:30 I got Jack on the phone and told him LBJ's decision as above. He said something to the general effect that he was in a general mess because some liberals were against LBJ. He said he was in a meeting with others right then and that people were urging that "no one had anything against Symington." He and I had discussed this earlier in the week and he made some reference conceding Symington had no affirmative qualifications. He then asked me to call back for a decision "in three minutes."

By this time Rowe and I were as calm as Chileans on top of an earthquake. I remember Jim's dropping a lighted cigarette on the rug and also pursuing me constantly with ash trays and lighted matches every time I took out a cigarette.

As a guide to Kennedy's request, I took off my wrist watch and placed it by the phone. We both agreed that "three minutes" in these circumstances meant ten minutes, and my memory is that about 2:40 or 2:45 I phoned back.

(In our earlier conversation I had—urged on by notes from Rowe—conveyed to Jack the South's opposition to LBJ's joining the ticket and the importance of LBJ in those areas.)

Each call, of course, was a telephonic obstacle course. To get Jack's secretary, Mrs. Lincoln, from the switchboard meant one delay, then from Mrs. Lincoln to Shriver or Steve Smith another, then insisting on Jack directly took more.

On the phone Jack was utterly calm. It's all set, he said. Tell Lyndon I want him and will have Lawrence nominate him, etc. He said he'd be busy getting Lawrence and the seconders and preparing his statement and also putting out fires of opposition, and so he asked me to call Adlai and tell him the decision and ask him for full support.

My memory is vague, but I believe I called Adlai first because of the need to get his support in such a short time. In any event, I talked to him first and then gave the phone to Jim Rowe and went down the hall to LBJ's bedroom. He and Lady Bird naturally quizzed me in detail about the conversation with Jack. Also Lady Bird was especially curious about Adlai's reaction. I remember shrugging and saying, Oh, you know, sort of as you'd expect.

I went back to the bedroom Rowe and I had been using to call Al Friendly and say I'd be down later. I told him I thought I'd better not tell him the decision as he'd be learning it soon.

At this point LBJ sent for me to say that Bobby Kennedy had been back down to see Rayburn some twenty minutes before (say, roughly, 3:00) and had said Jack would phone directly. No call had come and LBJ was considerably on edge. I wrote down the private numbers in his bedroom (his switchboard had long ago broken down) and said I'll call Jack.

About 3:30 I got Jack, who said he had assumed my message would suffice. I explained what Bobby had told Sam, and he said he'd call at once. He then again mentioned opposition to LBJ and asked for my judgment. I said something to the effect that the Southern gains would more than off-set liberal losses, and added that anyway it was too late to be mind-changing and that he should remember "You ain't no Adlai." He agreed about the finality of things and asked what Stevenson had said. I said he was wobbling about but would be all right and he asked me to call Stevenson back and ask him to issue a statement shortly after Jack issued his at 4:00.

I told LBJ Jack would be phoning him and then Rowe and I returned to the vacant bedroom to call Adlai. In our prior talk he had argued for Symington on pure expediency grounds and I had been a bit testy in pointing out that any VP was likely to be President. Rowe had continued that conversation while I had gone to LBJ's rooms and Rowe told me he had had the same argument but he'd told Adlai the decision was made and that he should line up support.

I got Adlai on the phone again after some delay and found him still brimming over with an account of the difficulties. Despite Rowe's and my assertions to him, he had spent the time since our last call in "canvassing" various people's reactions. After I listened to a recital of this for some time, I interrupted and with (I hope) polite firmness said "the nominee" had asked me to ask him to please issue a supporting statement as soon as Kennedy made his statement on TV, which should be very soon. Stevenson then quickly agreed.

At this point (approximately 3:40) John Connally burst into the room, saying he had to get Bobby Kennedy at once, as various Southern governors were objecting to some of the seconders and he wanted Bobby to add Almond or Ellington or Combs. Rowe and I urged Combs as the least objectionable and we spent a futile 10-15 minutes trying to reach Bobby.

Sometime around this point I went back into LBJ's bedroom and listened to Senator Kerr and Governor Daniels objecting to LBJ's taking the VPship. By now Rayburn, originally against the idea, was supporting Lyndon's decision and Kerr was soon able, very articulately, to change his mind. But Daniels remained opposed—and very loudly so. On the assumption that our "Yankee faces" might create more furor, Rowe and I left.

Connally, Rowe and I were still trying to reach Bobby (shortly after 4:00) when LBJ's appointment secretary, Bill Moyers, rushed into our room to say Lyndon wanted me at once. "I'll be along in just a minute." "That won't do," Moyers yelled, and, grabbing my arm, dragged me down the hall through a solid jam of press people and into the entrance hall of the suite with Rowe and Connally close behind.

LBJ was in a state of high nerves and said we must talk alone at once. His bedroom was still packed with Daniels, *et al*, so we headed into the adjoining room, which in proper pure farce setting, was full of about fifteen Hawaiian delegates. Johnson called out that he was sorry he had to have the room and they solemnly filed out, bowing their serious Oriental faces in turn to all of us at the door (LBJ, Lady Bird, Rayburn, Connally, Rowe and me) with LBJ loudly chanting, "Thank you, boys, thank you. Thank you for all you did."

In a minute they were gone and the six of us swept into the bedroom, joined by Bobby Baker. LBJ seemed about to jump out of his skin. He shouted at me that Bobby Kennedy had just come in and told Rayburn and him that there was much opposition and that Lyndon should withdraw for the sake of the Party.

There was considerable milling about and hubbub, and finally Mr. Rayburn said, "Phil, call Jack." The only phone was a regular Biltmore extension and it took a minute which seemed an hour to get the operator, then another series of hour-like minutes as we got Kennedy's switchboard, then his secretary, and finally Kennedy. "Jack," I said, "Bobby is down here and is telling the Speaker and Lyndon that there is opposition and that Lyndon should withdraw."

"Oh," said Jack, as calmly as though we were discussing the weather, "that's all right; Bobby's been out of touch and doesn't know what's been happening."

"Well, what do you want Lyndon to do?" I asked.

"I want him to make a statement right away; I've just finished making mine." (I believe he was reported to have made his statement downstairs in the Biltmore Bowl at 4:05.)

"You'd better speak to Lyndon," I said.

"Okay," he answered, "but I want to talk to you again when we're through."

I was standing between twin beds, and as I handed the phone to LBJ he sprawled out across the bed in front of me, lay on his side, and said, "Yes . . . Yes . . . Yes . . ." and then "Okay, here's Phil," as he handed the phone back to me.

At this point Kennedy chatted along to me as though we were discussing someone else's problems. He said Alex Rose was threatening not to list him on the Liberal Party ticket in New York because of Johnson and that "this is a problem we'll just have to solve." I heard myself saying, "Oh, don't worry, we'll solve that." And then, returning closer to sanity, I said, "You'd better speak to Bobby." So Baker dashed out to fetch Bobby, who walked in looking dead tired.

"Bobby, your brother wants to speak to you," said I (in what at once seemed to me the silliest line in the whole play). Bobby took the phone, and as I walked out of the room I heard him say, "Well, it's too late now," and half slam down the phone. [NOTE: Teddy—Since then I've heard from UAW V-P Leonard Woodcox that Alex Rose was in a white fury when Dubinsky called from N.Y. and said: the greatest political masterstroke I've ever seen.]

In the entrance hall of the suite LBJ and Lady Bird were standing, looking as though they had just survived an airplane crash, with Lyndon holding a typed statement accepting the VPship. "I was just going to read this on TV when Bobby came in, and now I don't know what I ought to do."

With more ham than I ever suspected myself of, I suddenly blurted: "Of course you know what you're going to do. Throw your shoulders back and your chin out and go out and make that announcement. And then go on and win. Everything's wonderful."

This soap-opera thrust was somehow wonderfully appropriate, and young Bill Moyers echoed loud approval while swinging open the hall doors and pushing Johnson out into the TV lights and the explosion of flashbulbs. Someone else propelled Lady Bird, and from the hall I could watch them rising to stand on some chairs, and as they rose their faces metamorphosed into enthusiasm and confidence.

The news spread while Johnson remained before the cameras, and more and more people began crawling into the living room through the back door of a bedroom down the hall. It had been barely over two hours since I'd got involved, yet it seemed that we'd all been under strain for weeks. I suddenly felt quite wildly elated and in a few minutes found myself talking to Elizabeth Guest (!) and being enthusiastically introduced by a Florida delegate to an 18-year-old blonde with a large ribbon across her ample front emblazoned "Miss Miami Beach."

At this point I felt withdrawal to a saner environment urgently needed, so I fled to an elevator and joined my wife for a ride to the Convention Hall and a late lunch.

(1) One other incident I cannot place exactly in time and so have left to a postscript, but I think it occurred just before Johnson asked me to call Kennedy (i.e., about 2:15). Rayburn told me the details of Bobby's visit to offer the VP (which must have been around 1:30). He said Bobby came in and sat down with Rayburn (and I believe Connally), with Rayburn waiting for the obvious. Whereupon Bobby said he wondered if Johnson would like to be National Chairman. "Shit," answered Rayburn, whereupon Bobby offered the Vice-Presidency.

(2) Did Jack offer the VP hoping LBJ would turn it down? Did LBJ really want it? [NOTE: *Teddy—Over the phone I'll tell you why LBJ wanted the VP. I did not know it when I wrote this.*] Did Bobby try to sabotage the offer? And if so, did he do so on his own or with Jack's approval? I have no confident answer to any of those questions. I am quite certain that Jack ultimately made up his mind in the ten minutes between my first two phone calls. As for Bobby, I later learned he had (in the hours just before the Presidential balloting) assured several liberal delegates it would *not* be Johnson. My guess is that he made that assurance on his own and tried to bring it about on his own during his dealings with Johnson and Rayburn.[1]

[1] Because of various things I have learned since, I know that this guess in Mr. Graham's superb account was based on incomplete facts and thus was in error. I am certain that Robert Kennedy was *not* acting on his own, but had been snarled by a confusion of communications and messages in a turbulent day. T.H.W.

(3) In explanation of Stevenson's conduct, I learned later that night (Thursday) while I was dining with him that he had talked over the VP thing with Jack at lunchtime Thursday and later heard that Jack had made a definite "deal" with Johnson at 7:30 that morning. Consequently, he had felt Jack had made a fool of him by discussing the matter at lunch as though it were unsettled. I explained to Adlai just when Jack had first definitely assured Johnson of his choice.

INDEX

THE AWARD-WINNING SERIES THAT CHANGED AMERICAN POLITICS FOREVER—NOW BACK IN PRINT FOR THE FIRST TIME IN DECADES

THE MAKING OF THE PRESIDENT 1960

With a New Foreword by Robert Dallek

ISBN 978-0-06-190060-0 (paperback)

The groundbreaking national bestseller and Pulitzer Prize-winning account of the 1960 presidential campaign and the election of John F. Kennedy.

"A notable achievement. White has written a fascinating story of a fascinating campaign." —*Time* magazine

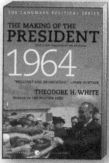

THE MAKING OF THE PRESIDENT 1964

With a New Foreword by Jon Meacham

ISBN 978-0-06-190061-7 (paperback)

The criticatlly-acclaimed account of the 1964 presidential campaign—from the assassination of JFK through the battle for power between Lyndon B. Johnson and Barry Goldwater.

THE MAKING OF THE PRESIDENT 1968

With a New Foreword by Chris Matthews

ISBN 978-0-06-190064-8 (paperback)

The compelling story of the turbulent 1968 presidential campaign, the assassinations of Martin Luther King, Jr. and presidential candidate Robert F. Kennedy, and the election of Richard Nixon.

THE MAKING OF THE PRESIDENT 1972

With a New Foreword by Cokie Roberts

ISBN 978-0-06-190067-9 (paperback)

In this fourth book in the landmark The Making of the President series, White recounts the 1972 presidential campaign, the landslide victory of Richard Nixon, and the seeds of the Watergate scandal.

Available wherever books are sold, or call 1-800-331-3761 to order.